Lecture Notes in Mathematics

For further volumes:
http://www.springer.com/series/304

Vladimir Maz'ya • Alexander Movchan
Michael Nieves

Green's Kernels and Meso-Scale Approximations in Perforated Domains

Springer

Vladimir Maz'ya
Department of Mathematics
Linköping University
Linköping, Sweden

Alexander Movchan
Department of Mathematical Sciences
University of Liverpool
Liverpool, United Kingdom

Michael Nieves
School of Engineering
Liverpool John Moores University
Liverpool, United Kingdom

ISBN 978-3-319-00356-6 ISBN 978-3-319-00357-3 (eBook)
DOI 10.1007/978-3-319-00357-3
Springer Heidelberg New York Dordrecht London

Lecture Notes in Mathematics ISSN print edition: 0075-8434
ISSN electronic edition: 1617-9692

Library of Congress Control Number: 2013939619

Mathematics Subject Classification (2010): 4E10, 35B40, 35J08

Printed on acid-free paper

Springer is part of Springer Science+Business Media (www.springer.com)

Preface

The book is based on the authors' results on asymptotic approximations of Green's kernels for elliptic boundary value problems in perforated domains. A new feature is the uniformity of the asymptotics with respect to the independent variables. Formal asymptotic approximations are supplied with estimates of the remainder terms. For the case when the number of perforations or inclusions becomes large, a novel method of meso-scale asymptotic approximations is introduced, and uniform asymptotic approximations of Green's kernels as well as solutions of boundary value problems in multiply perforated domains are presented. Such approximations do not require periodicity or other typical constraints attributed to homogenization approximations.

Applications are considered for problems of linear elasticity in planar and three-dimensional domains containing multiple small holes or inclusions. Illustrative computational examples are included to compare asymptotic approximations with accurate finite element numerical simulations, which demonstrate the advantages of the asymptotic method.

This book is addressed to mathematicians, physicists and engineers, as well as research students, interested in asymptotic analysis and numerical computations for solutions to partial differential equations. The required background includes a basic theory of partial differential equations and elements of functional analysis.

Acknowledgements The authors would like to thank the University of Liverpool for providing excellent academic facilities throughout the duration of the project, which has led to this book. The support of the UK Engineering and Physical Sciences Research Council via the grant EP/F005563/1 is gratefully acknowledged.

Linköping, Sweden — V. Maz'ya
Liverpool, UK — A. Movchan
Liverpool, UK — M. Nieves

Preface

Contents

Introduction

There is a wide range of applications in physics and structural mechanics involving domains with singular perturbations of the boundary. Examples include perforated domains and bodies with defects of different types. Accurate direct numerical treatment of such problems is challenging. As alternative means of efficient solution one can use asymptotic approximations. In particular, the multi-scale asymptotic approximations and their justification have been developed for homogenization problems by Marchenko and Khruslov [21], Sánchez-Palencia [39–41], Zhikov [44], Zhikov, Kozlov and Oleinik [43], Allaire [1], Chechkin [4], and Cioranescu and Murat [5].

A comprehensive asymptotic theory of boundary value problems in singularly perturbed domains was developed during the last three decades (see the monographs by Bakhvalov and Panasenko [2], Il'in [15], Kozlov, Maz'ya and Movchan [19], Maz'ya, Nazarov and Plamenevskii [30] and the bibliography therein). This theory includes a general methodology of asymptotic analysis of solutions to boundary value problems, eigenvalues of the corresponding operators and other set functions, such as energy, capacity and stress intensity factors.

In this book, we deal with the analysis of Green's functions and matrices, i.e. kernels of the integral operators representing solutions to elliptic boundary value problems. The exposition is based on the recent work by Maz'ya and Movchan [23–27] and Maz'ya, Movchan and Nieves [28, 29, 32, 33].

The first results on asymptotic approximations of Green's kernels $G_\varepsilon(\mathbf{x}, \mathbf{y})$ for certain classical boundary value problems under small variations of a domain are due to Hadamard [13], who considered regular perturbations of a planar domain with smooth boundary. In connection with our work, it is appropriate to mention that asymptotic approximations in [13] are not uniform with respect to the position of $\mathbf{x}$ and $\mathbf{y}$.

The importance of Green's functions is paramount. Important applications of asymptotic analysis of Green's kernels include extremal problems in the complex function theory in Julia [16], Barnard, Pearce and Campbell [3], shape sensitivity analysis in Fremiot and Sokolowski [11], free boundary problems in Palmerio and Dervieux [37] and theory of reproducing kernels in Englis et al. [7], and Komatsu [17].

Green's function $G_\varepsilon(\mathbf{x}, \mathbf{y})$ is considered here as the main object for study rather than a tool to generate solutions of specific boundary value problems. Singular perturbations occur while both $\mathbf{x}$ and $\mathbf{y}$ approach the boundary, even in the cases when the boundary itself is smooth. The uniformity of the asymptotic approximations is the principal point of attention. We also show non-trivial links between Green's functions and solutions of boundary value problems for meso-scale structures. Such systems involve a large number of small inclusions, so that a small parameter, the relative size of an inclusion, may compete with a large parameter, represented as an overall number of inclusions.

The main focus of this text is on two topics: (a) asymptotics of Green's functions and tensors for the Laplace and Lamé operators in domains with *singularly perturbed boundaries* and (b) meso-scale asymptotic approximations of physical fields in non-periodic domains with many inclusions. The novel feature of these asymptotic approximations is their *uniformity* with respect to the independent variables.

The book consists of three parts.

The derivation and analysis of the uniform asymptotics of Green's kernels in singularly perturbed domains for the Laplace operator is the main focus of Part I.

To give an impression of such approximations we show the following typical example. Let $G_\varepsilon(\mathbf{x}, \mathbf{y})$ be Green's function of the Dirichlet problem for the operator $-\Delta$ in a two-dimensional domain Ω_ε with a small Jordan inclusion $F_\varepsilon = \{\mathbf{x} : \varepsilon^{-1}\mathbf{x} \in F\}$ (see Fig. 1). We find the asymptotic approximation of G_ε in the form

$$G_\varepsilon(\mathbf{x}, \mathbf{y}) = G(\mathbf{x}, \mathbf{y}) + g(\frac{\mathbf{x}}{\varepsilon}, \frac{\mathbf{y}}{\varepsilon}) - g(\frac{\mathbf{x}}{\varepsilon}, \infty) - g(\infty, \frac{\mathbf{y}}{\varepsilon}) + \frac{1}{2\pi} \log \frac{|\mathbf{x} - \mathbf{y}|}{\varepsilon r_F}$$
$$- \frac{2\pi}{\log(\varepsilon r_F R_\Omega{}^{-1})} \left(G(\mathbf{x}, 0) + \frac{1}{2\pi} \log \frac{|\mathbf{x}|}{\varepsilon r_F} - g(\frac{\mathbf{x}}{\varepsilon}, \infty) \right)$$
$$\times \left(G(0, \mathbf{y}) + \frac{1}{2\pi} \log \frac{|\mathbf{y}|}{\varepsilon r_F} - g(\infty, \frac{\mathbf{y}}{\varepsilon}) \right) + O(\varepsilon),$$

where G and g are Green's functions of "model" interior and exterior Dirichlet problems in "limit" domains Ω and $\mathbb{R}^2 \setminus F$, independent of ε; r_F and R_Ω are the inner and outer conformal radii of F and Ω, respectively, as defined in Appendix G of Pólya and Szegö [38]. We emphasize that the estimate of the error term in the above asymptotic formula is uniform with respect to $\mathbf{x}$ and $\mathbf{y}$.

Furthermore, we obtain uniform asymptotics of Green's kernels for mixed boundary value problems in domains containing a small hole or a rigid inclusion. We address the Neumann condition on the hole and the Dirichlet condition on the exterior boundary, as well as the Neumann condition on the exterior boundary and Dirichlet condition on the defect. We also derive uniform asymptotics of the Neumann function in the perforated domain. Then, the asymptotic approximations of Green's kernels are constructed in a domain with several small inclusions.

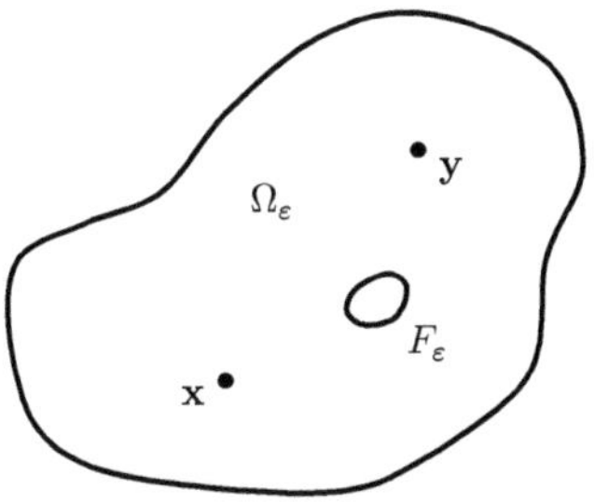

Fig. 1 A domain Ω_ε containing a small hole F_ε

Other examples of asymptotic approximations of Green's functions in singularly perturbed domains include a domain with the singular perturbation of the exterior smooth boundary, a truncated cone and a thin cylindrical body.

Part II is focused on the uniform asymptotic approximations of Green's tensors for linear elasticity in domains with small defects. We obtain uniform asymptotics of Green's tensor in a planar domain and a three-dimensional body containing a small rigid inclusion. This is followed by the construction of uniform asymptotics for Green's tensors in domains with multiple rigid inclusions. Here, instead of the capacitary potential used in approximations of Green's functions for clamped perforated domains in Part I, we introduce the matrix of the *elastic capacity* and study its properties. It will also be shown that this matrix plays an important role in the asymptotic algorithm.

Once the uniform asymptotic approximations for Green's tensor in a domain with multiple small inclusions have been obtained, we consider the asymptotics of Green's tensor in a planar body containing a single small void and furthermore extend this analysis to the case when the body contains several voids. Since the traction conditions are set on the boundary of small defects, we use the dipole fields of linear elasticity in the description of the boundary layer fields.

In Part III, we consider the case when the perforated geometries contain many inclusions or voids of different sizes, as illustrated in Fig. 2, and introduce a novel method of meso-scale asymptotic approximations. First, we deal with asymptotics of solutions to Dirichlet problems for the Poisson equation $-\Delta u = f$ in a three-dimensional body with many perforations. An example of the formal asymptotic representation for the solution of such a boundary value problem is

$$u(\mathbf{x}) \sim v_f(\mathbf{x}) + \sum_{j=1}^{N} C_j \Big(P^{(j)}(\mathbf{x}) - 4\pi \operatorname{cap}(F^{(j)})\, H(\mathbf{x}, \mathbf{O}^{(j)}) \Big), \qquad (1)$$

where

- v_f is the solution of the same equation in a domain Ω without inclusions
- $P^{(j)}$ is the harmonic capacitary potential of the inclusion $F^{(j)}$

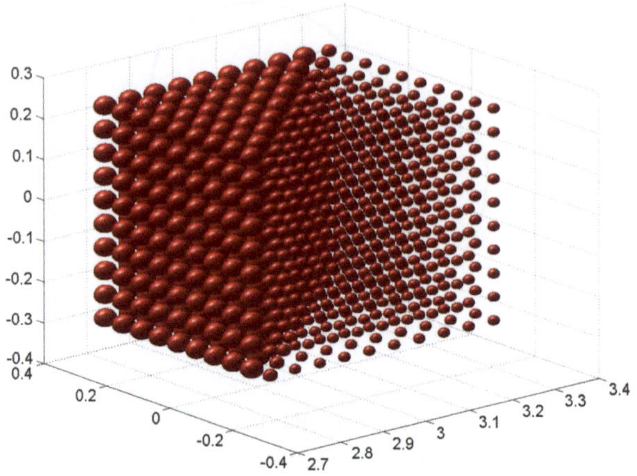

Fig. 2 The non-periodic cloud of 1,000 voids of different sizes. The method of meso-scale asymptotic approximations is applied to obtain a uniform approximation to the solution of the Dirichlet problem in the multiply perforated domain

- $\text{cap}(F^{(j)})$ is the harmonic capacity of $F^{(j)}$
- H is the regular part of Green's function G of Ω

The coefficients C_j satisfy a certain algebraic system, which includes the information about the positions, size and shapes of inclusions.

Furthermore, the text includes meso-scale approximations of Green's function for the Dirichlet problem in a multiply perforated body in $\mathbb{R}^3$:

$$G_N(\mathbf{x}, \mathbf{y}) = G(\mathbf{x}, \mathbf{y}) - \sum_{j=1}^{N} \Big\{ h^{(j)}(\mathbf{x}, \mathbf{y}) - P^{(j)}(\mathbf{y}) H(\mathbf{x}, \mathbf{O}^{(j)})$$

$$- P^{(j)}(\mathbf{x}) H(\mathbf{O}^{(j)}, \mathbf{y}) + 4\pi \, \text{cap}(F^{(j)}) H(\mathbf{x}, \mathbf{O}^{(j)}) H(\mathbf{O}^{(j)}, \mathbf{y})$$

$$+ H(\mathbf{O}^{(j)}, \mathbf{O}^{(j)}) \, T^{(j)}(\mathbf{x}) T^{(j)}(\mathbf{y}) - \sum_{i=1}^{N} \mathcal{C}_{ij} T^{(i)}(\mathbf{x}) T^{(j)}(\mathbf{y}) \Big\} + O(\varepsilon d^{-2}).$$

Here, d is another small parameter characterizing the minimum distance between each inclusion,

$$T^{(j)}(\mathbf{y}) = P^{(j)}(\mathbf{y}) - 4\pi \, \text{cap}(F^{(j)}) H(\mathbf{O}^{(j)}, \mathbf{y}),$$

and again the entries of the matrix $\mathcal{C} = (\mathcal{C}_{ij})_{i,j=1}^{N}$ are solutions of a certain algebraic system containing information about the inclusions.

Moreover, in addition to the meso-scale treatment of Dirichlet problems in domains with many inclusions, we present uniform asymptotic formulae for solutions to mixed boundary value problems in a body with a cloud of many voids, whose boundaries are subjected to Neumann boundary conditions. Important components of the asymptotic algorithm are the boundary layers near individual voids, whose formal description incorporates the dipole fields characterizing the shape of the voids and their orientation. A model algebraic problem is introduced and solved to evaluate the coefficients in the meso-scale asymptotic approximations. The energy estimates are obtained for the remainder terms.

In particular, for solids containing non-uniform clouds of many spherical voids, the asymptotic approximation takes a form where all boundary layer terms can be written explicitly; this makes such an approximation extremely simple and appealing for numerical implementation in practical problems, where traditional computational approaches like FEM become inefficient.

This book provides an exposition of novel asymptotic approximations, highly efficient for physical problems in multiply perforated domains with non-uniform distribution of defects such as voids or inclusions. The book would be of interest for a mathematician as well as for a physicist or an engineer, who can use the advantage of powerful methods of multi-scale asymptotic approximations in challenging physical problems for composite densely perforated media.

Part I
Green's Functions in Singularly Perturbed Domains

Chapter 1
Uniform Asymptotic Formulae for Green's Functions for the Laplacian in Domains with Small Perforations

We derive here uniform asymptotic formulae for Green's functions of the Dirichlet problem for the operator $-\Delta$ in n-dimensional domains with small holes, first for $n > 2$ and then for $n = 2$. We also show that these formulae can be simplified under certain constraints on the independent variables.

Now, we list several notations adopted here and throughout the text of the book. Let Ω be a domain in $\mathbb{R}^n$, $n \geq 2$, with compact closure $\overline{\Omega}$ and boundary $\partial\Omega$. By F we denote a contractible compact set of positive harmonic capacity in $\mathbb{R}^n$; its complement is $F^c = \mathbb{R}^n \setminus F$. We suppose that both Ω and F contain the origin $\mathbf{O}$ as an interior point. Without loss of generality, it is assumed that the minimum distance between $\mathbf{O}$ and the points of $\partial\Omega$ is equal to 1. Also, the maximum distance between $\mathbf{O}$ and the points of ∂F^c will be taken as 1. We introduce the set $F_\varepsilon = \{\mathbf{x} : \varepsilon^{-1}\mathbf{x} \in F\}$, where ε is a small positive parameter, and the open set $\Omega_\varepsilon = \Omega \setminus F_\varepsilon$. The notation B_ρ stands for the open ball centered at $\mathbf{O}$ with radius ρ.

Here, Green's function for the operator $-\Delta$ in Ω_ε, will be denoted by G_ε. In the sequel, along with $\mathbf{x}$ and $\mathbf{y}$, we use the scaled variables $\boldsymbol{\xi} = \varepsilon^{-1}\mathbf{x}$ and $\boldsymbol{\eta} = \varepsilon^{-1}\mathbf{y}$. By Const we always mean different positive constants independent of ε. Finally, the notation $f = O(g)$ is equivalent to the inequality $|f| \leq \text{Const}\, g$.

1.1 Green's Function for a Multi-dimensional Domain with a Small Hole

We assume here that $n > 2$. Let G and g denote Green's functions of the Dirichlet problem for the operator $-\Delta$ in the sets Ω and $F^c = \mathbb{R}^n \setminus F$. We make use of the regular parts of G and g, respectively:

$$\mathcal{H}(\mathbf{x}, \mathbf{y}) = (n-2)^{-1}|S^{n-1}|^{-1}|\mathbf{x} - \mathbf{y}|^{2-n} - G(\mathbf{x}, \mathbf{y}), \tag{1.1}$$

V. Maz'ya et al., *Green's Kernels and Meso-Scale Approximations in Perforated Domains*, Lecture Notes in Mathematics 2077, DOI 10.1007/978-3-319-00357-3_1,

and

$$h(\xi,\eta) = (n-2)^{-1}|S^{n-1}|^{-1}|\xi-\eta|^{2-n} - g(\xi,\eta), \tag{1.2}$$

where $|S^{n-1}|$ denotes the $(n-1)$-dimensional measure of the unit sphere S^{n-1}.

By $P(\xi)$ we mean the capacitary potential of F defined as a unique solution of the following Dirichlet problem in F^c

$$\Delta_\xi P(\xi) = 0 \ \text{ in } F^c, \tag{1.3}$$

$$P(\xi) = 1 \ \text{ on } \partial F^c, \tag{1.4}$$

$$P(\xi) \to 0 \ \text{ as } \ |\xi| \to \infty, \tag{1.5}$$

where the boundary condition (1.4) is interpreted in the sense of the Sobolev space H^1 (see Maz'ya [22]).

The following auxiliary assertion is classical.

Lemma 1.1. *(i) The potential P satisfies the estimate*

$$0 < P(\xi) \le \min\left\{1, |\xi|^{2-n}\right\}. \tag{1.6}$$

(ii) If $|\xi| \ge 2$, then

$$\left|P(\xi) - \frac{\operatorname{cap}(F)}{(n-2)|S^{n-1}|}|\xi|^{2-n}\right| \le \text{Const}\, |\xi|^{1-n} \tag{1.7}$$

Proof. *(i)* Inequalities (1.6) follow from the maximum principle for variational solutions of Laplace's equation.

(ii) Inequality (1.7) results from the expansion of P in spherical harmonics. □

Lemma 1.2. *For all $\eta \in F^c$ and for ξ with $|\xi| > 2$ the estimate holds:*

$$|h(\xi,\eta) - P(\eta)(n-2)^{-1}|S^{n-1}|^{-1}|\xi|^{2-n}| \le \text{Const}\, |\xi|^{1-n} P(\eta). \tag{1.8}$$

Proof. By (1.2), h satisfies the Dirichlet problem

$$\Delta_\xi h(\xi,\eta) = 0, \ \ \xi,\eta \in F^c, \tag{1.9}$$

$$h(\xi,\eta) = (n-2)^{-1}|S^{n-1}|^{-1}|\xi-\eta|^{2-n}, \quad \xi \in \partial F^c \text{ and } \eta \in F^c, \tag{1.10}$$

$$h(\xi,\eta) \to 0 \ \text{ as } |\xi| \to \infty \text{ and } \eta \in F^c. \tag{1.11}$$

We fix $\eta \in F^c$. By the series expansion of g in spherical harmonics,

$$|\xi|^{n-2}\Big(g(\xi,\eta) - \frac{C(\eta)}{(n-2)|S^{n-1}||\xi|^{n-2}}\Big) \to 0 \text{ as } |\xi| \to \infty. \tag{1.12}$$

We apply Green's formula to the functions $g(\xi,\eta)$ and $1 - P(\xi)$ restricted to the domain $B_R \setminus F$, where $B_R = \{\xi : |\xi| < R\}$ is the ball of a sufficiently large radius R. Taking into account that $P(\xi) = 1$ and $g(\xi,\eta) = 0$ when $\xi \in \partial(F^c)$ we deduce

$$\int_{B_R\setminus F} \nabla_\xi g(\xi,\eta)\cdot\nabla_\xi P(\xi)d\xi = P(\eta)-1-\int_{\partial B_R}(1-P(\xi))\frac{\partial}{\partial|\xi|}g(\xi,\eta)ds_\xi, \tag{1.13}$$

and

$$\int_{B_R\setminus F} \nabla_\xi g(\xi,\eta)\cdot\nabla_\xi P(\xi)d\xi = \int_{\partial B_R} g(\xi,\eta)\frac{\partial}{\partial|\xi|}P(\xi)ds_\xi. \tag{1.14}$$

Hence,

$$1 - P(\eta) = -\int_{\partial B_R}\Big(g(\xi,\eta)\frac{\partial}{\partial|\xi|}P(\xi) + (1-P(\xi))\frac{\partial}{\partial|\xi|}g(\xi,\eta)\Big)ds_\xi. \tag{1.15}$$

It follows from (1.12) that

$$1 - P(\eta) = -\lim_{R\to\infty}\int_{\partial B_R}\frac{\partial}{\partial|\xi|}\frac{C(\eta)}{(n-2)|S^{n-1}||\xi|^{n-2}}ds_\xi = C(\eta).$$

Let $|\xi| > 2$. Then for $\eta \in \partial F^c$

$$|h(\xi,\eta)-(n-2)^{-1}|S^{n-1}|^{-1}|\xi|^{2-n}P(\eta)| = (n-2)^{-1}|S^{n-1}|^{-1}\Big||\xi-\eta|^{2-n}-|\xi|^{2-n}\Big|$$

$$\le \text{Const}\,|\eta||\xi|^{1-n} \le \text{Const}\,|\xi|^{1-n}. \tag{1.16}$$

In the above estimate, we used the assumption of the maximum distance between the origin and the points of ∂F^c being equal to 1. From (1.16) and the maximum principle for functions harmonic in η, we deduce

$$|h(\xi,\eta) - \Big((n-2)|S^{n-1}|\Big)^{-1}|\xi|^{2-n}P(\eta)| \le \text{Const}\,|\xi|^{1-n}P(\eta),$$

for all $\eta \in F^c$ and $|\xi| > 2$. □

Our main result concerning the uniform approximation of Green's function G_ε in the multi-dimensional case is given by

Theorem 1.1. *Green's function $G_\varepsilon(\mathbf{x},\mathbf{y})$ admits the representation*

$$\begin{aligned}G_\varepsilon(\mathbf{x},\mathbf{y}) &= G(\mathbf{x},\mathbf{y}) + \varepsilon^{2-n} g(\varepsilon^{-1}\mathbf{x}, \varepsilon^{-1}\mathbf{y}) - ((n-2)|S^{n-1}||\mathbf{x}-\mathbf{y}|^{n-2})^{-1} \\ &+\mathcal{H}(0,\mathbf{y})P(\varepsilon^{-1}\mathbf{x}) + \mathcal{H}(\mathbf{x},0)P(\varepsilon^{-1}\mathbf{y}) - \mathcal{H}(0,0)P(\varepsilon^{-1}\mathbf{x})P(\varepsilon^{-1}\mathbf{y}) \\ &-\varepsilon^{n-2}\ \mathrm{cap}(F)\ \mathcal{H}(\mathbf{x},0)\mathcal{H}(0,\mathbf{y}) + O\Big(\varepsilon^{n-1}(\min\{|\mathbf{x}|,|\mathbf{y}|\}+\varepsilon)^{2-n}\Big), \qquad (1.17)\end{aligned}$$

uniformly with respect to $\mathbf{x},\mathbf{y} \in \Omega_\varepsilon$. *Here, $\mathcal{H}$ and h are regular parts of Green's functions G and g, respectively* (*see* (1.1), (1.2)), *and P is the capacitary potential of F.*

Before presenting a proof of this theorem, we give a *plausible formal argument* leading to (1.17). We note the form of the remainder as

$$O\Big(\varepsilon^{n-1}(\min\{|\mathbf{x}|,|\mathbf{y}|\}+\varepsilon)^{2-n}\Big),$$

which is valid uniformly through Ω_ε. Further in the text, we also present corollaries that include simplified asymptotic representations of Green's kernel subject to certain constraints on $\mathbf{x}$ and $\mathbf{y}$.

Let G_ε be represented in the form

$$G_\varepsilon(\mathbf{x},\mathbf{y}) = \Big((n-2)|S^{n-1}|\Big)^{-1}|\mathbf{x}-\mathbf{y}|^{2-n} - \mathcal{H}_\varepsilon(\mathbf{x},\mathbf{y}) - h_\varepsilon(\mathbf{x},\mathbf{y}), \qquad (1.18)$$

where $\mathcal{H}_\varepsilon$ and h_ε are solutions of the Dirichlet problems

$$\begin{aligned}\Delta_x\mathcal{H}_\varepsilon(\mathbf{x},\mathbf{y}) &= 0, \ \ \mathbf{x},\mathbf{y}\in\Omega_\varepsilon, \\ \mathcal{H}_\varepsilon(\mathbf{x},\mathbf{y}) &= \Big((n-2)|S^{n-1}|\Big)^{-1}|\mathbf{x}-\mathbf{y}|^{2-n}, \ \ \mathbf{x}\in\partial\Omega, \ \mathbf{y}\in\Omega_\varepsilon, \\ \mathcal{H}_\varepsilon(\mathbf{x},\mathbf{y}) &= 0, \ \ \mathbf{x}\in\partial F_\varepsilon^c, \ \mathbf{y}\in\Omega_\varepsilon,\end{aligned}$$

and

$$\begin{aligned}\Delta_x h_\varepsilon(\mathbf{x},\mathbf{y}) &= 0, \ \ \mathbf{x},\mathbf{y}\in\Omega_\varepsilon, \\ h_\varepsilon(\mathbf{x},\mathbf{y}) &= \Big((n-2)|S^{n-1}|\Big)^{-1}|\mathbf{x}-\mathbf{y}|^{2-n}, \ \ \mathbf{x}\in\partial F_\varepsilon^c, \ \mathbf{y}\in\Omega_\varepsilon, \\ h_\varepsilon(\mathbf{x},\mathbf{y}) &= 0, \ \ \mathbf{x}\in\partial\Omega, \ \mathbf{y}\in\Omega_\varepsilon.\end{aligned} \qquad (1.19)$$

By (1.18), it suffices to find asymptotic formulae for $\mathcal{H}_\varepsilon$ and h_ε.

Function $\mathcal{H}_\varepsilon$. Obviously, $\mathcal{H}_\varepsilon(\mathbf{x},\mathbf{y}) - \mathcal{H}(\mathbf{x},\mathbf{y})$ is harmonic in Ω_ε, and $\mathcal{H}_\varepsilon(\mathbf{x},\mathbf{y}) - \mathcal{H}(\mathbf{x},\mathbf{y}) = 0$ for $\mathbf{x}\in\partial\Omega$. On the other hand, for $\mathbf{x}\in\partial F_\varepsilon^c$ the leading part of $\mathcal{H}_\varepsilon(\mathbf{x},\mathbf{y}) - \mathcal{H}(\mathbf{x},\mathbf{y})$ is equal to the function $-\mathcal{H}(0,\mathbf{y})$. This function can be extended onto F_ε^c, harmonically in $\mathbf{x}$, as $-\mathcal{H}(0,\mathbf{y})P(\varepsilon^{-1}\mathbf{x})$, whose leading-order part is equal to $-\varepsilon^{n-2}\mathrm{cap}(F)\,\mathcal{H}(\mathbf{x},0)\mathcal{H}(0,\mathbf{y})$ for $\mathbf{x}\in\partial\Omega$. Hence,

$$\begin{aligned}\mathcal{H}_\varepsilon(\mathbf{x},\mathbf{y})-\mathcal{H}(\mathbf{x},\mathbf{y}) &\sim -\mathcal{H}(0,\mathbf{y})P(\varepsilon^{-1}\mathbf{x})\\ &+\varepsilon^{n-2}\mathrm{cap}(F)\,\mathcal{H}(\mathbf{x},0)\mathcal{H}(0,\mathbf{y}) \text{ for all } \mathbf{x},\mathbf{y}\in\Omega_\varepsilon.\end{aligned} \tag{1.20}$$

Function h_ε. By definitions (1.2) and (1.19) of h and h_ε,

$$h_\varepsilon(\mathbf{x},\mathbf{y})-\varepsilon^{2-n}h(\varepsilon^{-1}\mathbf{x},\varepsilon^{-1}\mathbf{y})=0 \text{ for } \mathbf{x}\in\partial F_\varepsilon^c.$$

Furthermore, by Lemma 1.2

$$\begin{aligned}&h_\varepsilon(\mathbf{x},\mathbf{y})-\varepsilon^{2-n}h(\varepsilon^{-1}\mathbf{x},\varepsilon^{-1}\mathbf{y})\\ &\sim -\Big((n-2)|S^{n-1}|\Big)^{-1}|\mathbf{x}|^{2-n}P(\varepsilon^{-1}\mathbf{y}) \text{ for } \mathbf{x}\in\partial\Omega.\end{aligned}$$

The harmonic function in $\mathbf{x}\in\Omega$, with the Dirichlet data

$$-\Big((n-2)|S^{n-1}|\Big)^{-1}|\mathbf{x}|^{2-n}P(\varepsilon^{-1}\mathbf{y})$$

on $\partial\Omega$, is $-\mathcal{H}(\mathbf{x},0)P(\varepsilon^{-1}\mathbf{y})$, and it is asymptotically equal to $-\mathcal{H}(0,0)P(\varepsilon^{-1}\mathbf{y})$ on ∂F_ε^c, which is not necessarily small. The harmonic in $\mathbf{x}$ extension of $\mathcal{H}(0,0)P(\varepsilon^{-1}\mathbf{y})$ onto F_ε^c is given by $\mathcal{H}(0,0)P(\varepsilon^{-1}\mathbf{y})P(\varepsilon^{-1}\mathbf{x})$. Since this function is small for $\mathbf{x}\in\partial\Omega$, one may assume the asymptotic representation

$$\begin{aligned}h_\varepsilon(\mathbf{x},\mathbf{y}) &-\varepsilon^{2-n}h(\varepsilon^{-1}\mathbf{x},\varepsilon^{-1}\mathbf{y})+\mathcal{H}(\mathbf{x},0)P(\varepsilon^{-1}\mathbf{y})\\ &\sim \mathcal{H}(0,0)P(\varepsilon^{-1}\mathbf{x})P(\varepsilon^{-1}\mathbf{y}) \qquad \text{for all } \mathbf{x},\mathbf{y}\in\Omega_\varepsilon.\end{aligned} \tag{1.21}$$

Substituting (1.20) and (1.21) into (1.18), we deduce

$$\begin{aligned}G_\varepsilon(\mathbf{x},\mathbf{y}) &\sim \Big((n-2)|S^{n-1}|\Big)^{-1}|\mathbf{x}-\mathbf{y}|^{2-n}-\mathcal{H}(\mathbf{x},\mathbf{y})-\varepsilon^{2-n}h(\varepsilon^{-1}\mathbf{x},\varepsilon^{-1}\mathbf{y})\\ &+\mathcal{H}(0,\mathbf{y})P(\varepsilon^{-1}\mathbf{x})+\mathcal{H}(\mathbf{x},0)P(\varepsilon^{-1}\mathbf{y})-\mathcal{H}(0,0)P(\varepsilon^{-1}\mathbf{x})P(\varepsilon^{-1}\mathbf{y})\\ &-\varepsilon^{n-2}\mathrm{cap}(F)\,\mathcal{H}(\mathbf{x},0)\mathcal{H}(0,\mathbf{y}),\end{aligned}$$

which is equivalent to

$$\begin{aligned}G_\varepsilon(\mathbf{x},\mathbf{y}) &\sim G(\mathbf{x},\mathbf{y})+\varepsilon^{2-n}g(\varepsilon^{-1}\mathbf{x},\varepsilon^{-1}\mathbf{y})-((n-2)|S^{n-1}|)^{-1}|\mathbf{x}-\mathbf{y}|^{2-n}\\ &+\mathcal{H}(0,\mathbf{y})P(\varepsilon^{-1}\mathbf{x})+\mathcal{H}(\mathbf{x},0)P(\varepsilon^{-1}\mathbf{y})-\mathcal{H}(0,0)P(\varepsilon^{-1}\mathbf{x})P(\varepsilon^{-1}\mathbf{y})\\ &-\varepsilon^{n-2}\mathrm{cap}(F)\,\mathcal{H}(\mathbf{x},0)\mathcal{H}(0,\mathbf{y}).\end{aligned}$$

Now, we give a rigorous proof of (1.17).

Proof of Theorem 1.1.
The remainder $r_\varepsilon(\mathbf{x}, \mathbf{y})$ in (1.17) is a solution of the boundary value problem

$$\Delta_x r_\varepsilon(\mathbf{x}, \mathbf{y}) = 0, \quad \mathbf{x}, \mathbf{y} \in \Omega_\varepsilon, \tag{1.22}$$

$$\begin{aligned} r_\varepsilon(\mathbf{x}, \mathbf{y}) &= \mathcal{H}(\mathbf{x}, \mathbf{y}) - \mathcal{H}(0, \mathbf{y}) \\ &-(\mathcal{H}(\mathbf{x}, 0) - \mathcal{H}(0,0)) P(\varepsilon^{-1}\mathbf{y}) \\ &+\varepsilon^{n-2}\mathrm{cap}(F)\, \mathcal{H}(\mathbf{x}, 0)\mathcal{H}(0, \mathbf{y}), \quad \mathbf{x} \in \partial F_\varepsilon^c, \ \mathbf{y} \in \Omega_\varepsilon, \end{aligned} \tag{1.23}$$

$$\begin{aligned} r_\varepsilon(\mathbf{x}, \mathbf{y}) &= \varepsilon^{2-n} h(\varepsilon^{-1}\mathbf{x}, \varepsilon^{-1}\mathbf{y}) - \mathcal{H}(0, \mathbf{y}) P(\varepsilon^{-1}\mathbf{x}) \\ &-\mathcal{H}(\mathbf{x}, 0) P(\varepsilon^{-1}\mathbf{y}) + \mathcal{H}(0,0) P(\varepsilon^{-1}\mathbf{x}) P(\varepsilon^{-1}\mathbf{y}) \\ &+\varepsilon^{n-2}\mathrm{cap}(F)\, \mathcal{H}(\mathbf{x}, 0)\mathcal{H}(0, \mathbf{y}), \ \mathbf{x} \in \partial\Omega, \ \mathbf{y} \in \Omega_\varepsilon. \end{aligned} \tag{1.24}$$

The functions $\mathcal{H}(\mathbf{x}, 0)$ and $\mathcal{H}(0, \mathbf{y})$ are harmonic in Ω and are bounded by Const on $\partial\Omega$. Hence, they are bounded by Const for $\mathbf{x} \in \partial F_\varepsilon^c$, $\mathbf{y} \in \Omega_\varepsilon$ and for $\mathbf{x} \in \partial\Omega$, $\mathbf{y} \in \Omega_\varepsilon$, respectively. The terms $\varepsilon^{n-2}\mathrm{cap}(F)\mathcal{H}(\mathbf{x}, 0)\mathcal{H}(0, \mathbf{y})$ in the right-hand sides of (1.23) and (1.24) are bounded by Const ε^{n-2}.

By definition (1.1), $\nabla_x \mathcal{H}(\mathbf{x}, \mathbf{y})$ is bounded by Const uniformly with respect to $\mathbf{y} \in \Omega$ for every $\mathbf{x} \in B_{1/2}$. Hence, by (1.23) and the inequalities $0 < P(\mathbf{x}) \le 1$,

$$\begin{aligned} &|\mathcal{H}(\mathbf{x}, \mathbf{y}) - \mathcal{H}(0, \mathbf{y}) - (\mathcal{H}(\mathbf{x}, 0) - \mathcal{H}(0,0)) P(\varepsilon^{-1}\mathbf{y})| \\ &\le \text{Const}\, \varepsilon \sup_{\mathbf{z} \in B_\varepsilon} |\nabla_z \mathcal{H}(\mathbf{z}, \mathbf{y})| \le \text{Const}\, \varepsilon, \end{aligned}$$

for $\mathbf{x} \in \partial F_\varepsilon^c$, $\mathbf{y} \in \Omega_\varepsilon$. Thus, the following estimate holds when $\mathbf{x} \in \partial F_\varepsilon^c$ and $\mathbf{y} \in \Omega_\varepsilon$

$$|r_\varepsilon(\mathbf{x}, \mathbf{y})| \le \text{Const}\, \varepsilon \sup_{\mathbf{z} \in B_\varepsilon} |\nabla_z \mathcal{H}(\mathbf{z}, \mathbf{y})| \le \text{Const}\, \varepsilon. \tag{1.25}$$

Next, we estimate $|r_\varepsilon(\mathbf{x}, \mathbf{y})|$ for $\mathbf{x} \in \partial\Omega$ and $\mathbf{y} \in \Omega_\varepsilon$. By Lemma 1.1, the capacitary potential $P(\varepsilon^{-1}\mathbf{x})$ satisfies the inequalities

$$0 \le P(\varepsilon^{-1}\mathbf{x}) \le \text{Const} \frac{\varepsilon^{n-2}}{(|\mathbf{x}| + \varepsilon)^{n-2}}, \tag{1.26}$$

for $\mathbf{x} \in \Omega_\varepsilon$, and

$$\begin{aligned} &\left| P(\varepsilon^{-1}\mathbf{x}) - \frac{\varepsilon^{n-2}\mathrm{cap}(F)}{(n-2)|S^{n-1}||\mathbf{x}|^{n-2}} \right| \\ &\le \text{Const} \left(\varepsilon/|\mathbf{x}|\right)^{n-1} \le \text{Const}\, \varepsilon^{n-1}, \end{aligned} \tag{1.27}$$

for $\mathbf{x} \in \partial\Omega$. Now, (1.27) and the definition of $\mathcal{H}(\mathbf{x}, \mathbf{y})$ imply

$$|\varepsilon^{n-2}\text{cap}(F)\mathcal{H}(\mathbf{x}, 0)\mathcal{H}(0, \mathbf{y}) - \mathcal{H}(0, \mathbf{y})P(\varepsilon^{-1}\mathbf{x})| \leq \text{Const}\, \varepsilon^{n-1}. \tag{1.28}$$

Also, we have the estimate

$$\begin{aligned}
&|\varepsilon^{2-n}h(\varepsilon^{-1}\mathbf{x}, \varepsilon^{-1}\mathbf{y}) - \mathcal{H}(\mathbf{x}, 0)P(\varepsilon^{-1}\mathbf{y})| \\
&= \varepsilon^{2-n}\left|h(\varepsilon^{-1}\mathbf{x}, \varepsilon^{-1}\mathbf{y}) - \frac{P(\varepsilon^{-1}\mathbf{y})}{(n-2)|S^{n-1}||\mathbf{x}/\varepsilon|^{n-2}}\right| \\
&\leq \text{Const}\, \varepsilon|\mathbf{x}|^{1-n}P(\varepsilon^{-1}\mathbf{y}) \\
&\leq \text{Const}\, \frac{\varepsilon^{n-1}}{(|\mathbf{y}| + \varepsilon)^{n-2}}, \quad \mathbf{x} \in \partial\Omega, \ \mathbf{y} \in \Omega_\varepsilon,
\end{aligned} \tag{1.29}$$

which follows from the definition (1.1) of $\mathcal{H}(\mathbf{x}, \mathbf{y})$ and the estimates (1.8) and (1.26). Combining (1.26), (1.28) and (1.29) we obtain from (1.24) that the trace of the function $\mathbf{x} \to |r_\varepsilon(\mathbf{x}, \mathbf{y})|$ on $\partial\Omega$ does not exceed

$$\text{Const}\, \frac{\varepsilon^{n-1}}{(|\mathbf{y}| + \varepsilon)^{n-2}},$$

for $\mathbf{y} \in \Omega_\varepsilon$. Using this and (1.25), we deduce by the maximum principle that

$$|r_\varepsilon(\mathbf{x}, \mathbf{y})| \leq \text{Const}\left\{\varepsilon P\left(\frac{\mathbf{x}}{\varepsilon}\right) + \frac{\varepsilon^{n-1}}{(|\mathbf{y}| + \varepsilon)^{n-2}}\right\},$$

for all $\mathbf{x}, \mathbf{y} \in \Omega_\varepsilon$. Taking into account (1.26), we arrive at

$$|r_\varepsilon(\mathbf{x}, \mathbf{y})| \leq \text{Const}\, \frac{\varepsilon^{n-1}}{(\min\{|\mathbf{x}|, |\mathbf{y}|\} + \varepsilon)^{n-2}} \tag{1.30}$$

The proof is complete. □

1.2 Green's Function for the Dirichlet Problem in a Planar Domain with a Small Hole

In this section, we find an asymptotic approximation of G_ε in the two-dimensional case. We shall see that this approximation has new features in comparison with that in Theorem 1.1.

The notations $\Omega_\varepsilon, \Omega, F_\varepsilon, F$, introduced earlier, will be used here. As before, we assume that the minimum distance from the origin to $\partial\Omega$ and the maximum distance between the origin and the points of ∂F^c are equal to 1.

Green's function $G(\mathbf{x}, \mathbf{y})$ for the unperturbed domain Ω has the form

$$G(\mathbf{x}, \mathbf{y}) = (2\pi)^{-1} \log |\mathbf{x} - \mathbf{y}|^{-1} - H(\mathbf{x}, \mathbf{y}), \tag{1.31}$$

where H is its regular part satisfying

$$\Delta_x H(\mathbf{x}, \mathbf{y}) = 0, \quad \mathbf{x}, \mathbf{y} \in \Omega, \tag{1.32}$$

$$H(\mathbf{x}, \mathbf{y}) = (2\pi)^{-1} \log |\mathbf{x} - \mathbf{y}|^{-1}, \quad \mathbf{x} \in \partial\Omega, \ \mathbf{y} \in \Omega. \tag{1.33}$$

The scaled coordinates $\boldsymbol{\xi} = \varepsilon^{-1}\mathbf{x}$ and $\boldsymbol{\eta} = \varepsilon^{-1}\mathbf{y}$ will be used as in the multi-dimensional case. Similar to Sect. 1.1, $g(\boldsymbol{\xi}, \boldsymbol{\eta})$ and $h(\boldsymbol{\xi}, \boldsymbol{\eta})$ are Green's function and its regular part in F^c:

$$\Delta_{\xi} g(\boldsymbol{\xi}, \boldsymbol{\eta}) + \delta(\boldsymbol{\xi} - \boldsymbol{\eta}) = 0, \quad \boldsymbol{\xi}, \boldsymbol{\eta} \in F^c, \tag{1.34}$$

$$g(\boldsymbol{\xi}, \boldsymbol{\eta}) = 0, \quad \boldsymbol{\xi} \in \partial F, \ \boldsymbol{\eta} \in F^c, \tag{1.35}$$

$$g(\boldsymbol{\xi}, \boldsymbol{\eta}) \ \text{is bounded as} \ |\boldsymbol{\xi}| \to \infty \ \text{and} \ \boldsymbol{\eta} \in F^c, \tag{1.36}$$

and

$$h(\boldsymbol{\xi}, \boldsymbol{\eta}) = (2\pi)^{-1} \log |\boldsymbol{\xi} - \boldsymbol{\eta}|^{-1} - g(\boldsymbol{\xi}, \boldsymbol{\eta}). \tag{1.37}$$

Solvability of elliptic boundary value problems in domains of infinite extent is well studied, and we refer to the work by Oleinik and Yosifian [35] addressing boundary value problems for second order elliptic equations in unbounded domains and Saint-Venant's principle. The limit behaviour, as $|\boldsymbol{\xi}| \to \infty$, of a bounded solution of (1.34)–(1.36) depends on $\boldsymbol{\eta}$, corresponding to a position of the source. Hence, we introduce a function ζ by

$$\zeta(\boldsymbol{\eta}) = \lim_{|\boldsymbol{\xi}| \to \infty} g(\boldsymbol{\xi}, \boldsymbol{\eta}). \tag{1.38}$$

Furthermore, the model problem for the function ζ is also discussed in Sect. 2.1.1.

Lemma 1.3. *Let $|\boldsymbol{\xi}| > 2$. Then the regular part $h(\boldsymbol{\xi}, \boldsymbol{\eta})$ of Green's function g in F^c admits the asymptotic representation*

$$h(\boldsymbol{\xi}, \boldsymbol{\eta}) = -(2\pi)^{-1} \log |\boldsymbol{\xi}| - \zeta(\boldsymbol{\eta}) + O(|\boldsymbol{\xi}|^{-1}), \tag{1.39}$$

which is uniform with respect to $\boldsymbol{\eta} \in F^c$.

Proof. Following the inversion transformation, we use the variables:

$$\xi' = |\xi|^{-2}\xi, \quad \eta' = |\eta|^{-2}\eta,$$

and the identity

$$|\xi - \eta|^{-1}|\xi||\eta| = |\xi' - \eta'|^{-1}.$$

Then, the boundary values of $h(\xi, \eta)$, as $\xi \in \partial F^c, \eta \in F^c$, can be expressed in the form

$$h(\xi, \eta) = \mathfrak{H}(\xi', \eta') - (2\pi)^{-1} \log |\xi||\eta|, \tag{1.40}$$

where $\mathfrak{H}(\xi', \eta')$, $\xi' \in \partial(F^c)'$, is the boundary value of the regular part of Green's function in the bounded transformed set $(F^c)'$. Namely, the function $\mathfrak{H}(\xi', \eta')$ is defined as a solution of the Dirichlet problem

$$\Delta_{\xi'} \mathfrak{H}(\xi', \eta') = 0, \quad \xi', \eta' \in (F^c)', \tag{1.41}$$

$$\mathfrak{H}(\xi', \eta') = (2\pi)^{-1} \log |\xi' - \eta'|^{-1}, \quad \xi' \in \partial(F^c)'. \tag{1.42}$$

It follows from (1.40) that the harmonic extension of $h(\xi, \eta)$ is

$$h(\xi, \eta) = \mathfrak{H}(\xi', \eta') - (2\pi)^{-1} \log |\xi||\eta|, \quad \xi, \eta \in F^c. \tag{1.43}$$

Since $\mathfrak{H}(\xi', \eta')$ is smooth in $(F^c)' \times (F^c)'$, we deduce

$$h(\xi, \eta) = \mathfrak{H}(0, \eta') - (2\pi)^{-1} \log |\xi||\eta| + O(|\xi'|), \tag{1.44}$$

for $|\xi'| < 1/2$ and for all $\eta' \in (F^c)'$. Also, by (1.43) and the definition of $h(\xi, \eta)$,

$$\mathfrak{H}(\xi', \eta') = -g(\xi, \eta) + (2\pi)^{-1} \log |\xi||\eta| - (2\pi)^{-1} \log |\xi - \eta|. \tag{1.45}$$

Then, applying (1.38) and taking the limit in (1.45), as $|\xi'| \to 0$, we arrive at

$$\mathfrak{H}(0, \eta') = -\zeta(\eta) + (2\pi)^{-1} \lim_{|\xi| \to \infty} \log(|\xi - \eta|^{-1}|\xi|) + (2\pi)^{-1} \log |\eta|$$

$$= (2\pi)^{-1} \log |\eta| - \zeta(\eta).$$

Further substitution of $\mathfrak{H}(0, \eta')$ into (1.44) leads to

$$h(\xi, \eta) = -(2\pi)^{-1} \log |\xi| - \zeta(\eta) + O(|\xi|^{-1}),$$

for $|\xi| > 2$ and for all $\eta \in F^c$. The proof is complete. □

After comparing (1.37) and (1.39) we introduce the constant

$$\zeta_\infty = \lim_{|\boldsymbol{\eta}|\to\infty} \{\zeta(\boldsymbol{\eta}) - (2\pi)^{-1}\log|\boldsymbol{\eta}|\}, \tag{1.46}$$

which will be used in further asymptotic representations.

1.2.1 Asymptotic Approximation of the Capacitary Potential

The *capacitary potential* $P_\varepsilon(\mathbf{x})$ is introduced as a solution of the following Dirichlet problem in Ω_ε

$$\Delta P_\varepsilon(\mathbf{x}) = 0, \quad \mathbf{x} \in \Omega_\varepsilon, \tag{1.47}$$

$$P_\varepsilon(\mathbf{x}) = 0, \quad \mathbf{x} \in \partial\Omega, \tag{1.48}$$

$$P_\varepsilon(\mathbf{x}) = 1, \quad \mathbf{x} \in \partial F_\varepsilon^c. \tag{1.49}$$

Lemma 1.4. *The asymptotic approximation of* $P_\varepsilon(\mathbf{x})$ *is given by the formula*

$$P_\varepsilon(\mathbf{x}) = \frac{-G(\mathbf{x},0) + \zeta(\frac{\mathbf{x}}{\varepsilon}) - \frac{1}{2\pi}\log\frac{|\mathbf{x}|}{\varepsilon} - \zeta_\infty}{\frac{1}{2\pi}\log\varepsilon + H(0,0) - \zeta_\infty} + p_\varepsilon(\mathbf{x}), \tag{1.50}$$

where ζ_∞ *is defined by* (1.46)*, and* p_ε *is the remainder term such that*

$$|p_\varepsilon(\mathbf{x})| \le \text{Const}\,\varepsilon|\log\varepsilon|^{-1}$$

uniformly with respect to $\mathbf{x} \in \Omega_\varepsilon$.

Proof. Direct substitution of (1.50) into (1.47)–(1.49) yields the Dirichlet problem for the remainder term p_ε

$$\Delta p_\varepsilon(\mathbf{x}) = 0, \quad \mathbf{x} \in \Omega_\varepsilon, \tag{1.51}$$

$$p_\varepsilon(\mathbf{x}) = -\frac{\zeta(\varepsilon^{-1}\mathbf{x}) - \frac{1}{2\pi}\log(\varepsilon^{-1}|\mathbf{x}|) - \zeta_\infty}{\frac{1}{2\pi}\log\varepsilon + H(0,0) - \zeta_\infty}, \quad \mathbf{x} \in \partial\Omega, \tag{1.52}$$

$$p_\varepsilon(\mathbf{x}) = 1 - \frac{H(\mathbf{x},0) + \frac{1}{2\pi}\log\varepsilon - \zeta_\infty}{\frac{1}{2\pi}\log\varepsilon + H(0,0) - \zeta_\infty}, \quad \mathbf{x} \in \partial F_\varepsilon^c. \tag{1.53}$$

Using (1.46) and the expansion of $\zeta(\boldsymbol{\xi})$ in spherical harmonics, we deduce

$$\zeta(\varepsilon^{-1}\mathbf{x}) - (2\pi)^{-1}\log(\varepsilon^{-1}|\mathbf{x}|) - \zeta_\infty = O(\varepsilon),$$

as $\mathbf{x} \in \partial\Omega$, and hence the right-hand side in (1.52) is $O(\varepsilon|\log\varepsilon|^{-1})$. Since $H(\mathbf{x}, 0)$ is smooth in Ω, we have

$$H(\mathbf{x}, 0) - H(0, 0) = O(\varepsilon),$$

as $\mathbf{x} \in \partial F_\varepsilon^c$, and therefore the right-hand side in (1.53) is also $O(\varepsilon|\log\varepsilon|^{-1})$. Applying the maximum principle, we arrive at the result of Lemma. □

Remark. For the case when Ω is a Jordan domain and F is the closure of a Jordan domain, we can adopt the notions of [38]: the inner conformal radius r_F of F, with respect to $\mathbf{O}$, and the outer conformal radius R_Ω of Ω, with respect to $\mathbf{O}$, are defined as

$$r_F = \exp(-2\pi\zeta_\infty), \ R_\Omega = \exp(-2\pi H(0,0)),$$

respectively. In this case, the capacitary potential $P_\varepsilon(\mathbf{x})$ can be represented in the form

$$P_\varepsilon(\mathbf{x}) = \frac{-G(\mathbf{x}, 0) + \zeta(\frac{\mathbf{x}}{\varepsilon}) - \frac{1}{2\pi}\log\frac{|\mathbf{x}|}{\varepsilon r_F}}{\frac{1}{2\pi}\log\frac{\varepsilon r_F}{R_\Omega}} + p_\varepsilon(\mathbf{x}).$$

1.2.2 Uniform Asymptotic Approximation

Theorem 1.2. *Green's function G_ε for the operator $-\Delta$ in $\Omega_\varepsilon \subset \mathbb{R}^2$ admits the representation*

$$G_\varepsilon(\mathbf{x}, \mathbf{y}) = G(\mathbf{x}, \mathbf{y}) + g(\varepsilon^{-1}\mathbf{x}, \varepsilon^{-1}\mathbf{y}) + (2\pi)^{-1}\log(\varepsilon^{-1}|\mathbf{x} - \mathbf{y}|)$$
$$+\frac{\Big((2\pi)^{-1}\log\varepsilon + \zeta(\frac{\mathbf{x}}{\varepsilon}) - \zeta_\infty + H(\mathbf{x}, 0)\Big)\Big((2\pi)^{-1}\log\varepsilon + \zeta(\frac{\mathbf{y}}{\varepsilon}) - \zeta_\infty + H(0, \mathbf{y})\Big)}{(2\pi)^{-1}\log\varepsilon + H(0,0) - \zeta_\infty}$$
$$-\zeta(\varepsilon^{-1}\mathbf{x}) - \zeta(\varepsilon^{-1}\mathbf{y}) + \zeta_\infty + O(\varepsilon), \quad (1.54)$$

which is uniform with respect to $(\mathbf{x}, \mathbf{y}) \in \Omega_\varepsilon \times \Omega_\varepsilon$.

Proof. Let

$$G_\varepsilon(\mathbf{x}, \mathbf{y}) = (2\pi)^{-1}\log|\mathbf{x} - \mathbf{y}|^{-1} - H_\varepsilon(\mathbf{x}, \mathbf{y}) - h_\varepsilon(\mathbf{x}, \mathbf{y}), \quad (1.55)$$

where H_ε and h_ε are defined as solutions of the Dirichlet problems

$$\Delta_x H_\varepsilon(\mathbf{x}, \mathbf{y}) = 0, \quad \mathbf{x},\ \mathbf{y} \in \Omega_\varepsilon, \tag{1.56}$$

$$H_\varepsilon(\mathbf{x}, \mathbf{y}) = (2\pi)^{-1} \log |\mathbf{x} - \mathbf{y}|^{-1}, \quad \mathbf{x} \in \partial\Omega,\ \mathbf{y} \in \Omega_\varepsilon, \tag{1.57}$$

$$H_\varepsilon(\mathbf{x}, \mathbf{y}) = 0, \quad \mathbf{x} \in \partial F_\varepsilon,\ \mathbf{y} \in \Omega_\varepsilon, \tag{1.58}$$

and

$$\Delta_x h_\varepsilon(\mathbf{x}, \mathbf{y}) = 0, \quad \mathbf{x},\ \mathbf{y} \in \Omega_\varepsilon, \tag{1.59}$$

$$h_\varepsilon(\mathbf{x}, \mathbf{y}) = 0, \quad \mathbf{x} \in \partial\Omega,\ \mathbf{y} \in \Omega_\varepsilon, \tag{1.60}$$

$$h_\varepsilon(\mathbf{x}, \mathbf{y}) = (2\pi)^{-1} \log |\mathbf{x} - \mathbf{y}|^{-1}, \quad \mathbf{x} \in \partial F_\varepsilon,\ \mathbf{y} \in \Omega_\varepsilon. \tag{1.61}$$

The function H_ε is represented in the form

$$H_\varepsilon(\mathbf{x}, \mathbf{y}) = C(\mathbf{y}, \log\varepsilon) G(\mathbf{x}, 0) + H(\mathbf{x}, \mathbf{y}) + R_\varepsilon(\mathbf{x}, \mathbf{y}, \log\varepsilon), \tag{1.62}$$

where $C(\mathbf{y}, \log\varepsilon)$ is to be determined, G and H are defined by (1.31)–(1.33), and the third term R_ε satisfies the boundary value problem

$$\Delta_x R_\varepsilon(\mathbf{x}, \mathbf{y}, \log\varepsilon) = 0, \quad \mathbf{x},\ \mathbf{y} \in \Omega_\varepsilon, \tag{1.63}$$

$$R_\varepsilon(\mathbf{x}, \mathbf{y}, \log\varepsilon) = 0, \quad \mathbf{x} \in \partial\Omega,\ \mathbf{y} \in \Omega_\varepsilon, \tag{1.64}$$

$$R_\varepsilon(\mathbf{x}, \mathbf{y}, \log\varepsilon) = -C G(\mathbf{x}, 0) - H(\mathbf{x}, \mathbf{y}), \quad \mathbf{x} \in \partial F_\varepsilon,\ \mathbf{y} \in \Omega_\varepsilon, \tag{1.65}$$

and it is approximated by a function $R(\varepsilon^{-1}\mathbf{x}, \mathbf{y}, \log\varepsilon)$ defined in scaled coordinates in such a way that

$$\Delta_\xi R(\boldsymbol{\xi}, \mathbf{y}, \log\varepsilon) = 0, \quad \boldsymbol{\xi} \in F^c, \tag{1.66}$$

$$\begin{aligned} R(\boldsymbol{\xi}, \mathbf{y}, \log\varepsilon) = {} & C(2\pi)^{-1}(\log|\boldsymbol{\xi}| + \log\varepsilon) \\ & + CH(0,0) - H(0, \mathbf{y}), \quad \boldsymbol{\xi} \in \partial F^c, \end{aligned} \tag{1.67}$$

$$R(\boldsymbol{\xi}, \mathbf{y}, \log\varepsilon) \to 0 \text{ as } |\boldsymbol{\xi}| \to \infty, \tag{1.68}$$

where $\mathbf{y} \in \Omega_\varepsilon$. The solution of the above problem has the form

$$\begin{aligned} R(\boldsymbol{\xi}, \mathbf{y}, \log\varepsilon) = {} & -C\{(2\pi)^{-1} \log|\boldsymbol{\xi}|^{-1} + \zeta(\boldsymbol{\xi})\} \\ & + C\{(2\pi)^{-1} \log\varepsilon + H(0,0)\} - H(0, \mathbf{y}), \end{aligned} \tag{1.69}$$

with ζ defined by (1.38).

The condition (1.68) is satisfied provided

$$C(\mathbf{y}, \log\varepsilon) = \frac{H(0, \mathbf{y})}{H(0,0) + \frac{1}{2\pi}\log\varepsilon - \zeta_\infty}. \tag{1.70}$$

Combining (1.69), (1.70), and (1.62), we deduce

$$H_\varepsilon(\mathbf{x},\mathbf{y}) = -H(0,\mathbf{y})P_\varepsilon(\mathbf{x}) + H(\mathbf{x},\mathbf{y}) + \tilde{H}_\varepsilon(\mathbf{x},\mathbf{y}), \tag{1.71}$$

where $\tilde{H}_\varepsilon$ is the remainder term, such that

$$\Delta_x \tilde{H}_\varepsilon(\mathbf{x},\mathbf{y}) = 0, \quad \mathbf{x},\mathbf{y} \in \Omega_\varepsilon, \tag{1.72}$$

$$\tilde{H}_\varepsilon(\mathbf{x},\mathbf{y}) = 0, \quad \mathbf{x} \in \partial\Omega, \ \mathbf{y} \in \Omega_\varepsilon, \tag{1.73}$$

$$\tilde{H}_\varepsilon(\mathbf{x},\mathbf{y}) = H(0,\mathbf{y}) - H(\mathbf{x},\mathbf{y}), \quad \mathbf{x} \in \partial F_\varepsilon, \ \mathbf{y} \in \Omega_\varepsilon, \tag{1.74}$$

where the modulus of the right-hand side in (1.74) is estimated by Const ε, uniformly with respect to $\mathbf{x} \in \partial F_\varepsilon^c$ and $\mathbf{y} \in \Omega_\varepsilon$. The maximum principle leads to the estimate $|\tilde{H}(\mathbf{x},\mathbf{y})| \leq \text{Const}\, \varepsilon$, which is uniform for $\mathbf{x},\mathbf{y} \in \Omega_\varepsilon$.

The approximation of h_ε (see (1.59)–(1.61)) also involves the capacitary potential P_ε from Sect. 1.2.1. The harmonic function h_ε satisfies the homogeneous Dirichlet condition on $\partial\Omega$, and the boundary condition on ∂F_ε^c is rewritten as

$$h_\varepsilon(\mathbf{x},\mathbf{y}) = -(2\pi)^{-1}\log(\varepsilon^{-1}|\mathbf{x}-\mathbf{y}|) - (2\pi)^{-1}\log\varepsilon, \quad \mathbf{x} \in \partial F_\varepsilon^c, \mathbf{y} \in \Omega_\varepsilon.$$

Hence $h_\varepsilon(\mathbf{x},\mathbf{y})$ is sought in the form

$$h_\varepsilon(\mathbf{x},\mathbf{y}) = h(\varepsilon^{-1}\mathbf{x},\varepsilon^{-1}\mathbf{y}) - (2\pi)^{-1}\log\varepsilon + \tilde{h}_\varepsilon^{(1)}(\mathbf{x},\mathbf{y}), \tag{1.75}$$

where the harmonic function $\tilde{h}_\varepsilon^{(1)}$ vanishes when $\mathbf{x} \in \partial F_\varepsilon^c$, $\mathbf{y} \in \Omega_\varepsilon$, and

$$\tilde{h}_\varepsilon^{(1)}(\mathbf{x},\mathbf{y}) = (2\pi)^{-1}\log\varepsilon - h(\varepsilon^{-1}\mathbf{x},\varepsilon^{-1}\mathbf{y}), \quad \mathbf{x} \in \partial\Omega, \mathbf{y} \in \Omega_\varepsilon. \tag{1.76}$$

Representing the right-hand side in (1.76) according to Lemma 1.3, we obtain

$$\tilde{h}_\varepsilon^{(1)}(\mathbf{x},\mathbf{y}) = (2\pi)^{-1}\log|\mathbf{x}| + \zeta(\varepsilon^{-1}\mathbf{y}) + O(\varepsilon),$$

uniformly for $\mathbf{x} \in \partial\Omega, \mathbf{y} \in \Omega_\varepsilon$. Using the capacitary potential P_ε and the definition (1.33) of $H(\mathbf{x},\mathbf{y})$, we write $\tilde{h}_\varepsilon^{(1)}$ as

$$\tilde{h}_\varepsilon^{(1)}(\mathbf{x},\mathbf{y}) = -H(\mathbf{x},0) + \zeta(\varepsilon^{-1}\mathbf{y})(1 - P_\varepsilon(\mathbf{x})) + \tilde{h}_\varepsilon^{(2)}(\mathbf{x},\mathbf{y}), \tag{1.77}$$

where $\tilde{h}_\varepsilon^{(2)}$ is a harmonic function, which is $O(\varepsilon)$ for all $\mathbf{x} \in \partial\Omega, \mathbf{y} \in \Omega_\varepsilon$, and satisfies

$$\tilde{h}_\varepsilon^{(2)}(\mathbf{x},\mathbf{y}) = H(\mathbf{x},0) = H(0,0) + O(\varepsilon),$$

for all $\mathbf{x} \in \partial F_\varepsilon^c, \mathbf{y} \in \Omega_\varepsilon$. Hence,

$$\tilde{h}_\varepsilon^{(2)}(\mathbf{x}, \mathbf{y}) = H(0,0) P_\varepsilon(\mathbf{x}) + O(\varepsilon), \tag{1.78}$$

uniformly with respect to $\mathbf{x}, \mathbf{y} \in \Omega_\varepsilon$.

Combining (1.75), (1.77) and (1.78), we deduce

$$\begin{aligned} h_\varepsilon(\mathbf{x}, \mathbf{y}) = \ & h(\varepsilon^{-1}\mathbf{x}, \varepsilon^{-1}\mathbf{y}) - (2\pi)^{-1} \log \varepsilon - H(\mathbf{x}, 0) \\ & + \zeta(\varepsilon^{-1}\mathbf{y})(1 - P_\varepsilon(\mathbf{x})) + H(0,0) P_\varepsilon(\mathbf{x}) + O(\varepsilon), \end{aligned} \tag{1.79}$$

uniformly with respect to $\mathbf{x}, \mathbf{y} \in \Omega_\varepsilon$.

Furthermore, it follows from (1.55), (1.71) and (1.79) that Green's function G_ε admits the representation

$$\begin{aligned} G_\varepsilon(\mathbf{x}, \mathbf{y}) = \ & (2\pi)^{-1} \log |\mathbf{x} - \mathbf{y}|^{-1} - H(\mathbf{x}, \mathbf{y}) - h(\varepsilon^{-1}\mathbf{x}, \varepsilon^{-1}\mathbf{y}) \\ & + (2\pi)^{-1} \log \varepsilon - \zeta(\boldsymbol{\eta}) + H(\mathbf{x}, 0) \\ & - P_\varepsilon(\mathbf{x})(H(0,0) - H(0, \mathbf{y}) - \zeta(\varepsilon^{-1}\mathbf{y})) + O(\varepsilon), \end{aligned} \tag{1.80}$$

which is uniform with respect to $\mathbf{x}, \mathbf{y} \in \Omega_\varepsilon$.

By Lemma 1.4, (1.80) takes the form

$$\begin{aligned} G_\varepsilon(\mathbf{x}, \mathbf{y}) \ & = (2\pi)^{-1} \log |\mathbf{x} - \mathbf{y}|^{-1} - H(\mathbf{x}, \mathbf{y}) - h(\varepsilon^{-1}\mathbf{x}, \varepsilon^{-1}\mathbf{y}) \\ & + \frac{(H(0,0) - H(\mathbf{x}, 0) - \zeta(\varepsilon^{-1}\mathbf{x}))(H(0,0) - H(0, \mathbf{y}) - \zeta(\varepsilon^{-1}\mathbf{y}))}{\frac{1}{2\pi} \log \varepsilon + H(0,0) - \zeta_\infty} \\ & + (2\pi)^{-1} \log \varepsilon + H(\mathbf{x}, 0) + H(0, \mathbf{y}) - H(0,0) + O(\varepsilon). \end{aligned} \tag{1.81}$$

Also with the use of Lemma 1.4, for all $\mathbf{x}, \mathbf{y} \in \Omega_\varepsilon$, the above formula can be written as

$$\begin{aligned} G_\varepsilon(\mathbf{x}, \mathbf{y}) = \ & (2\pi)^{-1} \log |\mathbf{x} - \mathbf{y}|^{-1} - H(\mathbf{x}, \mathbf{y}) - h(\varepsilon^{-1}\mathbf{x}, \varepsilon^{-1}\mathbf{y}) \\ & + ((2\pi)^{-1} \log \varepsilon + H(0,0) - \zeta_\infty)(1 - P_\varepsilon(\mathbf{x}))(1 - P_\varepsilon(\mathbf{y})) \\ & + (2\pi)^{-1} \log \varepsilon + H(\mathbf{x}, 0) + H(0, \mathbf{y}) - H(0,0) + O(\varepsilon) \\ = \ & (2\pi)^{-1} \log |\mathbf{x} - \mathbf{y}|^{-1} - H(\mathbf{x}, \mathbf{y}) - h(\varepsilon^{-1}\mathbf{x}, \varepsilon^{-1}\mathbf{y}) \\ & + ((2\pi)^{-1} \log \varepsilon + H(0,0) - \zeta_\infty) P_\varepsilon(\mathbf{x}) P_\varepsilon(\mathbf{y}) \\ & - \zeta(\varepsilon^{-1}\mathbf{x}) - \zeta(\varepsilon^{-1}\mathbf{y}) + \zeta_\infty + O(\varepsilon), \end{aligned} \tag{1.82}$$

which is equivalent to (1.54). The proof is complete. □

1.3 Corollaries

The asymptotic formulae of Sects. 1.1 and 1.2 can be simplified under constraints on positions of the points $\mathbf{x}, \mathbf{y}$ within Ω_ε.

Corollary 1.1. *(a) Let* $\mathbf{x}$ *and* $\mathbf{y}$ *be points of* $\Omega_\varepsilon \subset \mathbb{R}^n, n > 2$, *such that*

$$\min\{|\mathbf{x}|, |\mathbf{y}|\} > 2\varepsilon. \tag{1.83}$$

Then

$$G_\varepsilon(\mathbf{x}, \mathbf{y}) = G(\mathbf{x}, \mathbf{y}) \ -\varepsilon^{n-2}\text{cap}(F)\, G(\mathbf{x}, 0)G(0, \mathbf{y}) + O\Big(\frac{\varepsilon^{n-1}}{(|\mathbf{x}||\mathbf{y}|)^{n-2}\min\{|\mathbf{x}|, |\mathbf{y}|\}}\Big). \tag{1.84}$$

(b) If $\max\{|\mathbf{x}|, |\mathbf{y}|\} < 1/2$, *then*

$$G_\varepsilon(\mathbf{x}, \mathbf{y}) = \varepsilon^{2-n} g(\varepsilon^{-1}\mathbf{x}, \varepsilon^{-1}\mathbf{y})$$

$$- \mathcal{H}(0, 0)(P(\varepsilon^{-1}\mathbf{x}) - 1)(P(\varepsilon^{-1}\mathbf{y}) - 1) + O(\max\{|\mathbf{x}|, |\mathbf{y}|\}). \tag{1.85}$$

Both (1.84) *and* (1.85) *are uniform with respect to* ε *and* $(\mathbf{x}, \mathbf{y}) \in \Omega_\varepsilon \times \Omega_\varepsilon$.

Proof. (a) The formula (1.17) is equivalent to

$$G_\varepsilon(\mathbf{x}, \mathbf{y}) = G(\mathbf{x}, \mathbf{y}) - \varepsilon^{2-n} h(\varepsilon^{-1}\mathbf{x}, \varepsilon^{-1}\mathbf{y}) \tag{1.86}$$

$$+\mathcal{H}(0, \mathbf{y}) P(\varepsilon^{-1}\mathbf{x}) + \mathcal{H}(\mathbf{x}, 0) P(\varepsilon^{-1}\mathbf{y}) - \mathcal{H}(0, 0) P(\varepsilon^{-1}\mathbf{x}) P(\varepsilon^{-1}\mathbf{y})$$

$$-\varepsilon^{n-2}\text{cap}(F)\, \mathcal{H}(\mathbf{x}, 0)\mathcal{H}(0, \mathbf{y}) + O\Big(\frac{\varepsilon^{n-1}}{(\min\{|\mathbf{x}|, |\mathbf{y}|\}^{n-2})}\Big).$$

By Lemmas 1.1 and 1.2

$$P(\varepsilon^{-1}\mathbf{x}) = \frac{\varepsilon^{n-2}\,\text{cap}(F)}{(n-2)|S^{n-1}||\mathbf{x}|^{n-2}} + O\Big(\frac{\varepsilon^{n-1}}{|\mathbf{x}|^{n-1}}\Big), \tag{1.87}$$

and

$$\varepsilon^{2-n} h(\varepsilon^{-1}\mathbf{x}, \varepsilon^{-1}\mathbf{y}) = \frac{P(\varepsilon^{-1}\mathbf{y})}{(n-2)|S^{n-1}||\mathbf{x}|^{n-2}} + O\Big(\frac{\varepsilon^{n-1}}{|\mathbf{x}|^{n-1}|\mathbf{y}|^{n-2}}\Big) \tag{1.88}$$

$$= \frac{\varepsilon^{n-2}\text{cap}(F)}{((n-2)|S^{n-1}|)^2|\mathbf{x}|^{n-2}|\mathbf{y}|^{n-2}} + O\Big(\frac{\varepsilon^{n-1}}{(|\mathbf{x}||\mathbf{y}|)^{n-2}\min\{|\mathbf{x}|, |\mathbf{y}|\}}\Big).$$

Direct substitution of (1.88) and (1.87) into (1.86) leads to

$$\begin{aligned}
G_\varepsilon(\mathbf{x},\mathbf{y}) = {} & G(\mathbf{x},\mathbf{y}) - \frac{\varepsilon^{n-2}\mathrm{cap}(F)}{(n-2)^2|S^{n-1}|^2|\mathbf{x}|^{n-2}|\mathbf{y}|^{n-2}} \\
& + \varepsilon^{n-2}\mathrm{cap}(F)\Big(\frac{\mathcal{H}(0,\mathbf{y})}{(n-2)|S^{n-1}||\mathbf{x}|^{n-2}} + \frac{\mathcal{H}(\mathbf{x},0)}{(n-2)|S^{n-1}||\mathbf{y}|^{n-2}} \\
& - \mathcal{H}(\mathbf{x},0)\mathcal{H}(0,\mathbf{y})\Big) + O\Big(\frac{\varepsilon^{n-1}}{(|\mathbf{x}||\mathbf{y}|)^{n-2}\min\{|\mathbf{x}|,|\mathbf{y}|\}}\Big) \\
= {} & G(\mathbf{x},\mathbf{y}) - \varepsilon^{n-2}\mathrm{cap}(F)\Big((n-2)^{-1}|S^{n-1}|^{-1}|\mathbf{x}|^{2-n} - \mathcal{H}(\mathbf{x},0)\Big) \\
& \times\Big((n-2)^{-1}|S^{n-1}|^{-1}|\mathbf{y}|^{2-n} - \mathcal{H}(0,\mathbf{y})\Big) \\
& + O\Big(\frac{\varepsilon^{n-1}}{(|\mathbf{x}||\mathbf{y}|)^{n-2}\min\{|\mathbf{x}|,|\mathbf{y}|\}}\Big),
\end{aligned}$$

which is equivalent to (1.84).

(b) Since $\mathcal{H}(\mathbf{x},\mathbf{y})$ is smooth in the vicinity of $(\mathbf{O},\mathbf{O})$ formula (1.17) can be presented in the form

$$\begin{aligned}
G_\varepsilon(\mathbf{x},\mathbf{y}) = {} & \varepsilon^{2-n} g(\varepsilon^{-1}\mathbf{x},\varepsilon^{-1}\mathbf{y}) - \mathcal{H}(0,0) \\
& + (\mathcal{H}(0,0) + O(|\mathbf{y}|))P(\varepsilon^{-1}\mathbf{x}) + (\mathcal{H}(0,0) + O(|\mathbf{x}|))P(\varepsilon^{-1}\mathbf{y}) \\
& - \mathcal{H}(0,0)P(\varepsilon^{-1}\mathbf{x})P(\varepsilon^{-1}\mathbf{y}) + O(\max\{|\mathbf{x}|,|\mathbf{y}|\}),
\end{aligned}$$

which is equivalent to (1.85). The proof is complete. □

Asymptotic formulae, similar to (1.84), are also presented in [36].

Next, we give an analogue of Corollary 1.1 for the planar case.

Corollary 1.2. *(a) Let* $\mathbf{x}$ *and* $\mathbf{y}$ *be points of* $\Omega_\varepsilon \subset \mathbb{R}^2$ *subject to* (1.83). *Then*

$$G_\varepsilon(\mathbf{x},\mathbf{y}) = G(\mathbf{x},\mathbf{y}) + \frac{G(\mathbf{x},0)G(0,\mathbf{y})}{\frac{1}{2\pi}\log\varepsilon + H(0,0) - \zeta_\infty} + O\Big(\frac{\varepsilon}{\min\{|\mathbf{x}|,|\mathbf{y}|\}}\Big). \tag{1.89}$$

(b) If $\max\{|\mathbf{x}|,|\mathbf{y}|\} < 1/2$, *then*

$$\begin{aligned}
G_\varepsilon(\mathbf{x},\mathbf{y}) = {} & g(\varepsilon^{-1}\mathbf{x},\varepsilon^{-1}\mathbf{y}) \\
& + \frac{\zeta(\varepsilon^{-1}\mathbf{x})\zeta(\varepsilon^{-1}\mathbf{y})}{\frac{1}{2\pi}\log\varepsilon + H(0,0) - \zeta_\infty} + O(\max\{|\mathbf{x}|,|\mathbf{y}|\}),
\end{aligned} \tag{1.90}$$

Both (1.89) *and* (1.90) *are uniform with respect to* ε *and* $(\mathbf{x},\mathbf{y}) \in \Omega_\varepsilon \times \Omega_\varepsilon$.

Proof. *(a)* Formula (1.54) can be written as

$$G_\varepsilon(\mathbf{x},\mathbf{y}) = (2\pi)^{-1}\log|\mathbf{x}-\mathbf{y}|^{-1} - H(\mathbf{x},\mathbf{y}) - h(\boldsymbol{\xi},\boldsymbol{\eta})$$
$$+\frac{(G(\mathbf{x},0)-\zeta(\boldsymbol{\xi})+\frac{1}{2\pi}\log|\boldsymbol{\xi}|+\zeta_\infty)(G(0,\mathbf{y})-\zeta(\boldsymbol{\eta})+\frac{1}{2\pi}\log|\boldsymbol{\eta}|+\zeta_\infty)}{\frac{1}{2\pi}\log\varepsilon+H(0,0)-\zeta_\infty}$$
$$-\zeta(\boldsymbol{\xi})-\zeta(\boldsymbol{\eta})+\zeta_\infty+O(\varepsilon). \tag{1.91}$$

It follows from Lemma 1.3 and definition (1.38) that

$$h(\boldsymbol{\xi},\boldsymbol{\eta}) = -(2\pi)^{-1}\log|\boldsymbol{\xi}| - \zeta(\boldsymbol{\eta}) + O(\varepsilon/|\mathbf{x}|), \tag{1.92}$$

and

$$\zeta(\boldsymbol{\xi}) = (2\pi)^{-1}\log|\boldsymbol{\xi}| + \zeta_\infty + O(\varepsilon/|\mathbf{x}|). \tag{1.93}$$

Direct substitution of (1.92) and (1.93) into (1.91) yields

$$G_\varepsilon(\mathbf{x},\mathbf{y}) = (2\pi)^{-1}\log|\mathbf{x}-\mathbf{y}|^{-1} - H(\mathbf{x},\mathbf{y})$$
$$+\frac{(-G(\mathbf{x},0)+O(\varepsilon/|\mathbf{x}|))(-G(0,\mathbf{y})+O(\varepsilon/|\mathbf{y}|))}{\frac{1}{2\pi}\log\varepsilon+H(0,0)-\zeta_\infty}$$
$$+O\left(\frac{\varepsilon}{\min\{|\mathbf{x}|,|\mathbf{y}|\}}\right), \tag{1.94}$$

and hence we arrive at (1.89).

(*b*) When $\max\{|\mathbf{x}|,|\mathbf{y}|\} < 1/2$, (1.54) is presented in the form:

$$G_\varepsilon(\mathbf{x},\mathbf{y}) = g(\varepsilon^{-1}\mathbf{x},\varepsilon^{-1}\mathbf{y}) - H(\mathbf{x},\mathbf{y})$$
$$+\frac{(H(0,0)-H(\mathbf{x},0)-\zeta(\varepsilon^{-1}\mathbf{x}))(H(0,0)-H(0,\mathbf{y})-\zeta(\varepsilon^{-1}\mathbf{y}))}{\frac{1}{2\pi}\log\varepsilon+H(0,0)-\zeta_\infty}$$
$$+H(\mathbf{x},0)+H(0,\mathbf{y})-H(0,0)+O(\varepsilon)$$

(compare with (1.82)). Since $H(\mathbf{x},\mathbf{y})$ is smooth in a vicinity of $(\mathbf{O},\mathbf{O})$, we obtain

$$G_\varepsilon(\mathbf{x},\mathbf{y}) = g(\varepsilon^{-1}\mathbf{x},\varepsilon^{-1}\mathbf{y}) + \frac{(-\zeta(\varepsilon^{-1}\mathbf{x})+O(|\mathbf{x}|))(-\zeta(\varepsilon^{-1}\mathbf{y})+O(|\mathbf{y}|))}{\frac{1}{2\pi}\log\varepsilon+H(0,0)-\zeta_\infty}$$
$$+O(\max\{|\mathbf{x}|,|\mathbf{y}|\})$$
$$= g(\varepsilon^{-1}\mathbf{x},\varepsilon^{-1}\mathbf{y})$$
$$+\frac{\zeta(\varepsilon^{-1}\mathbf{x})\zeta(\varepsilon^{-1}\mathbf{y})+O(|\mathbf{y}|\log(|\mathbf{x}|/\varepsilon))+O(|\mathbf{x}|\log(|\mathbf{y}|/\varepsilon))}{\frac{1}{2\pi}\log\varepsilon+H(0,0)-\zeta_\infty}$$
$$+O(\max\{|\mathbf{x}|,|\mathbf{y}|\}),$$

which implies (1.90). □

Chapter 2
Mixed and Neumann Boundary Conditions for Domains with Small Holes and Inclusions: Uniform Asymptotics of Green's Kernels

In this chapter, we derive and justify asymptotic approximations of Green's kernels for singularly perturbed domains whose boundary, or some part of it, supports the *Neumann boundary condition.* We also derive simpler asymptotic formulae, which become efficient when certain constraints are imposed on the independent variables.

Sections 2.1 and 2.2 deal with the Dirichlet–Neumann problems in two-dimensional domains with small holes, inclusions or cracks. Section 2.3 gives the uniform approximation of Green's function for the Neumann problem in the domain of the same type. Finally, in Sect. 2.4 we formulate similar asymptotic approximations of Green's kernels in three-dimensional domains with small holes or small inclusions.

2.1 Mixed Boundary Value Problem in a Planar Domain with a Small Hole or a Crack

Let Ω be a bounded domain in $\mathbb{R}^2$, which contains the origin $\mathbf{O}$, and let F be a compact set in $\mathbb{R}^2$, $\mathbf{O} \in F$. We suppose that the boundary $\partial\Omega$ is smooth. This constraint is not essential and can be considerably weakened. We assume, without loss of generality, that $\operatorname{diam} F = 1/2$, and that $\operatorname{dist}(\mathbf{O}, \partial\Omega) = 1$. We also introduce the set $F_\varepsilon = \{\mathbf{x} : \varepsilon^{-1}\mathbf{x} \in F\}$, with ε being a small positive parameter. The boundary ∂F is required to be piecewise smooth, with the angle openings from the side of $\mathbb{R}^2 \setminus F$ belonging to $(0, 2\pi]$. In the case of a crack, ∂F and ∂F_ε are treated as two-sided. We assume that $\Omega_\varepsilon = \Omega \setminus F_\varepsilon$ is connected, and in the sequel we refer to it as a domain with a small hole (or possibly a small crack).

Let $G_\varepsilon^{(N)}$ denote Green's function of the operator $-\Delta$, with the Neumann data on ∂F_ε and the Dirichlet data on $\partial\Omega$. In other words, $G_\varepsilon^{(N)}$ is a solution of the problem

V. Maz'ya et al., *Green's Kernels and Meso-Scale Approximations in Perforated Domains*, Lecture Notes in Mathematics 2077, DOI 10.1007/978-3-319-00357-3_2,

$$\Delta_x G_\varepsilon^{(N)}(\mathbf{x},\mathbf{y}) + \delta(\mathbf{x}-\mathbf{y}) = 0, \quad \mathbf{x},\mathbf{y} \in \Omega_\varepsilon, \tag{2.1}$$

$$G_\varepsilon^{(N)}(\mathbf{x},\mathbf{y}) = 0, \quad \mathbf{x} \in \partial\Omega, \ \mathbf{y} \in \Omega_\varepsilon, \tag{2.2}$$

$$\frac{\partial G_\varepsilon^{(N)}}{\partial n_x}(\mathbf{x},\mathbf{y}) = 0, \quad \mathbf{x} \in \partial F_\varepsilon, \ \mathbf{y} \in \Omega_\varepsilon. \tag{2.3}$$

Here and elsewhere *the Neumann condition is understood in the variational sense*, (see Sect. 4.10 in Courant and Hilbert [6]).

In this section, we construct an asymptotic approximation of $G_\varepsilon^{(N)}(\mathbf{x},\mathbf{y})$, uniform with respect to $\mathbf{x}$ and $\mathbf{y}$ in Ω_ε.

2.1.1 Special Solutions of Model Problems

While constructing the asymptotic approximation of $G_\varepsilon^{(N)}$, we use the variational solutions $G(\mathbf{x},\mathbf{y}), \mathcal{D}(\varepsilon^{-1}\mathbf{x}), \zeta(\varepsilon^{-1}\mathbf{x})$ and $\mathcal{N}(\varepsilon^{-1}\mathbf{x},\varepsilon^{-1}\mathbf{y})$ of certain model problems in the limit domains Ω and $\mathbb{R}^2 \setminus F$. It is standard that all solutions, introduced in this subsection, exist and are unique. We describe these solutions.

1. Let G be *Green's function for the Dirichlet problem in* Ω:

$$G(\mathbf{x},\mathbf{y}) = (2\pi)^{-1}\log|\mathbf{x}-\mathbf{y}|^{-1} - H(\mathbf{x},\mathbf{y}), \tag{2.4}$$

where H is the regular part of G, i.e. a unique solution of the Dirichlet problem

$$\Delta_x H(\mathbf{x},\mathbf{y}) = 0, \quad \mathbf{x},\mathbf{y} \in \Omega, \tag{2.5}$$

$$H(\mathbf{x},\mathbf{y}) = (2\pi)^{-1}\log|\mathbf{x}-\mathbf{y}|^{-1}, \quad \mathbf{x} \in \partial\Omega, \ \mathbf{y} \in \Omega. \tag{2.6}$$

2. We introduce the scaled coordinates $\boldsymbol{\xi} = \varepsilon^{-1}\mathbf{x}$ and $\boldsymbol{\eta} = \varepsilon^{-1}\mathbf{y}$. The notation ζ is used for a unique special solution of the Dirichlet problem:

$$\Delta\zeta(\boldsymbol{\xi}) = 0 \ \text{in} \ \mathbb{R}^2 \setminus F, \tag{2.7}$$

$$\zeta(\boldsymbol{\xi}) = 0 \ \text{for} \ \boldsymbol{\xi} \in \partial F, \tag{2.8}$$

$$\zeta(\boldsymbol{\xi}) = (2\pi)^{-1}\log|\boldsymbol{\xi}| + \zeta_\infty + O(|\boldsymbol{\xi}|^{-1}), \ \text{as} \ |\boldsymbol{\xi}| \to \infty, \tag{2.9}$$

where ζ_∞ is constant.

Also, it can be shown that ζ is the limit of Green's function $\mathcal{G}$ of the exterior Dirichlet problem in $\mathbb{R}^2 \setminus F$

$$\zeta(\boldsymbol{\eta}) = \lim_{|\boldsymbol{\xi}|\to\infty} \mathcal{G}(\boldsymbol{\xi},\boldsymbol{\eta}), \tag{2.10}$$

where

$$\Delta_\xi \mathcal{G}(\boldsymbol{\xi},\boldsymbol{\eta}) + \delta(\boldsymbol{\xi}-\boldsymbol{\eta}) = 0, \quad \boldsymbol{\xi},\boldsymbol{\eta} \in \mathbb{R}^2 \setminus F, \tag{2.11}$$

$$\mathcal{G}(\boldsymbol{\xi},\boldsymbol{\eta}) = 0, \quad \boldsymbol{\xi} \in \partial F, \ \boldsymbol{\eta} \in \mathbb{R}^2 \setminus F, \tag{2.12}$$

$$\mathcal{G}(\boldsymbol{\xi},\boldsymbol{\eta}) \ \text{is bounded as } |\boldsymbol{\xi}| \to \infty \text{ and } \boldsymbol{\eta} \in \mathbb{R}^2 \setminus F. \tag{2.13}$$

Representation (2.10) follows from Green's formula applied to ζ and $\mathcal{G}$. Here and elsewhere $B_R = \{\mathbf{X} \in \mathbb{R}^2 : \ |\mathbf{X}| < R\}$. We derive

$$\begin{aligned}
\zeta(\boldsymbol{\eta}) &= -\lim_{R\to\infty} \int_{B_R\setminus F} \zeta(\boldsymbol{\xi})\Delta_\xi \mathcal{G}(\boldsymbol{\xi},\boldsymbol{\eta})d\boldsymbol{\xi} \\
&= \lim_{R\to\infty} \int_{|\xi|=R} \left(\mathcal{G}(\boldsymbol{\xi},\boldsymbol{\eta}) \frac{\partial \zeta(\boldsymbol{\xi})}{\partial|\boldsymbol{\xi}|} - \zeta(\boldsymbol{\xi}) \frac{\partial \mathcal{G}(\boldsymbol{\xi},\boldsymbol{\eta})}{\partial |\boldsymbol{\xi}|} \right) dS_\xi \\
&= (2\pi)^{-1} \lim_{R\to\infty} \int_{|\boldsymbol{\xi}|=R} \mathcal{G}(\boldsymbol{\xi},\boldsymbol{\eta})|\boldsymbol{\xi}|^{-1} dS_\xi = \mathcal{G}(\infty,\boldsymbol{\eta}),
\end{aligned} \tag{2.14}$$

which yields (2.10).

3. Let $\mathcal{N}(\boldsymbol{\xi},\boldsymbol{\eta})$ be *the Neumann function* in $\mathbb{R}^2 \setminus F$ defined by

$$\mathcal{N}(\boldsymbol{\xi},\boldsymbol{\eta}) = (2\pi)^{-1} \log|\boldsymbol{\xi}-\boldsymbol{\eta}|^{-1} - h_N(\boldsymbol{\xi},\boldsymbol{\eta}), \tag{2.15}$$

where h_N is the regular part of $\mathcal{N}$ subject to

$$\Delta_\xi h_N(\boldsymbol{\xi},\boldsymbol{\eta}) = 0, \quad \boldsymbol{\xi},\boldsymbol{\eta} \in \mathbb{R}^2 \setminus F, \tag{2.16}$$

$$\frac{\partial h_N}{\partial n_\xi}(\boldsymbol{\xi},\boldsymbol{\eta}) = \frac{1}{2\pi}\frac{\partial}{\partial n_\xi}(\log|\boldsymbol{\xi}-\boldsymbol{\eta}|^{-1}), \quad \boldsymbol{\xi} \in \partial F, \ \boldsymbol{\eta} \in \mathbb{R}^2 \setminus F, \tag{2.17}$$

$$h_N(\boldsymbol{\xi},\boldsymbol{\eta}) \to 0, \ \text{ as } |\boldsymbol{\xi}| \to \infty, \ \boldsymbol{\eta} \in \mathbb{R}^2 \setminus F. \tag{2.18}$$

We note that the Neumann function $\mathcal{N}$ used here, is symmetric. This follows from Green's formula applied to $U(\mathbf{X}) := \mathcal{N}(\mathbf{X},\boldsymbol{\xi})$ and $V(\mathbf{X}) := \mathcal{N}(\mathbf{X},\boldsymbol{\eta})$, where $\boldsymbol{\xi}$ and $\boldsymbol{\eta}$ are arbitrary fixed points in $\mathbb{R}^2 \setminus F$. We have

$$\begin{aligned}
U(\boldsymbol{\eta}) - V(\boldsymbol{\xi}) &= \lim_{R\to\infty} \int_{B_R\setminus F} \Big\{ V(\mathbf{X})\Delta_X U(\mathbf{X}) - U(\mathbf{X})\Delta_X V(\mathbf{X}) \Big\} d\mathbf{X} \\
&= \lim_{R\to\infty} \int_{|\mathbf{X}|=R} \{ V(\mathbf{X}) \frac{\partial}{\partial|\mathbf{X}|} U(\mathbf{X}) - U(\mathbf{X}) \frac{\partial}{\partial|\mathbf{X}|} V(\mathbf{X}) \} dS_X \\
= -\lim_{R\to\infty} (4\pi^2 R)^{-1} & \int_{|\mathbf{X}|=R} \Big\{ (\log|\mathbf{X}-\boldsymbol{\eta}|^{-1} + O(R^{-1})) \Big(\frac{\mathbf{X}\cdot(\mathbf{X}-\boldsymbol{\xi})}{|\mathbf{X}-\boldsymbol{\xi}|^2} + O(R^{-2}) \Big) \\
&- (\log|\mathbf{X}-\boldsymbol{\xi}|^{-1} + O(R^{-1})) \Big(\frac{\mathbf{X}\cdot(\mathbf{X}-\boldsymbol{\eta})}{|\mathbf{X}-\boldsymbol{\eta}|^2} + O(R^{-2}) \Big) \Big\} dS_x = 0.
\end{aligned}$$

Thus,

$$0 = U(\boldsymbol{\eta}) - V(\boldsymbol{\xi}) = \mathcal{N}(\boldsymbol{\eta}, \boldsymbol{\xi}) - \mathcal{N}(\boldsymbol{\xi}, \boldsymbol{\eta}).$$

4. The *vector of dipole fields* $\boldsymbol{\mathcal{D}}(\boldsymbol{\xi}) = (\mathcal{D}_1(\boldsymbol{\xi}), \mathcal{D}_2(\boldsymbol{\xi}))^T$ is a solution of the exterior Neumann problem

$$\Delta \boldsymbol{\mathcal{D}}(\boldsymbol{\xi}) = 0 \text{ in } \mathbb{R}^2 \setminus F, \tag{2.19}$$

$$\frac{\partial \mathcal{D}_j}{\partial n}(\boldsymbol{\xi}) = n_j \text{ for } \boldsymbol{\xi} \in \partial F, \ j = 1, 2, \tag{2.20}$$

$$\mathcal{D}_j(\boldsymbol{\xi}) \to 0 \text{ as } |\boldsymbol{\xi}| \to \infty, \ j = 1, 2, \tag{2.21}$$

were n_1, n_2 are components of the unit normal on ∂F.

2.1.2 *The Dipole Matrix $\mathcal{P}$*

The dipole fields $\mathcal{D}_j$, $j = 1, 2$, defined in (2.19)–(2.21), allow for the asymptotic representation (see, for example, [38])

$$\mathcal{D}_j(\boldsymbol{\xi}) = \frac{1}{2\pi} \sum_{k=1}^{2} \frac{\mathcal{P}_{jk} \xi_k}{|\boldsymbol{\xi}|^2} + O(|\boldsymbol{\xi}|^{-2}), \tag{2.22}$$

where $|\boldsymbol{\xi}| > 2$, and $\mathcal{P} = (\mathcal{P}_{jk})_{j,k=1}^2$ is the *dipole matrix*.

The symmetry of $\mathcal{P}$ can be verified as follows. Let B_R be a disk of sufficiently large radius R, centered at the origin. We apply Green's formula to $\xi_j - \mathcal{D}_j(\boldsymbol{\xi})$ and $\mathcal{D}_k(\boldsymbol{\xi})$ in $B_R \setminus F$, and deduce

$$\int_{\partial B_R} \Big\{ (\xi_j - \mathcal{D}_j(\boldsymbol{\xi})) \frac{\partial \mathcal{D}_k(\boldsymbol{\xi})}{\partial |\boldsymbol{\xi}|} - \mathcal{D}_k(\boldsymbol{\xi}) \frac{\partial}{\partial |\boldsymbol{\xi}|} (\xi_j - \mathcal{D}_j(\boldsymbol{\xi})) \Big\} dS$$
$$= - \int_{\partial F} (\xi_j - \mathcal{D}_j(\boldsymbol{\xi})) \frac{\partial \mathcal{D}_k(\boldsymbol{\xi})}{\partial n} dS, \tag{2.23}$$

where $\partial/\partial n$ is the normal derivative in the direction of the interior normal with respect to F. In the limit, as $R \to \infty$, the integral in the left-hand side of (2.23) tends to $-\mathcal{P}_{kj}$, whereas the integral in the right-hand side becomes

$$- \int_{\partial F} \xi_j \frac{\partial \xi_k}{\partial n} dS + \int_{\partial F} \mathcal{D}_j(\boldsymbol{\xi}) \frac{\partial \mathcal{D}_k(\boldsymbol{\xi})}{\partial n} dS$$
$$= \delta_{jk} \text{meas}(F) + \int_{\mathbb{R}^2 \setminus F} \nabla \mathcal{D}_j(\boldsymbol{\xi}) \cdot \nabla \mathcal{D}_k(\boldsymbol{\xi}) \, d\boldsymbol{\xi},$$

where meas(F) stands for the two-dimensional Lebesgue measure of the set F. Thus, the representation for components of the dipole matrix takes the form

$$\mathcal{P}_{kj} = -\delta_{jk}\text{meas}(F) - \int_{\mathbb{R}^2\setminus F} \nabla\mathcal{D}_j(\boldsymbol{\xi}) \cdot \nabla\mathcal{D}_k(\boldsymbol{\xi})\, d\boldsymbol{\xi}, \tag{2.24}$$

which implies that the *dipole matrix* $\boldsymbol{\mathcal{P}}$ *for the hole* F *is symmetric and negative definite.*

2.1.3 Pointwise Estimate of a Solution to the Exterior Neumann Problem

In this subsection, we make use of the function spaces $L_2^1(\mathbb{R}^2 \setminus F)$, $W_p^1(\mathbb{R}^2 \setminus F)$ and $W_p^{-1/p}(\partial F)$. The first of them is the space of distributions whose gradients belong to $L_2(\mathbb{R}^2 \setminus F)$. The second one is the usual Sobolev's space consisting of functions in $L_p(\mathbb{R}^2 \setminus F)$ with distributional first derivatives in $L_p(\mathbb{R}^2 \setminus F)$. Finally, $W_p^{-1/p}(\partial F)$ stands for the dual of the space of traces on ∂F of functions in $W_{p'}^1(\mathbb{R}^2 \setminus F)$, $p + p' = pp'$.

The following pointwise estimate will be used repeatedly in the sequel.

Lemma 2.1. *Let* $U \in L_2^1(\mathbb{R}^2 \setminus F)$ *be a solution of the exterior Neumann problem*

$$\Delta U(\boldsymbol{\xi}) = 0, \quad \boldsymbol{\xi} \in \mathbb{R}^2 \setminus F, \tag{2.25}$$

$$\frac{\partial U}{\partial n}(\boldsymbol{\xi}) = \varphi(\boldsymbol{\xi}), \quad \boldsymbol{\xi} \in \partial F, \tag{2.26}$$

$$U(\boldsymbol{\xi}) \to 0 \ \textit{as} \ |\boldsymbol{\xi}| \to \infty, \tag{2.27}$$

where $\partial/\partial n$ *is the normal derivative on* ∂F*, outward with respect to* $\mathbb{R}^2 \setminus F$*, and* $\varphi \in L_\infty(\partial F)$*,*

$$\int_{\partial F} \varphi(\boldsymbol{\xi}) dS = 0. \tag{2.28}$$

We also assume that

$$\int_{\partial F} U(\boldsymbol{\xi}) \frac{\partial \zeta}{\partial n}(\boldsymbol{\xi}) dS = 0, \tag{2.29}$$

where ζ *is the same as in* (2.10). *Then*

$$\sup_{\boldsymbol{\xi}\in\mathbb{R}^2\setminus F} \{(|\boldsymbol{\xi}| + 1)|U(\boldsymbol{\xi})|\} \le C\, \|\varphi\|_{L_\infty(\partial F)}, \tag{2.30}$$

where C *is a constant depending on* ∂F.

Proof. Let B_r denote the disk of radius r centered at $\mathbf{O}$ and let $W_2^1(B_r \setminus F)$ be the space of restrictions of functions in $W_2^1(\mathbb{R}^2 \setminus F)$ to $B_r \setminus F$. By the W_p^1 local coercivity result by Maz'ya and Plamenevskii [31], $U \in W_p^1(B_2 \setminus F)$ for any $p \in (1, 4)$, and

$$\|U\|_{W_p^1(B_2\setminus F)} \le C\Big(\|\varphi\|_{W_p^{-1/p}(\partial F)} + \|U\|_{L_2(B_3\setminus F)}\Big). \tag{2.31}$$

The first term in the right-hand side of (2.31) satisfies

$$\|\varphi\|_{W_p^{-1/p}(\partial F)} \le C\|\varphi\|_{L_\infty(\partial F)}. \tag{2.32}$$

It follows from (2.25) and (2.26) that

$$\|\nabla U\|^2_{L_2(\mathbb{R}^2\setminus F)} = \int_{\partial F} U(\xi)\varphi(\xi)dS \le \|U\|_{L_2(\partial F)}\|\varphi\|_{L_2(\partial F)}. \tag{2.33}$$

Note that by Sobolev's trace theorem

$$\|U\|_{L_q(\partial F)} \le C\|U\|_{W_2^1(B_2\setminus F)} \tag{2.34}$$

for any $q < \infty$ (see, for instance, Theorem 1.4.5 in [22]). It follows from our assumptions on F that

$$\left|\frac{\partial\zeta(\xi)}{\partial n}\right| \le C(\delta(\xi))^{-1/2}, \tag{2.35}$$

where $\delta(\xi)$ is the distance from $\xi \in \partial F$ to the nearest angle vertex on ∂F. Hence

$$\left|\int_{\partial F} U(\xi)\frac{\partial\zeta(\xi)}{\partial n}dS\right| \le C\|U\|_{L_q(\partial F)} \tag{2.36}$$

for any $q > 2$. This inequality, together with (2.34), shows that the left-hand side in (2.36) is a semi-norm, continuous in $W_2^1(B_2 \setminus F)$. Besides,

$$\int_{\partial F} \frac{\partial\zeta}{\partial n}(\xi)dS = \lim_{R\to\infty}(2\pi)^{-1}\int_{|\xi|=R}\frac{\partial}{\partial|\xi|}\log|\xi|\,dS = 1.$$

Now, Sobolev's equivalent normalizations theorem (see Sect. 1.1.15 in [22]) implies that the norm in $W_2^1(B_2 \setminus F)$ is equivalent to the norm

$$\|\nabla U\|_{L_2(B_2\setminus F)} + \left|\int_{\partial F} U(\xi)\frac{\partial\zeta}{\partial n}(\xi)dS\right|.$$

Combining this fact with (2.34) and using (2.29), we arrive at

$$\|U\|_{L_2(\partial F)} \le C\|\nabla U\|_{L_2(\mathbb{R}^2\setminus F)}. \tag{2.37}$$

Then, (2.33) and (2.37) yield

$$\|\nabla U\|_{L_2(\mathbb{R}^2\setminus F)} + \|U\|_{L_2(\partial F)} \le C\,\|\varphi\|_{L_2(\partial F)}. \tag{2.38}$$

By (2.34), the norm in $W_2^1(B_3 \setminus F)$ is equivalent to the norm

$$\|\nabla U\|_{L_2(B_3\setminus F)} + \|U\|_{L_2(\partial F)}.$$

Hence

$$\|U\|_{L_2(B_3\setminus F)} \le C\Big(\|\nabla U\|_{L_2(\mathbb{R}^2\setminus F)} + \|U\|_{L_2(\partial F)}\Big), \tag{2.39}$$

which, together with (2.38), gives

$$\|U\|_{L_2(B_3\setminus F)} \le C\,\|\varphi\|_{L_2(\partial F)}. \tag{2.40}$$

Substituting estimates (2.32) and (2.40) into (2.31), we arrive at

$$\|U\|_{W_p^1(B_2\setminus F)} \le C\,\|\varphi\|_{L_\infty(\partial F)}. \tag{2.41}$$

Recalling that $W_p^1(B_2 \setminus F)$ is embedded into $C(\overline{B_2 \setminus F})$ for $p > 2$, by another Sobolev's theorem (see Theorem 1.4.5 in [22]), we obtain

$$\sup_{B_2\setminus F} |U| \le C\,\|\varphi\|_{L_\infty(\partial F)}. \tag{2.42}$$

Since $U(\boldsymbol{\xi}) \to 0$ as $|\boldsymbol{\xi}| \to \infty$ (see (2.28) and (2.29)), we have the Poisson's formula

$$U(\boldsymbol{\xi}) = \frac{1}{\pi} Re \int_0^{2\pi} \frac{U(1,\theta')}{\rho e^{i(\theta-\theta')} - 1} d\theta', \;\; \boldsymbol{\xi} = \rho e^{i\theta}, \tag{2.43}$$

which, together with (2.42), implies for $|\boldsymbol{\xi}| > 1$ that

$$\big(1 + |\boldsymbol{\xi}|\big)|U(\boldsymbol{\xi})| \le C \max_{\boldsymbol{\xi}\in\partial B_1} |U(\boldsymbol{\xi})| \le C\,\|\varphi\|_{L_\infty(\partial F)}. \tag{2.44}$$

Applying (2.42) once more, we complete the proof. □

2.1.4 Asymptotic Properties of the Regular Part of the Neumann Function in $\mathbb{R}^2 \setminus F$

Lemma 2.1 proved in the previous section enables one to describe the asymptotic behaviour of the function h_N defined in (2.16)–(2.18).

Lemma 2.2. *The solution $h_N(\boldsymbol{\xi}, \boldsymbol{\eta})$ of problem* (2.16)–(2.18) *satisfies the estimate*

$$\left| h_N(\boldsymbol{\xi}, \boldsymbol{\eta}) - \frac{\mathcal{D}(\boldsymbol{\eta}) \cdot \boldsymbol{\xi}}{2\pi |\boldsymbol{\xi}|^2} \right| \leq \text{Const}\ (1 + |\boldsymbol{\eta}|)^{-1} |\boldsymbol{\xi}|^{-2} \tag{2.45}$$

as $|\boldsymbol{\xi}| > 2$ and $\boldsymbol{\eta} \in \mathbb{R}^2 \setminus F$.

Proof. The leading-order approximation of the harmonic function $h_N(\boldsymbol{\xi}, \boldsymbol{\eta})$, as $|\boldsymbol{\xi}| \to \infty$, is sought in the form

$$(2\pi)^{-1} |\boldsymbol{\xi}|^{-2} (C_1 \xi_1 + C_2 \xi_2).$$

Applying Green's formula in $B_R \setminus F$ to $h_N(\boldsymbol{\xi}, \boldsymbol{\eta})$ and $\mathcal{D}_j(\boldsymbol{\xi}) - \xi_j$, and taking the limit, as $R \to \infty$, we obtain

$$\lim_{R \to \infty} \int_{|\mathbf{x}|=R} \Big\{ h_N(\boldsymbol{\xi}, \boldsymbol{\eta}) \frac{\partial (\mathcal{D}_j(\boldsymbol{\xi}) - \xi_j)}{\partial |\boldsymbol{\xi}|} + (\xi_j - \mathcal{D}_j(\boldsymbol{\xi})) \frac{\partial h_N(\boldsymbol{\xi}, \boldsymbol{\eta})}{\partial |\boldsymbol{\xi}|} \Big\} dS_\xi$$

$$= \int_{\partial F} (\mathcal{D}_j(\boldsymbol{\xi}) - \xi_j) \frac{\partial h_N(\boldsymbol{\xi}, \boldsymbol{\eta})}{\partial n} dS_\xi, \tag{2.46}$$

where $\partial/\partial n$ is the normal derivative in the direction of the inward normal with respect to F. As $R \to \infty$, the left-hand side of (2.46) becomes

$$\frac{1}{2\pi} \lim_{R \to +\infty} \int_{|\mathbf{x}|=R} \Big\{ -2 \frac{(C_1 \xi_1 + C_2 \xi_2) \xi_j}{R^3} \Big\} dS_\xi$$

$$= -\frac{1}{\pi} \lim_{R \to +\infty} \int_0^{2\pi} (C_1 \cos\theta + C_2 \sin\theta) R^{-1} \xi_j \, d\theta = -C_j. \tag{2.47}$$

Taking into account the definition of the dipole fields $\mathcal{D}_j$ (see (2.19)–(2.21)) and the definition of the regular part h_N of Neumann's function (see (2.16)–(2.18)) in $\mathbb{R}^2 \setminus F$, we can reduce the integral $\mathcal{I}$ in the right-hand side of (2.46) to the form

$$\mathcal{I} = \frac{1}{2\pi} \Big\{ \int_{\partial F} \Big(\mathcal{D}_j(\boldsymbol{\xi}) \frac{\partial}{\partial n_\xi} \Big(\log |\boldsymbol{\xi} - \boldsymbol{\eta}|^{-1} \Big) - \log |\boldsymbol{\xi} - \boldsymbol{\eta}|^{-1} \frac{\partial}{\partial n_\xi} \mathcal{D}_j(\boldsymbol{\xi}) \Big) dS_\xi$$

$$+ \int_{\partial F} \Big(n_j \log |\boldsymbol{\xi} - \boldsymbol{\eta}|^{-1} - \xi_j \frac{\partial}{\partial n_\xi} \Big(\log |\boldsymbol{\xi} - \boldsymbol{\eta}|^{-1} \Big) \Big) dS_\xi \Big\}. \tag{2.48}$$

The second integral in (2.48) equals zero. Applying Green's formula to the first integral in (2.48) we obtain

$$\frac{1}{2\pi} \int_{\partial F} \Big(\mathcal{D}_j(\boldsymbol{\xi}) \frac{\partial}{\partial n_\xi} \Big(\log |\boldsymbol{\xi} - \boldsymbol{\eta}|^{-1} \Big)$$

$$- \log |\boldsymbol{\xi} - \boldsymbol{\eta}|^{-1} \frac{\partial}{\partial n_\xi} \mathcal{D}_j(\boldsymbol{\xi}) \Big) dS_\xi = -\mathcal{D}_j(\boldsymbol{\eta}). \tag{2.49}$$

Hence, it follows from (2.47)–(2.49) that

$$C_j = \mathcal{D}_j(\boldsymbol{\eta}),\ j = 1, 2. \tag{2.50}$$

We note that the function

$$h_N(\boldsymbol{\xi}, \boldsymbol{\eta}) + \boldsymbol{\mathcal{D}}(\boldsymbol{\eta}) \cdot \nabla_\xi \Big(\frac{1}{2\pi} \log |\boldsymbol{\xi}|^{-1}\Big) \tag{2.51}$$

is harmonic in $\mathbb{R}^2 \setminus F$, both in $\boldsymbol{\xi}$ and $\boldsymbol{\eta}$, and it vanishes at infinity. Using (2.17) and (2.20), we obtain

$$\begin{aligned}
&\frac{\partial}{\partial n_\eta}\Big(h_N(\boldsymbol{\xi}, \boldsymbol{\eta}) + \boldsymbol{\mathcal{D}}(\boldsymbol{\eta}) \cdot \nabla_\xi \big(\frac{1}{2\pi} \log |\boldsymbol{\xi}|^{-1}\big)\Big) \\
&\quad = \frac{\partial}{\partial n_\eta} h_N(\boldsymbol{\xi}, \boldsymbol{\eta}) + \mathbf{n} \cdot \nabla_\xi \big(\frac{1}{2\pi} \log |\boldsymbol{\xi}|^{-1}\big) \\
&\quad = -\mathbf{n} \cdot \nabla_\xi \Big\{\frac{1}{2\pi} \log(|\boldsymbol{\xi}||\boldsymbol{\xi} - \boldsymbol{\eta}|^{-1})\Big\} \\
&\quad = -\frac{1}{2\pi |\boldsymbol{\xi}|^2} \mathbf{n} \cdot \Big\{\boldsymbol{\eta} - \frac{2\boldsymbol{\xi} \cdot \boldsymbol{\eta}}{|\boldsymbol{\xi}|^2} \boldsymbol{\xi} + O(|\boldsymbol{\xi}|^{-1})\Big\}
\end{aligned} \tag{2.52}$$

as $\boldsymbol{\eta} \in \partial F$ and $|\boldsymbol{\xi}| > 2$. We also note that

$$\int_{\partial F} \frac{\partial}{\partial n_\eta}\Big(h_N(\boldsymbol{\xi}, \boldsymbol{\eta}) + \boldsymbol{\mathcal{D}}(\boldsymbol{\eta}) \cdot \nabla_\xi \big(\frac{1}{2\pi} \log |\boldsymbol{\xi}|^{-1}\big)\Big) dS_\eta = 0.$$

Consider the problem (2.25)–(2.27) in the formulation of Lemma 2.1, where the variable $\boldsymbol{\xi}$ is replaced by $\boldsymbol{\eta}$, the differentiation is taken with respect to components of $\boldsymbol{\eta}$, and the function U is changed for (2.51), with fixed $\boldsymbol{\xi}$. In this case, the right-hand side φ in (2.26) is replaced by

$$\frac{\partial}{\partial n_\eta} h_N(\boldsymbol{\xi}, \boldsymbol{\eta}) + \mathbf{n} \cdot \nabla_\xi \Big(\frac{1}{2\pi} \log |\boldsymbol{\xi}|^{-1}\Big).$$

Then using (2.52) and applying Lemma 2.1, we obtain (2.45). □

Using the notion of the dipole matrix, from (2.22) and Lemma 2.2 we derive the following asymptotic representation of h_N.

Corollary 2.1. *Let* $|\boldsymbol{\xi}| > 2$, *and* $|\boldsymbol{\eta}| > 2$. *Then*

$$h_N(\boldsymbol{\xi}, \boldsymbol{\eta}) = \frac{1}{4\pi^2} \sum_{j,k=1}^{2} \frac{\mathcal{P}_{jk} \xi_j \eta_k}{|\boldsymbol{\xi}|^2 |\boldsymbol{\eta}|^2} + O\Big(\frac{|\boldsymbol{\xi}| + |\boldsymbol{\eta}|}{|\boldsymbol{\xi}|^2 |\boldsymbol{\eta}|^2}\Big). \tag{2.53}$$

2.1.5 *Maximum Modulus Estimate for Solutions to the Mixed Problem in Ω_ε, with the Neumann Data on ∂F_ε*

In the sequel, when estimating the remainder term in the asymptotic representation of $G_\varepsilon(\mathbf{x}, \mathbf{y})$, we use the following assertion.

Lemma 2.3. *Let u be a solution of the mixed boundary value problem*

$$\Delta u(\mathbf{x}) = 0, \quad \mathbf{x} \in \Omega_\varepsilon, \tag{2.54}$$

$$u(\mathbf{x}) = \varphi(\mathbf{x}), \quad \mathbf{x} \in \partial\Omega, \tag{2.55}$$

$$\frac{\partial u}{\partial n}(\mathbf{x}) = \psi_\varepsilon(\mathbf{x}), \quad \mathbf{x} \in \partial F_\varepsilon, \tag{2.56}$$

where $\varphi \in C(\partial\Omega)$, $\psi_\varepsilon \in L_\infty(\partial F_\varepsilon)$, and

$$\int_{\partial F_\varepsilon} \psi_\varepsilon(\mathbf{x}) ds = 0. \tag{2.57}$$

The solution u is sought in $C(\overline{\Omega}_\varepsilon)$, and it is also assumed that ∇u is square integrable in a neighbourhood of ∂F_ε. Then there exists a positive constant C, independent of ε and such that

$$\|u\|_{C(\overline{\Omega}_\varepsilon)} \leq \|\varphi\|_{C(\partial\Omega)} + \varepsilon C \, \|\psi_\varepsilon\|_{L_\infty(\partial F_\varepsilon)}. \tag{2.58}$$

Proof. (a) We introduce the inverse operator

$$\mathfrak{N} : \psi \to v \tag{2.59}$$

for the boundary value problem

$$\Delta v(\boldsymbol{\xi}) = 0, \ \boldsymbol{\xi} \in \mathbb{R}^2 \setminus F, \tag{2.60}$$

$$\frac{\partial v}{\partial n}(\boldsymbol{\xi}) = \psi(\boldsymbol{\xi}), \ \boldsymbol{\xi} \in \partial F, \tag{2.61}$$

$$v(\boldsymbol{\xi}) \to 0, \text{ as } |\boldsymbol{\xi}| \to \infty, \tag{2.62}$$

where $\psi \in L_\infty(\partial F)$, and

$$\int_{\partial F} \psi(\boldsymbol{\xi}) ds_\xi = 0. \tag{2.63}$$

In the scaled coordinates $\boldsymbol{\xi} = \varepsilon^{-1}\mathbf{x}$, the operator $\mathfrak{N}_\varepsilon$ is defined by

$$(\mathfrak{N}_\varepsilon \psi_\varepsilon)(\mathbf{x}) = (\mathfrak{N}\psi)(\boldsymbol{\xi}), \tag{2.64}$$

where $\psi_\varepsilon(\mathbf{x}) = \varepsilon^{-1}\psi(\varepsilon^{-1}\mathbf{x})$.

(b) We look for the solution u of (2.54)–(2.57) in the form

$$u = V(\mathbf{x}) + W(\mathbf{x}), \tag{2.65}$$

where $V = \mathfrak{N}_\varepsilon \psi_\varepsilon$, and the function W satisfies the problem

$$\Delta W(\mathbf{x}) = 0, \ \ \mathbf{x} \in \Omega_\varepsilon, \tag{2.66}$$

$$\frac{\partial W}{\partial n}(\mathbf{x}) = 0, \ \ \mathbf{x} \in \partial F_\varepsilon, \tag{2.67}$$

$$W(\mathbf{x}) = \varphi(\mathbf{x}) - V(\mathbf{x}), \ \ \mathbf{x} \in \partial\Omega. \tag{2.68}$$

By Lemma 2.1, we have

$$\max_{\overline{\Omega}_\varepsilon} |V| = \max_{\overline{\Omega}_\varepsilon} |\mathfrak{N}_\varepsilon \psi_\varepsilon| \le \varepsilon C \, \|\psi_\varepsilon\|_{L_\infty(\partial F_\varepsilon)}. \tag{2.69}$$

Hence, as follows from (2.68) and (2.69)

$$\max_{\partial\Omega} |W| \le \|\varphi\|_{C(\partial\Omega)} + \varepsilon C \, \|\psi_\varepsilon\|_{L_\infty(\partial F_\varepsilon)}, \tag{2.70}$$

and by the weak maximum principle for variational solutions (see, for example, Gilbarg and Trudinger [12], pages 215–216) of (2.66)–(2.68) we obtain

$$\max_{\overline{\Omega}_\varepsilon} |W| \le \|\varphi\|_{C(\partial\Omega)} + \varepsilon C \, \|\psi_\varepsilon\|_{L_\infty(\partial F_\varepsilon)}. \tag{2.71}$$

The result follows from (2.69), (2.71) combined with (2.65). □

2.1.6 Approximation of Green's Function $G_\varepsilon^{(N)}$

The required approximation of $G_\varepsilon^{(N)}$ is given in the next Theorem.

Theorem 2.1. *Green's function $G_\varepsilon^{(N)}(\mathbf{x}, \mathbf{y})$ for the boundary value problem* (2.1)–(2.3), *with the Neumann data on ∂F_ε and the Dirichlet data on $\partial\Omega$, has the asymptotic representation*

$$G_\varepsilon^{(N)}(\mathbf{x}, \mathbf{y}) = G(\mathbf{x}, \mathbf{y}) + \mathcal{N}(\varepsilon^{-1}\mathbf{x}, \varepsilon^{-1}\mathbf{y}) + (2\pi)^{-1} \log(\varepsilon^{-1}|\mathbf{x} - \mathbf{y}|)$$
$$+ \varepsilon \boldsymbol{\mathcal{D}}(\varepsilon^{-1}\mathbf{x}) \cdot \nabla_x H(0, \mathbf{y}) + \varepsilon \boldsymbol{\mathcal{D}}(\varepsilon^{-1}\mathbf{y}) \cdot \nabla_y H(\mathbf{x}, 0) + r_\varepsilon(\mathbf{x}, \mathbf{y}), \tag{2.72}$$

where

$$|r_\varepsilon(\mathbf{x}, \mathbf{y})| \le \text{Const}\, \varepsilon^2 \tag{2.73}$$

uniformly with respect to $\mathbf{x}, \mathbf{y} \in \Omega_\varepsilon$. Here, G, $\mathcal{N}$, $\boldsymbol{\mathcal{D}}$ and H are the same as in Sect. 2.1.1.

Proof. We begin with the formal argument leading to (2.72). First, we note that

$$N(\varepsilon^{-1}\mathbf{x}, \varepsilon^{-1}\mathbf{y}) + (2\pi)^{-1}\log(\varepsilon^{-1}|\mathbf{x}-\mathbf{y}|) = -h_N(\varepsilon^{-1}\mathbf{x}, \varepsilon^{-1}\mathbf{y}),$$

and then represent $G_\varepsilon^{(N)}(\mathbf{x}, \mathbf{y})$ in the form

$$G_\varepsilon^{(N)}(\mathbf{x}, \mathbf{y}) = G(\mathbf{x}, \mathbf{y}) - h_N(\varepsilon^{-1}\mathbf{x}, \varepsilon^{-1}\mathbf{y}) + \rho_\varepsilon(\mathbf{x}, \mathbf{y}). \tag{2.74}$$

By the direct substitution of (2.74) into (2.1)–(2.3) and using Lemma 2.2, we deduce that $\rho_\varepsilon(\mathbf{x}, \mathbf{y})$ satisfies the boundary value problem

$$\begin{aligned} \Delta_x \rho_\varepsilon(\mathbf{x}, \mathbf{y}) &= 0, \quad \mathbf{x}, \mathbf{y} \in \Omega_\varepsilon, \\ \rho_\varepsilon(\mathbf{x}, \mathbf{y}) &= h_N(\varepsilon^{-1}\mathbf{x}, \varepsilon^{-1}\mathbf{y}) \\ &= \frac{\varepsilon}{2\pi}\boldsymbol{\mathcal{D}}\Big(\frac{\mathbf{y}}{\varepsilon}\Big) \cdot \frac{\mathbf{x}}{|\mathbf{x}|^2} + O(\varepsilon^2), \quad \text{for } \mathbf{x} \in \partial\Omega, \mathbf{y} \in \Omega_\varepsilon, \end{aligned} \tag{2.75}$$

and

$$\begin{aligned} \frac{\partial \rho_\varepsilon}{\partial n_x}(\mathbf{x}, \mathbf{y}) &= \frac{\partial}{\partial n_x} H(\mathbf{x}, \mathbf{y}) \\ &= \mathbf{n} \cdot \nabla_x H(0, \mathbf{y}) + O(\varepsilon), \quad \text{for } \mathbf{x} \in \partial F_\varepsilon, \mathbf{y} \in \Omega_\varepsilon. \end{aligned} \tag{2.76}$$

Hence, by (2.5), (2.6) and (2.19)–(2.21), the leading-order approximation of ρ_ε is

$$\varepsilon\boldsymbol{\mathcal{D}}(\varepsilon^{-1}\mathbf{x}) \cdot \nabla_x H(0, \mathbf{y}) + \varepsilon\boldsymbol{\mathcal{D}}(\varepsilon^{-1}\mathbf{y}) \cdot \nabla_y H(\mathbf{x}, 0),$$

which, together with (2.74), leads to (2.72).

Now, we prove the remainder estimate (2.73). The direct substitution of (2.72) into (2.1)–(2.3) yields the boundary value problem for r_ε:

$$\Delta_x r_\varepsilon(\mathbf{x}, \mathbf{y}) = 0, \quad \text{for } \mathbf{x}, \mathbf{y} \in \Omega_\varepsilon, \tag{2.77}$$

$$\begin{aligned} r_\varepsilon(\mathbf{x}, \mathbf{y}) &= h_N(\varepsilon^{-1}\mathbf{x}, \varepsilon^{-1}\mathbf{y}) \\ &\quad -\varepsilon\boldsymbol{\mathcal{D}}(\varepsilon^{-1}\mathbf{x}) \cdot \nabla_x H(0, \mathbf{y}) - \varepsilon\boldsymbol{\mathcal{D}}(\varepsilon^{-1}\mathbf{y}) \cdot \nabla_y H(\mathbf{x}, 0), \\ &\quad \text{for } \mathbf{x} \in \partial\Omega, \ \mathbf{y} \in \Omega_\varepsilon, \end{aligned} \tag{2.78}$$

$$\begin{aligned} \frac{\partial r_\varepsilon(\mathbf{x}, \mathbf{y})}{\partial n_x} &= \mathbf{n} \cdot \nabla_x H(\mathbf{x}, \mathbf{y}) - \varepsilon\frac{\partial}{\partial n_x}\Big(\boldsymbol{\mathcal{D}}(\varepsilon^{-1}\mathbf{x}) \cdot \nabla_x H(0, \mathbf{y})\Big) \\ &\quad -\varepsilon\frac{\partial}{\partial n_x}\Big(\boldsymbol{\mathcal{D}}(\varepsilon^{-1}\mathbf{y}) \cdot \nabla_y H(\mathbf{x}, 0)\Big), \\ &\quad \text{for } \mathbf{x} \in \partial F_\varepsilon, \ \mathbf{y} \in \Omega_\varepsilon. \end{aligned} \tag{2.79}$$

We note that every term in the right-hand side of (2.79) has zero average on ∂F_ε, and hence

$$\int_{\partial F_\varepsilon} \frac{\partial r_\varepsilon(\mathbf{x},\mathbf{y})}{\partial n_x} dS_x = 0. \tag{2.80}$$

It follows from Lemma 2.2 that

$$|h_N(\varepsilon^{-1}\mathbf{x}, \varepsilon^{-1}\mathbf{y}) - \varepsilon \boldsymbol{\mathcal{D}}(\varepsilon^{-1}\mathbf{y}) \cdot \nabla_y H(\mathbf{x}, 0)| \le \text{Const}\, \varepsilon^2, \tag{2.81}$$

uniformly with respect to $\mathbf{x} \in \partial\Omega$ and $\mathbf{y} \in \Omega_\varepsilon$. Since $|\boldsymbol{\mathcal{D}}(\boldsymbol{\xi})| \le \text{Const}\, |\boldsymbol{\xi}|^{-1}$, as $|\boldsymbol{\xi}| \to \infty$, and $\nabla_x H(0, \mathbf{y})$ is smooth on Ω_ε, we deduce

$$|\varepsilon \boldsymbol{\mathcal{D}}(\varepsilon^{-1}\mathbf{x}) \cdot \nabla_x H(0, \mathbf{y})| \le \text{Const}\, \varepsilon^2 \tag{2.82}$$

uniformly with respect to $\mathbf{x} \in \partial\Omega$ and $\mathbf{y} \in \Omega_\varepsilon$. By (2.81) and (2.82), the modulus of the right-hand side in (2.78) is bounded by Const ε^2, uniformly in $\mathbf{x} \in \partial\Omega$ and $\mathbf{y} \in \Omega_\varepsilon$.

It also follows from the definition of the dipole fields $\mathcal{D}_j(\boldsymbol{\xi})$, $j = 1, 2$, and the smoothness of the function $H(\mathbf{x}, \mathbf{y})$ for all $\mathbf{x} \in \partial F_\varepsilon$, $\mathbf{y} \in \Omega_\varepsilon$ that

$$\left| \mathbf{n} \cdot \nabla_x H(\mathbf{x}, \mathbf{y}) - \varepsilon \frac{\partial}{\partial n_x} \Big(\boldsymbol{\mathcal{D}}(\varepsilon^{-1}\mathbf{x}) \cdot \nabla_x H(0, \mathbf{y}) \Big) \right| \le \text{Const}\, \varepsilon, \tag{2.83}$$

and

$$\left| \varepsilon \frac{\partial}{\partial n_x} \Big(\boldsymbol{\mathcal{D}}(\varepsilon^{-1}\mathbf{y}) \cdot \nabla_y H(\mathbf{x}, 0) \Big) \right| \le \text{Const}\, \varepsilon, \tag{2.84}$$

uniformly with respect to $\mathbf{x} \in \partial F_\varepsilon$, $\mathbf{y} \in \Omega_\varepsilon$. These estimates imply that the modulus of the right-hand side in (2.79) is bounded by Const ε, uniformly in $\mathbf{x} \in \partial F_\varepsilon$ and $\mathbf{y} \in \Omega_\varepsilon$.

Using the estimates on ∂F_ε and $\partial\Omega$, just obtained, together with the orthogonality condition (2.80), we deduce that the right-hand sides of problem (2.77)–(2.79) satisfy the conditions of Lemma 2.3. Applying Lemma 2.3, we obtain that $\|r_\varepsilon\|_{L_\infty(\Omega_\varepsilon)}$ is dominated by Const ε^2, which completes the proof. □

2.1.7 *Simpler Asymptotic Formulae for Green's Function* $G_\varepsilon^{(N)}$

Here we formulate two corollaries of Theorem 2.1. They contain simpler asymptotic formulae, which are efficient for the cases when both **x** and **y** are distant from F_ε or both **x** and **y** are sufficiently close to F_ε.

Corollary 2.2. *Let* $\min\{|\mathbf{x}|, |\mathbf{y}|\} > 2\varepsilon$. *Then the asymptotic formula holds*

$$G_\varepsilon^{(N)}(\mathbf{x}, \mathbf{y}) = G(\mathbf{x}, \mathbf{y}) - \frac{\varepsilon^2}{4\pi^2} \frac{\mathbf{x}^T}{|\mathbf{x}|^2} \mathcal{P} \frac{\mathbf{y}}{|\mathbf{y}|^2}$$
$$+ \frac{\varepsilon^2}{2\pi} \Big\{ \frac{\mathbf{x}^T}{|\mathbf{x}|^2} \mathcal{P} \nabla_x H(0, \mathbf{y}) + \frac{\mathbf{y}^T}{|\mathbf{y}|^2} \mathcal{P} \nabla_y H(\mathbf{x}, 0) \Big\}$$
$$+ \varepsilon^2 O(|\mathbf{x}|^{-2} + |\mathbf{y}|^{-2}), \tag{2.85}$$

where H *is the regular part of Green's function* G *in* Ω, *and* $\mathcal{P}$ *is the dipole matrix for* F, *as defined in* (2.22).

Proof. Using (2.53) for the regular part h_N of the Neumann function in $\mathbb{R}^2 \setminus F$, together with the asymptotic representation (2.22) of the dipole fields $\mathcal{D}_j$ in $\mathbb{R}^2 \setminus F$, we obtain

$$G_\varepsilon^{(N)}(\mathbf{x}, \mathbf{y}) = G(\mathbf{x}, \mathbf{y}) - \frac{\varepsilon^2}{4\pi^2} \sum_{j,k=1}^{2} \frac{\mathcal{P}_{jk} x_j y_k}{|\mathbf{x}|^2 |\mathbf{y}|^2} + O\Big(\varepsilon^3 \frac{|\mathbf{x}| + |\mathbf{y}|}{|\mathbf{x}|^2 |\mathbf{y}|^2}\Big)$$
$$+ \frac{1}{2\pi} \sum_{j,k=1}^{2} \Big\{ \varepsilon^2 \mathcal{P}_{jk} \Big(\frac{x_k}{|\mathbf{x}|^2} \frac{\partial H}{\partial x_j}(0, \mathbf{y}) + \frac{y_k}{|\mathbf{y}|^2} \frac{\partial H}{\partial y_j}(\mathbf{x}, 0) \Big)$$
$$+ \varepsilon^2 O(|\mathbf{x}|^{-2} + |\mathbf{y}|^{-2}) \Big\} + O(\varepsilon^2). \tag{2.86}$$

Combining the remainder terms and adopting the matrix representation involving the dipole matrix $\mathcal{P}$, we arrive at (2.85). □

The formula (2.85) becomes efficient when both $\mathbf{x}$ and $\mathbf{y}$ are sufficiently distant from the small hole F_ε. Compared to (2.72), formula (2.85) does not involve special solutions of model problems in $\mathbb{R}^2 \setminus F$, while the influence of the hole F is seen through the dipole matrix $\mathcal{P}$.

Corollary 2.3. *The following asymptotic formula for Green's function* $G_\varepsilon^{(N)}$ *of the boundary value problem* (2.1)–(2.3) *holds:*

$$G_\varepsilon^{(N)}(\mathbf{x}, \mathbf{y}) = (2\pi)^{-1} \log |\mathbf{x} - \mathbf{y}|^{-1} - h_N(\varepsilon^{-1}\mathbf{x}, \varepsilon^{-1}\mathbf{y}) - H(0, 0)$$
$$- (\mathbf{x} - \varepsilon \mathcal{D}(\varepsilon^{-1}\mathbf{x})) \cdot \nabla_x H(0, \mathbf{y}) - (\mathbf{y} - \varepsilon \mathcal{D}(\varepsilon^{-1}\mathbf{y})) \cdot \nabla_y H(\mathbf{x}, 0)$$
$$+ O(\varepsilon^2 + |\mathbf{x}|^2 + |\mathbf{y}|^2), \tag{2.87}$$

for $\mathbf{x}, \mathbf{y} \in \Omega_\varepsilon$.

Proof. Using the Taylor expansion of $H(\mathbf{x}, \mathbf{y})$ in a neighbourhood of the origin, we obtain

$$G_\varepsilon^{(N)}(\mathbf{x}, \mathbf{y}) = -H(0, 0) - \mathbf{x} \cdot \nabla_x H(0, \mathbf{y}) - \mathbf{y} \cdot \nabla_y H(\mathbf{x}, 0) + O(|\mathbf{x}|^2 + |\mathbf{y}|^2)$$
$$+ \mathcal{N}(\varepsilon^{-1}\mathbf{x}, \varepsilon^{-1}\mathbf{y}) - (2\pi)^{-1} \log \varepsilon$$

$$
\begin{aligned}
&+ \varepsilon \boldsymbol{\mathcal{D}}(\varepsilon^{-1}\mathbf{x}) \cdot \nabla_x H(0, \mathbf{y}) \\
&+ \varepsilon \boldsymbol{\mathcal{D}}(\varepsilon^{-1}\mathbf{y}) \cdot \nabla_y H(\mathbf{x}, 0) + O(\varepsilon^2).
\end{aligned} \tag{2.88}
$$

By substituting

$$
\mathcal{N}(\varepsilon^{-1}\mathbf{x}, \varepsilon^{-1}\mathbf{y}) = (2\pi)^{-1} \log |\mathbf{x} - \mathbf{y}|^{-1} + (2\pi)^{-1} \log \varepsilon - h_N(\varepsilon^{-1}\mathbf{x}, \varepsilon^{-1}\mathbf{y})
$$

into (2.88) and rearranging the terms, we arrive at (2.87). □

2.2 Mixed Boundary Value Problem with the Dirichlet Condition on ∂F_ε

In the present section, the meaning of the notations Ω, F and F_ε, already used in Sect. 2.1, will be slightly altered. Let Ω be a bounded domain with smooth boundary, and let F stand for an arbitrary compact set in $\mathbb{R}^2$ of positive logarithmic capacity (see Landkof [20]). As in Sect. 2.1, it is assumed that diam $F = 1/2$, and that dist$(\mathbf{O}, \partial\Omega) = 1$. We also set $F_\varepsilon = \{\mathbf{x} : \varepsilon^{-1}\mathbf{x} \in F\}$.

We consider the mixed boundary value problem in a two-dimensional domain $\Omega_\varepsilon = \Omega \setminus F_\varepsilon$, with the Dirichlet data on ∂F_ε and the Neumann data on $\partial\Omega$.

Green's function $G_\varepsilon^{(D)}$ of this problem is a weak solution of

$$
\Delta_x G_\varepsilon^{(D)}(\mathbf{x}, \mathbf{y}) + \delta(\mathbf{x} - \mathbf{y}) = 0, \quad \mathbf{x}, \mathbf{y} \in \Omega_\varepsilon, \tag{2.89}
$$

$$
G_\varepsilon^{(D)}(\mathbf{x}, \mathbf{y}) = 0, \quad \mathbf{x} \in \partial F_\varepsilon, \ \mathbf{y} \in \Omega_\varepsilon, \tag{2.90}
$$

$$
\frac{\partial G_\varepsilon^{(D)}}{\partial n_x}(\mathbf{x}, \mathbf{y}) = 0, \quad \mathbf{x} \in \partial\Omega, \ \mathbf{y} \in \Omega_\varepsilon. \tag{2.91}
$$

Before deriving an asymptotic approximation of $G_\varepsilon^{(D)}(\mathbf{x}, \mathbf{y})$, uniform with respect to $\mathbf{x}, \mathbf{y} \in \Omega_\varepsilon$, we outline the properties of solutions of auxiliary model problems in limit domains.

2.2.1 *Special Solutions of Model Problems*

1. Let $N(\mathbf{x}, \mathbf{y})$ be the Neumann function in Ω, i.e.

$$
\Delta N(\mathbf{x}, \mathbf{y}) + \delta(\mathbf{x} - \mathbf{y}) = 0, \quad \mathbf{x}, \mathbf{y} \in \Omega, \tag{2.92}
$$

$$
\frac{\partial}{\partial n_x}\Big(N(\mathbf{x}, \mathbf{y}) + (2\pi)^{-1} \log |\mathbf{x}|\Big) = 0, \quad \mathbf{x} \in \partial\Omega, \ \mathbf{y} \in \Omega, \tag{2.93}
$$

and

$$\int_{\partial\Omega} N(\mathbf{x},\mathbf{y}) \frac{\partial}{\partial n_x} \log|\mathbf{x}| dS_x = 0. \tag{2.94}$$

Condition (2.94) implies the symmetry of $N(\mathbf{x},\mathbf{y})$. In fact, let $U(\mathbf{x}) = N(\mathbf{x},\mathbf{z})$ and $V(\mathbf{x}) = N(\mathbf{x},\mathbf{y})$, where $\mathbf{z}$ and $\mathbf{y}$ are fixed points in Ω. Then applying Green's formula to U and V and using (2.92)–(2.94) we deduce

$$U(\mathbf{y}) - V(\mathbf{z}) = \int_{\Omega} \Big(V(\mathbf{x})\Delta_x U(\mathbf{x}) - U(\mathbf{x})\Delta_x V(\mathbf{x})\Big) d\mathbf{x}$$

$$= \frac{1}{2\pi} \int_{\partial\Omega} \Big(U(\mathbf{x}) \frac{\partial}{\partial n_x}(\log|\mathbf{x}|) - V(\mathbf{x}) \frac{\partial}{\partial n_x}(\log|\mathbf{x}|)\Big) dS_x$$

$$= \frac{1}{2\pi} \Big\{ \int_{\partial\Omega} N(\mathbf{x},\mathbf{z}) \frac{\partial}{\partial n_x}(\log|\mathbf{x}|) dS_x - \int_{\partial\Omega} N(\mathbf{x},\mathbf{y}) \frac{\partial}{\partial n_x}(\log|\mathbf{x}|) dS_x \Big\} = 0,$$

where $\partial/\partial n_x$ is the normal derivative in the direction of the outward normal on $\partial\Omega$. Hence $N(\mathbf{y},\mathbf{z}) = N(\mathbf{z},\mathbf{y})$.

The regular part of the Neumann function is defined by

$$R(\mathbf{x},\mathbf{y}) = (2\pi)^{-1} \log|\mathbf{x}-\mathbf{y}|^{-1} - N(\mathbf{x},\mathbf{y}). \tag{2.95}$$

Note that

$$R(0,\mathbf{y}) = -(2\pi)^{-2} \int_{\partial\Omega} \log|\mathbf{x}| \frac{\partial}{\partial n} \log|\mathbf{x}| ds_x, \tag{2.96}$$

which is verified by applying Green's formula to $R(\mathbf{x},\mathbf{y})$ and $(2\pi)^{-1}\log|\mathbf{x}|$ as follows:

$$R(0,\mathbf{y}) = \frac{1}{2\pi} \int_{\Omega} R(\mathbf{x},\mathbf{y}) \Delta_x (\log|\mathbf{x}|) d\mathbf{x}$$

$$= \frac{1}{2\pi} \int_{\partial\Omega} \Big(R(\mathbf{x},\mathbf{y}) \frac{\partial}{\partial n_x}(\log|\mathbf{x}|) - \log|\mathbf{x}| \frac{\partial}{\partial n_x} R(\mathbf{x},\mathbf{y})\Big) ds_x, \tag{2.97}$$

where $\partial/\partial n_x$ is the normal derivative in the outward direction on $\partial\Omega$. Taking into account (2.93)–(2.95), we can write (2.97) in the form

$$R(0,\mathbf{y}) = \frac{1}{4\pi^2} \int_{\partial\Omega} \Big(\log|\mathbf{x}-\mathbf{y}|^{-1} \frac{\partial}{\partial n_x}(\log|\mathbf{x}|) - \log|\mathbf{x}| \frac{\partial}{\partial n_x}(\log|\mathbf{x}-\mathbf{y}|^{-1})\Big) ds_x$$

$$+ \frac{1}{2\pi} \int_{\partial\Omega} \log|\mathbf{x}| \frac{\partial}{\partial n_x}(N(\mathbf{x},\mathbf{y})) ds_x. \tag{2.98}$$

The first integral in (2.98) is equal to zero, while the second integral in (2.98) is reduced to (2.96) because of the boundary condition (2.93).

As in Sect. 2.1, the notations $\boldsymbol{\xi}$ and $\boldsymbol{\eta}$ will be used for the scaled coordinates $\boldsymbol{\xi} = \varepsilon^{-1}\mathbf{x}$ and $\boldsymbol{\eta} = \varepsilon^{-1}\mathbf{y}$. The corresponding limit domain is $\mathbb{R}^2 \setminus F$.

2. Green's function $\mathcal{G}(\boldsymbol{\xi}, \boldsymbol{\eta})$ for the Dirichlet problem in $\mathbb{R}^2 \setminus F$ is a unique solution to the problem (2.11)–(2.13). The regular part $h(\boldsymbol{\xi}, \boldsymbol{\eta})$ of Green's function $\mathcal{G}(\boldsymbol{\xi}, \boldsymbol{\eta})$ is

$$h(\boldsymbol{\xi}, \boldsymbol{\eta}) = (2\pi)^{-1} \log |\boldsymbol{\xi} - \boldsymbol{\eta}|^{-1} - \mathcal{G}(\boldsymbol{\xi}, \boldsymbol{\eta}). \tag{2.99}$$

3. Here and in the sequel, $\mathbf{D}(\boldsymbol{\xi})$ denotes a vector function, whose components $D_j,\ j = 1, 2$, satisfy the model problems

$$\Delta D_j(\boldsymbol{\xi}) = 0, \ \ \boldsymbol{\xi} \in \mathbb{R}^2 \setminus F, \tag{2.100}$$

$$D_j(\boldsymbol{\xi}) = \xi_j, \ \ \boldsymbol{\xi} \in \partial F, \tag{2.101}$$

$$D_j(\boldsymbol{\xi}) \text{ is bounded as } |\boldsymbol{\xi}| \to \infty. \tag{2.102}$$

We use the notations $D_j^\infty = \lim_{|\boldsymbol{\xi}|\to\infty} D_j(\boldsymbol{\xi})$ and $\mathbf{D}^\infty = (D_1^\infty, D_2^\infty)^T$.

Application of Green's formula to D_j and the function ζ, defined in (2.7)–(2.9), gives

$$D_j^\infty = -\int_{\partial F} \xi_j \frac{\partial \zeta(\boldsymbol{\xi})}{\partial n} dS_\xi. \tag{2.103}$$

Here and in other derivations of this section, $\partial/\partial n$ on ∂F is the normal derivative in the direction of the inward normal with respect to F.

We also find an additional connection between D_j and ζ by analyzing the asymptotic formula (compare with (2.9))

$$\zeta(\boldsymbol{\xi}) = (2\pi)^{-1} \log |\boldsymbol{\xi}| + \zeta_\infty + \frac{1}{2\pi} \sum_{k=1}^{2} \frac{\alpha_k \xi_k}{|\boldsymbol{\xi}|^2} + O(|\boldsymbol{\xi}|^{-2}), \ \ |\boldsymbol{\xi}| \to \infty, \tag{2.104}$$

and showing that

$$\alpha_k = -D_k^\infty. \tag{2.105}$$

Let us apply Green's formula to ξ_j and ζ:

$$\begin{aligned}
\int_{\partial F} \xi_j \frac{\partial \zeta(\boldsymbol{\xi})}{\partial n} dS_\xi &= \int_{\partial F} \Big\{ \xi_j \frac{\partial \zeta(\boldsymbol{\xi})}{\partial n} - \zeta(\boldsymbol{\xi}) \frac{\partial \xi_j}{\partial n} \Big\} dS_\xi \\
&= -\lim_{R\to\infty} \int_{|\xi|=R} \Big\{ \xi_j \frac{\partial \zeta(\boldsymbol{\xi})}{\partial |\boldsymbol{\xi}|} - \zeta(\boldsymbol{\xi}) \frac{\partial \xi_j}{\partial |\boldsymbol{\xi}|} \Big\} dS_\xi \\
&= \frac{1}{\pi} \lim_{R\to\infty} \int_{|\xi|=R} \sum_{k=1}^{2} \frac{\alpha_k \xi_k \xi_j}{|\boldsymbol{\xi}|^3} dS_\xi = \alpha_j.
\end{aligned} \tag{2.106}$$

Then formulae (2.106) and (2.103) lead to (2.105).

2.2.2 *Asymptotic Property of the Regular Part of Green's Function in* $\mathbb{R}^2 \setminus F$

Asymptotic representation at infinity for the regular part of Green's function in $\mathbb{R}^2 \setminus F$ is given by the following Lemma.

Lemma 2.4. *The regular part* (2.99) *of* $\mathcal{G}$ *satisfies the estimate*

$$\left| h(\xi,\eta) - (2\pi)^{-1}\log|\xi|^{-1} + \zeta(\eta) - \frac{1}{2\pi}\sum_{j=1}^{2}\frac{D_j(\eta)\xi_j}{|\xi|^2}\right| \le \frac{\text{Const}}{|\xi|^2}, \tag{2.107}$$

as $|\xi| > 2$, *and* $\eta \in \mathbb{R}^2 \setminus F$.

Proof. Let

$$\beta(\xi,\eta) = h(\xi,\eta) - (2\pi)^{-1}\log|\xi|^{-1} + \zeta(\eta) - \frac{1}{2\pi}\sum_{j=1}^{2}\frac{D_j(\eta)\xi_j}{|\xi|^2}.$$

We have

$$\Delta_\eta \beta(\xi,\eta) = 0, \quad \eta \in \mathbb{R}^2 \setminus F,$$

and

$$\begin{aligned}\beta(\xi,\eta) &= -\frac{1}{4\pi}\log\left(1 - 2\frac{\xi\cdot\eta}{|\xi|^2} + \frac{|\eta|^2}{|\xi|^2}\right) - \frac{\xi\cdot\eta}{2\pi|\xi|^2}\\ &= -\frac{1}{4\pi|\xi|^2}\left\{|\eta|^2 - 2\frac{(\xi\cdot\eta)^2}{|\xi|^2} + O(|\xi|^{-1})\right\}\end{aligned} \tag{2.108}$$

as $\eta \in \partial F$. By (2.7)–(2.9) and Green's formula

$$\beta(\xi,\infty) = -\int_{\partial F}\beta(\xi,\eta)\frac{\partial\zeta(\eta)}{\partial n_\eta}dS_\eta,$$

which together with (2.108) and (2.35) implies

$$|\beta(\xi,\infty)| \le C\,|\xi|^{-2}.$$

Hence the maximum principle gives (2.107). □

2.2.3 Maximum Modulus Estimate for Solutions to the Mixed Problem in Ω_ε, with the Dirichlet Data on ∂F_ε

Lemma 2.5. *Let u be a solution of the mixed problem*

$$\Delta u(\mathbf{x}) = 0, \quad \mathbf{x} \in \Omega_\varepsilon, \tag{2.109}$$

$$\frac{\partial u}{\partial n}(\mathbf{x}) = \psi(\mathbf{x}), \quad \mathbf{x} \in \partial\Omega, \tag{2.110}$$

$$u(\mathbf{x}) = \varphi_\varepsilon(\mathbf{x}), \quad \mathbf{x} \in \partial F_\varepsilon, \tag{2.111}$$

where $\psi \in C(\partial\Omega)$, $\varphi_\varepsilon \in C(\partial F_\varepsilon)$, *and*

$$\int_{\partial\Omega} \psi(\mathbf{x}) ds = 0. \tag{2.112}$$

The solution u is sought in $C(\overline{\Omega}_\varepsilon)$, *and it is also assumed that* ∇u *is square integrable in a neighbourhood of* $\partial\Omega$. *Then there exists a positive constant C such that*

$$\|u\|_{C(\overline{\Omega}_\varepsilon)} \le \|\varphi_\varepsilon\|_{C(\partial F_\varepsilon)} + C\,\|\psi\|_{C(\partial\Omega)}. \tag{2.113}$$

Proof. (a) First, we introduce the inverse operator

$$\mathfrak{N}_\Omega : \psi \to w \tag{2.114}$$

for the interior Neumann problem in Ω

$$\Delta w(\mathbf{x}) = 0, \quad \mathbf{x} \in \Omega, \tag{2.115}$$

$$\frac{\partial w}{\partial n}(\mathbf{x}) = \psi(\mathbf{x}), \quad \mathbf{x} \in \partial\Omega, \tag{2.116}$$

with $\psi \in C(\partial\Omega)$ and

$$\int_{\partial\Omega} \psi(\mathbf{x}) dS_x = 0 \text{ and } \int_{\partial\Omega} w(\mathbf{x}) \frac{\partial}{\partial n}\big(\log|\mathbf{x}|\big) dS_x = 0. \tag{2.117}$$

Applying Green's formula to $w(\mathbf{x})$ and $N(\mathbf{x},\mathbf{y})$ in Ω we obtain

$$w(\mathbf{y}) = \int_{\partial\Omega} \Big(N(\mathbf{x},\mathbf{y})\psi(\mathbf{x}) + \frac{1}{2\pi} w(\mathbf{x}) \frac{\partial}{\partial n_x}(\log|\mathbf{x}|)\Big) dS_x.$$

Then the unique solution of (2.115)–(2.117) is given by

$$w(\mathbf{x}) = \int_{\partial\Omega} N(\mathbf{x},\mathbf{y})\psi(\mathbf{y})dS_y, \tag{2.118}$$

and

$$\max_{\overline{\Omega}} |w| \leq C\,\|\psi\|_{C(\partial\Omega)}. \tag{2.119}$$

(b) The solution u of (2.109)–(2.111) is sought in the form

$$u(\mathbf{x}) = w(\mathbf{x}) + v(\mathbf{x}), \tag{2.120}$$

where $w = \mathfrak{N}_\Omega\psi$ is defined by (2.118), whereas the second term v satisfies the problem

$$\Delta v(\mathbf{x}) = 0, \quad \mathbf{x} \in \Omega_\varepsilon, \tag{2.121}$$

$$\frac{\partial v}{\partial n}(\mathbf{x}) = 0, \quad \mathbf{x} \in \partial\Omega, \tag{2.122}$$

$$v(\mathbf{x}) = \varphi_\varepsilon(\mathbf{x}) - w(\mathbf{x}), \quad \mathbf{x} \in \partial F_\varepsilon. \tag{2.123}$$

According to the estimate (2.119) and the maximum principle for variational solutions of (2.121)–(2.123) (see, for example, Gilbarg and Trudinger [12]) we have

$$\max_{\overline{\Omega}_\varepsilon} |v| \leq \|\varphi_\varepsilon\|_{C(\partial F_\varepsilon)} + C\,\|\psi\|_{C(\partial\Omega)}. \tag{2.124}$$

Finally, using the representation (2.120), together with the estimates (2.119) and (2.124), we obtain the result (2.113). This completes the proof. □

2.2.4 *Approximation of Green's Function* $G_\varepsilon^{(D)}$

We give a uniform asymptotic formula for Green's function solving the problem (2.89)–(2.91).

Theorem 2.2. *Green's function* $G_\varepsilon^{(D)}(\mathbf{x},\mathbf{y})$ *for problem* (2.89)–(2.91) *admits the asymptotic representation*

$$G_\varepsilon^{(D)}(\mathbf{x},\mathbf{y}) = \mathcal{G}(\varepsilon^{-1}\mathbf{x},\varepsilon^{-1}\mathbf{y}) + N(\mathbf{x},\mathbf{y}) - (2\pi)^{-1}\log|\mathbf{x}-\mathbf{y}|^{-1} + R(0,0)$$
$$+ \varepsilon\mathbf{D}(\varepsilon^{-1}\mathbf{y})\cdot\nabla_y R(\mathbf{x},0) + \varepsilon\mathbf{D}(\varepsilon^{-1}\mathbf{x})\cdot\nabla_x R(0,\mathbf{y}) + r_\varepsilon(\mathbf{x},\mathbf{y}), \tag{2.125}$$

where $\mathcal{G}, N, R, \mathbf{D}$ *are defined in* (2.11)–(2.13), (2.92)–(2.94), (2.95), (2.100)–(2.102), *and*

$$|r_\varepsilon(\mathbf{x},\mathbf{y})| \le \text{Const}\,\varepsilon^2,$$

which is uniform with respect to $\mathbf{x},\mathbf{y} \in \Omega_\varepsilon$.

Proof. First, we describe the formal argument leading to (2.125). Let $\rho_\varepsilon(\mathbf{x},\mathbf{y}) = G_\varepsilon^{(D)}(\mathbf{x},\mathbf{y}) - \mathcal{G}(\varepsilon^{-1}\mathbf{x}, \varepsilon^{-1}\mathbf{y})$. This function satisfies the problem

$$\Delta_x \rho_\varepsilon(\mathbf{x},\mathbf{y}) = 0, \quad \mathbf{x},\mathbf{y} \in \Omega_\varepsilon, \tag{2.126}$$

$$\rho_\varepsilon(\mathbf{x},\mathbf{y}) = 0 \ \text{ when } \mathbf{x} \in \partial F_\varepsilon,\ \mathbf{y} \in \Omega_\varepsilon, \tag{2.127}$$

and

$$\begin{aligned}\frac{\partial \rho_\varepsilon}{\partial n_x}(\mathbf{x},\mathbf{y}) &= -\frac{\partial}{\partial n_x}\Big(\frac{1}{2\pi}\log|\mathbf{x}-\mathbf{y}|^{-1} - h(\varepsilon^{-1}\mathbf{x}, \varepsilon^{-1}\mathbf{y})\Big) \\ &= -\frac{\partial}{\partial n_x}\Big(\frac{1}{2\pi}\log|\mathbf{x}-\mathbf{y}|^{-1} - N(\mathbf{x},\mathbf{y})\Big) \\ &\quad + \frac{\partial}{\partial n_x}\Big(\frac{1}{2\pi}\log|\mathbf{x}| + h(\varepsilon^{-1}\mathbf{x}, \varepsilon^{-1}\mathbf{y})\Big),\end{aligned} \tag{2.128}$$

where $\mathbf{x} \in \partial\Omega$, $\mathbf{y} \in \Omega_\varepsilon$. Here $h(\boldsymbol{\xi}, \boldsymbol{\eta})$ is the regular part of Green's function $\mathcal{G}$ in $\mathbb{R}^2 \setminus F$. Taking into account (2.95), we deduce that

$$\rho_\varepsilon(\mathbf{x},\mathbf{y}) = -R(\mathbf{x},\mathbf{y}) + R(0,0) + \mathcal{R}_\varepsilon(\mathbf{x},\mathbf{y}), \tag{2.129}$$

where $R(\mathbf{x},\mathbf{y})$ is the regular part of the Neumann function $N(\mathbf{x},\mathbf{y})$ in Ω, and $\mathcal{R}_\varepsilon$ is harmonic in Ω_ε and satisfies the boundary conditions

$$\frac{\partial \mathcal{R}_\varepsilon}{\partial n_x}(\mathbf{x},\mathbf{y}) = \frac{\partial}{\partial n_x}\Big(\frac{1}{2\pi}\log|\mathbf{x}| + h(\varepsilon^{-1}\mathbf{x}, \varepsilon^{-1}\mathbf{y})\Big) \ \text{ as } \ \mathbf{x} \in \partial\Omega,\ \mathbf{y} \in \Omega_\varepsilon, \tag{2.130}$$

$$\mathcal{R}_\varepsilon(\mathbf{x},\mathbf{y}) = \mathbf{x}\cdot\nabla_x R(0,\mathbf{y}) + O(\varepsilon^2) \ \text{ as } \ \mathbf{x} \in \partial F_\varepsilon,\ \mathbf{y} \in \Omega_\varepsilon. \tag{2.131}$$

The asymptotics of $h(\boldsymbol{\xi}, \boldsymbol{\eta})$ given by Lemma 2.4, can be used in evaluation of the right-hand side in (2.130).

The boundary condition (2.131) can be written as

$$\mathcal{R}_\varepsilon(\mathbf{x},\mathbf{y}) - \varepsilon\mathbf{D}(\boldsymbol{\xi})\cdot\nabla_x R(0,\mathbf{y}) = O(\varepsilon^2),$$

for $\mathbf{x} \in \partial F_\varepsilon$, $\mathbf{y} \in \Omega_\varepsilon$. In turn, the boundary condition (2.130) is reduced to

$$\frac{\partial}{\partial n_x}\Big\{\mathcal{R}_\varepsilon(\mathbf{x},\mathbf{y}) - \varepsilon\mathbf{D}(\boldsymbol{\eta})\cdot\nabla_y R(\mathbf{x},0)\Big\} = O(\varepsilon^2),$$

when $\mathbf{x} \in \partial\Omega,\ \ \mathbf{y} \in \Omega_\varepsilon$. Hence, representation (2.129) of ρ_ε can be updated to the form

$$\rho_\varepsilon(\mathbf{x},\mathbf{y}) = -R(\mathbf{x},\mathbf{y}) + R(0,0) \\ + \varepsilon\mathbf{D}(\boldsymbol{\xi}) \cdot \nabla_x R(0,\mathbf{y}) + \varepsilon\mathbf{D}(\boldsymbol{\eta}) \cdot \nabla_y R(\mathbf{x},0) + \mathcal{R}_\varepsilon^{(1)}(\mathbf{x},\mathbf{y}), \tag{2.132}$$

where the principal part of $\mathcal{R}_\varepsilon^{(1)}(\mathbf{x},\mathbf{y})$ compensates for the leading term of the discrepancy $\varepsilon^2\boldsymbol{\xi} \cdot \nabla_x\big(\mathbf{D}(\boldsymbol{\eta}) \cdot \nabla_y R(\mathbf{x},0)\big)\big|_{\mathbf{x}=0}$ brought by the term $\varepsilon\mathbf{D}(\boldsymbol{\eta}) \cdot \nabla_y R(\mathbf{x},0)$ into the boundary condition (2.127) on ∂F_ε. This leads to the required formula (2.125).

For the remainder $r_\varepsilon(\mathbf{x},\mathbf{y})$ in the asymptotic formula (2.125), we verify by the direct substitution that

$$\Delta_x r_\varepsilon(\mathbf{x},\mathbf{y}) = 0, \ \ \mathbf{x},\mathbf{y} \in \Omega_\varepsilon, \tag{2.133}$$

and that the boundary condition (2.90) implies

$$r_\varepsilon(\mathbf{x},\mathbf{y}) = R(0,\mathbf{y}) - R(0,0) + \mathbf{x} \cdot \nabla_x R(0,\mathbf{y}) \\ -\varepsilon\mathbf{D}(\mathbf{x}/\varepsilon) \cdot \nabla_x R(0,\mathbf{y}) + O(\varepsilon^2) = O(\varepsilon^2) \text{ as } \mathbf{x} \in \partial\omega_\varepsilon,\ \mathbf{y} \in \Omega_\varepsilon, \tag{2.134}$$

where $\mathbf{D}(\mathbf{x}/\varepsilon) = \varepsilon^{-1}\mathbf{x}$ for $\mathbf{x} \in \partial\omega_\varepsilon$, and formula (2.96) was used to state that $R(0,\mathbf{y})$ is independent of $\mathbf{y}$. In turn, the second boundary condition (2.91), together with formula (2.107), yields

$$\frac{\partial r_\varepsilon}{\partial n_x}(\mathbf{x},\mathbf{y}) = \frac{\partial}{\partial n_x}\Big(h(\varepsilon^{-1}\mathbf{x},\varepsilon^{-1}\mathbf{y}) - \frac{1}{2\pi}\log|\mathbf{x}|^{-1}\Big) \\ -\varepsilon\mathbf{D}(\varepsilon^{-1}\mathbf{y}) \cdot \frac{\partial}{\partial n_x}\Big(\nabla_y R(\mathbf{x},0)\Big) + O(\varepsilon^2) \\ = \varepsilon\sum_{j=1}^{2} D_j(\varepsilon^{-1}\mathbf{y})\frac{\partial}{\partial n_x}\Big(\frac{x_j}{2\pi|\mathbf{x}|^2}\Big) \\ -\varepsilon\mathbf{D}(\varepsilon^{-1}\mathbf{y}) \cdot \frac{\partial}{\partial n_x}\Big(\nabla_y R(\mathbf{x},0)\Big) + O(\varepsilon^2) = O(\varepsilon^2), \tag{2.135}$$

as $\mathbf{x} \in \partial\Omega,\ \mathbf{y} \in \Omega_\varepsilon$.

It can also be verified that $\int_{\partial\Omega} \frac{\partial}{\partial n_x} r_\varepsilon(\mathbf{x},\mathbf{y})dS_x = 0$. Indeed,

$$-\int_{\partial\Omega} \frac{\partial}{\partial n_x} r_\varepsilon(\mathbf{x},\mathbf{y})dS_x = \int_{\partial\Omega} \frac{\partial}{\partial n_x}\Big\{\mathcal{G}(\varepsilon^{-1}\mathbf{x},\varepsilon^{-1}\mathbf{y}) + \frac{1}{2\pi}\log\frac{|\mathbf{x}-\mathbf{y}|}{|\mathbf{x}|} \\ +\varepsilon\mathbf{D}(\varepsilon^{-1}\mathbf{y}) \cdot \nabla_y R(\mathbf{x},0) + \varepsilon\mathbf{D}(\varepsilon^{-1}\mathbf{x}) \cdot \nabla_x R(0,\mathbf{y})\Big\}dS_x$$

$$= \varepsilon \int_{\partial\Omega} \frac{\partial}{\partial n_x} \Big\{ \mathbf{D}(\varepsilon^{-1}\mathbf{y}) \cdot \nabla_y \Big((2\pi)^{-1} \log |\mathbf{x}-\mathbf{y}|^{-1} - N(\mathbf{x},\mathbf{y}) \Big) \Big|_{\mathbf{y}=0} \Big\} dS_x$$

$$= \frac{\varepsilon}{2\pi} \int_{\partial\Omega} \frac{\partial}{\partial n_x} \Big\{ \mathbf{D}(\varepsilon^{-1}\mathbf{y}) \cdot \frac{\mathbf{x}}{|\mathbf{x}|^2} \Big\} dS_x = 0.$$

Using (2.134), (2.135), together with Lemma 2.5, we complete the proof. □

2.2.5 *Simpler Asymptotic Representation of Green's Function $G_\varepsilon^{(D)}$*

Two corollaries, which will be formulated here, follow from Theorem 2.2. They include simplified asymptotic formulae for the Green's function, which are efficient for the cases when both $\mathbf{x}$ and $\mathbf{y}$ are distant from F_ε or both $\mathbf{x}$ and $\mathbf{y}$ are sufficiently close to F_ε.

Corollary 2.4. *Let* $\min\{|\mathbf{x}|, |\mathbf{y}|\} > 2\varepsilon$. *Then the asymptotic formula* (2.125) *is simplified to the form*

$$\begin{aligned} G_\varepsilon^{(D)}(\mathbf{x},\mathbf{y}) &= N(\mathbf{x},\mathbf{y}) - (2\pi)^{-1}\log\varepsilon + \zeta_\infty + R(0,0) \\ &+ (2\pi)^{-1}\log(|\mathbf{x}||\mathbf{y}|) - \frac{\varepsilon}{2\pi}\mathbf{D}^\infty \cdot \Big(\mathbf{x}|\mathbf{x}|^{-2} + \mathbf{y}|\mathbf{y}|^{-2} \Big) \\ &+ \varepsilon \mathbf{D}^\infty \cdot \Big(\nabla_x R(0,\mathbf{y}) + \nabla_y R(\mathbf{x},0) \Big) \\ &+ O(\varepsilon^2 |\mathbf{x}|^{-1}|\mathbf{y}|^{-1}), \end{aligned} \tag{2.136}$$

where R is the regular part of Neumann's function N in Ω.

Proof. Estimate (2.107) can be written in the form

$$\begin{aligned} h(\boldsymbol{\xi},\boldsymbol{\eta}) &= (2\pi)^{-1}\log(|\boldsymbol{\xi}||\boldsymbol{\eta}|)^{-1} - \zeta_\infty \\ &+ \frac{\varepsilon}{2\pi}\sum_{j=1}^{2} D_j^\infty \Big(\frac{x_j}{|\mathbf{x}|^2} + \frac{y_j}{|\mathbf{y}|^2} \Big) + O(\varepsilon^2|\mathbf{x}|^{-1}|\mathbf{y}|^{-1}). \end{aligned} \tag{2.137}$$

Using (2.99), (2.125) and (2.137) we obtain

$$\begin{aligned} G_\varepsilon^{(D)}(\mathbf{x},\mathbf{y}) &= -\frac{1}{2\pi}\log\varepsilon + \frac{1}{2\pi}\log\frac{|\mathbf{x}||\mathbf{y}|}{|\mathbf{x}-\mathbf{y}|} + \zeta_\infty \\ &- \frac{\varepsilon}{2\pi}\sum_{j=1}^{2} D_j^\infty \Big(\frac{x_j}{|\mathbf{x}|^2} + \frac{y_j}{|\mathbf{x}|^2} \Big) + O(\varepsilon^2|\mathbf{x}|^{-1}|\mathbf{y}|^{-1}) \end{aligned}$$

$$
\begin{aligned}
&+ N(\mathbf{x}, \mathbf{y}) - (2\pi)^{-1} \log |\mathbf{x} - \mathbf{y}|^{-1} + R(0, 0) \\
&+ \varepsilon \mathbf{D}^{\infty} \cdot \Big(\nabla_y R(\mathbf{x}, 0) + \nabla_x R(0, \mathbf{y}) \Big) \\
&+ \varepsilon^2 O(|\mathbf{x}|^{-1} + |\mathbf{y}|^{-1}).
\end{aligned} \tag{2.138}
$$

Rearranging the terms in (2.138) and taking into account that the remainder terms in the above formula are $O(\varepsilon^2 |\mathbf{x}|^{-1} |\mathbf{y}|^{-1})$, we arrive at (2.136). □

Formula (2.136) is efficient when both **x** and **y** are sufficiently distant from F_ε.

The next corollary of Theorem 2.2 gives the representation of $G_\varepsilon^{(D)}$, which is effective for the case when both **x** and **y** are sufficiently close to F_ε.

Corollary 2.5. *The following asymptotic formula for Green's function* $G_\varepsilon^{(D)}$ *of the boundary value problem* (2.89)–(2.91) *holds*

$$
\begin{aligned}
G_\varepsilon^{(D)}(\mathbf{x}, \mathbf{y}) = {} & \mathcal{G}(\varepsilon^{-1}\mathbf{x}, \varepsilon^{-1}\mathbf{y}) - (\mathbf{x} - \varepsilon \mathbf{D}(\varepsilon^{-1}\mathbf{x})) \cdot \nabla_x R(0, \mathbf{y}) \\
& - (\mathbf{y} - \varepsilon \mathbf{D}(\varepsilon^{-1}\mathbf{y})) \cdot \nabla_y R(\mathbf{x}, 0) \\
& + O(|\mathbf{x}|^2 + |\mathbf{y}|^2 + \varepsilon^2),
\end{aligned} \tag{2.139}
$$

for $\mathbf{x}, \mathbf{y} \in \Omega_\varepsilon$. *(The term* ε^2 *in the remainder can be omitted if the interior of* F *is nonempty and contains the origin.)*

Proof. Using the Taylor expansion of $R(\mathbf{x}, \mathbf{y})$ in a neighbourhood of the origin we reduce the formula (2.125) to the form

$$
\begin{aligned}
G_\varepsilon^{(D)}(\mathbf{x}, \mathbf{y}) = {} & \mathcal{G}(\varepsilon^{-1}\mathbf{x}, \varepsilon^{-1}\mathbf{y}) - R(\mathbf{x}, \mathbf{y}) + R(0, 0) \\
& + \varepsilon \mathbf{D}(\varepsilon^{-1}\mathbf{y}) \cdot \nabla_y R(\mathbf{x}, 0) + \varepsilon \mathbf{D}(\varepsilon^{-1}\mathbf{x}) \cdot \nabla_x R(0, \mathbf{y}) + O(\varepsilon^2) \\
= {} & \mathcal{G}(\varepsilon^{-1}\mathbf{x}, \varepsilon^{-1}\mathbf{y}) \\
& - \mathbf{x} \cdot \nabla_x R(0, \mathbf{y}) - \mathbf{y} \cdot \nabla_y R(\mathbf{x}, 0) + O(|\mathbf{x}|^2 + |\mathbf{y}|^2) \\
& + \varepsilon \mathbf{D}(\varepsilon^{-1}\mathbf{y}) \cdot \nabla_y R(\mathbf{x}, 0) + \varepsilon \mathbf{D}(\varepsilon^{-1}\mathbf{x}) \cdot \nabla_x R(0, \mathbf{y}) + O(\varepsilon^2).
\end{aligned} \tag{2.140}
$$

By rearranging the terms in the above formula, we arrive at (2.139). □

2.3 The Neumann Function for a Planar Domain with a Small Hole or Crack

It is noted that in the previous sections, boundary conditions of the Dirichlet type were set at a part of the boundary of Ω_ε. Now, we consider the case when $\partial\Omega_\varepsilon$ is subject to the Neumann boundary conditions. Here, the set F_ε is the same as in Sect. 2.1.

The *Neumann function* $N_\varepsilon(\mathbf{x}, \mathbf{y})$ for $\Omega_\varepsilon \subset \mathbb{R}^2$ is defined as a solution of the boundary value problem

$$\Delta_x N_\varepsilon(\mathbf{x}, \mathbf{y}) + \delta(\mathbf{x} - \mathbf{y}) = 0, \quad \mathbf{x}, \mathbf{y} \in \Omega_\varepsilon, \tag{2.141}$$

$$\frac{\partial}{\partial n_x}\Big(N_\varepsilon(\mathbf{x}, \mathbf{y}) + (2\pi)^{-1} \log|\mathbf{x}| \Big) = 0, \quad \mathbf{x} \in \partial\Omega, \ \mathbf{y} \in \Omega_\varepsilon, \tag{2.142}$$

$$\frac{\partial N_\varepsilon}{\partial n_x}(\mathbf{x}, \mathbf{y}) = 0, \quad \mathbf{x} \in \partial F_\varepsilon, \ \mathbf{y} \in \Omega_\varepsilon. \tag{2.143}$$

In addition, we require the orthogonality condition, which provides the symmetry of $N_\varepsilon(\mathbf{x}, \mathbf{y})$

$$\int_{\partial\Omega} N_\varepsilon(\mathbf{x}, \mathbf{y}) \frac{\partial}{\partial n} \log|\mathbf{x}| dS_x = 0. \tag{2.144}$$

The regular part $R_\varepsilon(\mathbf{x}, \mathbf{y})$ of the Neumann function is defined by

$$R_\varepsilon(\mathbf{x}, \mathbf{y}) = \frac{1}{2\pi} \log|\mathbf{x} - \mathbf{y}|^{-1} - N_\varepsilon(\mathbf{x}, \mathbf{y}).$$

2.3.1 Special Solutions of Model Problems

As in the previous sections, we consider two limit domains independent of the small parameter ε: the domain Ω (with no hole), and the unbounded domain $\mathbb{R}^2 \setminus F$ that represents the scaled exterior of the small hole. As always, the scaled coordinates $\boldsymbol{\xi} = \varepsilon^{-1}\mathbf{x}$ and $\boldsymbol{\eta} = \varepsilon^{-1}\mathbf{y}$ will be used.

The Neumann function $N(\mathbf{x}, \mathbf{y})$ of Ω is defined by (2.92)–(2.94), and the regular part $R(\mathbf{x}, \mathbf{y})$ of $N(\mathbf{x}, \mathbf{y})$ is the same as in (2.95).

We shall use the vector function $\mathcal{D}$ already defined in Sect. 2.1.

Another model field to be used is the Neumann function $\mathcal{N}(\boldsymbol{\xi}, \boldsymbol{\eta})$ in $\mathbb{R}^2 \setminus F$, as in (2.15), whose regular part h_N satisfies the problem (2.16)–(2.18).

2.3.2 Maximum Modulus Estimate for Solutions to the Neumann Problem in Ω_ε

First, we formulate and prove the auxiliary Lemma required for the forthcoming estimate of the remainder term in the approximation of N_ε.

Lemma 2.6. *Let u be a solution of the Neumann boundary value problem*

$$\Delta u(\mathbf{x}) = 0, \quad \mathbf{x} \in \Omega_\varepsilon, \tag{2.145}$$

$$\frac{\partial u}{\partial n}(\mathbf{x}) = \psi(\mathbf{x}), \quad \mathbf{x} \in \partial\Omega, \tag{2.146}$$

$$\frac{\partial u}{\partial n}(\mathbf{x}) = \varphi_\varepsilon(\mathbf{x}), \quad \mathbf{x} \in \partial F_\varepsilon, \tag{2.147}$$

where $\psi \in C(\partial\Omega)$, $\varphi_\varepsilon \in L_\infty(\partial F_\varepsilon)$, *and*

$$\int_{\partial F_\varepsilon} \varphi_\varepsilon(\mathbf{x}) ds = 0 \text{ and } \int_{\partial\Omega} \psi(\mathbf{x}) ds = 0. \tag{2.148}$$

The solution u is sought in $C(\overline{\Omega}_\varepsilon)$, *and it is also assumed that* ∇u *is square integrable in a neighbourhood of* $\partial\Omega_\varepsilon$, *and*

$$\left| \int_{\partial\Omega} u(\mathbf{x}) \frac{\partial}{\partial n} \big(\log |\mathbf{x}|\big) ds \right| \leq \text{Const} \,\{\|\psi\|_{C(\partial\Omega)} + \varepsilon \|\varphi_\varepsilon\|_{L_\infty(\partial F_\varepsilon)}\}. \tag{2.149}$$

Then there exists a positive constant C, independent of ε *and such that*

$$\|u\|_{C(\overline{\Omega}_\varepsilon)} \leq C \big\{ \|\psi\|_{C(\partial\Omega)} + \varepsilon \|\varphi_\varepsilon\|_{L_\infty(\partial F_\varepsilon)} \big\}. \tag{2.150}$$

Proof. (a) We use the operators $\mathfrak{N}$ and $\mathfrak{N}_\Omega$ of model problems (2.60)–(2.62) and (2.115)–(2.117) introduced in Sects. 2.1 and 2.2.

(b) We begin with the case of the homogeneous boundary condition on $\partial\Omega$, i.e.

$$\Delta u_1(\mathbf{x}) = 0, \quad \mathbf{x} \in \Omega_\varepsilon, \tag{2.151}$$

$$\frac{\partial u_1}{\partial n}(\mathbf{x}) = 0, \quad \mathbf{x} \in \partial\Omega, \tag{2.152}$$

$$\frac{\partial u_1}{\partial n}(\mathbf{x}) = \varphi_\varepsilon(\mathbf{x}), \quad \mathbf{x} \in \partial F_\varepsilon, \tag{2.153}$$

where the right-hand side φ_ε is such that

$$\int_{\partial F_\varepsilon} \varphi_\varepsilon(\mathbf{x}) ds = 0.$$

The operator $\mathfrak{N}_\varepsilon$ is defined as in (2.64), so that

$$(\mathfrak{N}_\varepsilon \varphi_\varepsilon)(\mathbf{x}) = (\mathfrak{N}\varphi)(\boldsymbol{\xi}),$$

where $\boldsymbol{\xi} = \varepsilon^{-1}\mathbf{x}$ and $\varphi_\varepsilon(\mathbf{x}) = \varepsilon^{-1}\varphi(\varepsilon^{-1}\mathbf{x})$.

The solution u_1 is sought in the form

$$u_1 = \mathfrak{N}_\varepsilon g_\varepsilon - \mathfrak{N}_\Omega \Big(\frac{\partial}{\partial n} (\mathfrak{N}_\varepsilon g_\varepsilon)_{\partial\Omega} \Big), \tag{2.154}$$

where g_ε is an unknown function such that

$$\int_{\partial F} g(\boldsymbol{\xi}) ds_\xi = 0.$$

By Lemma 2.1, we have

$$|\mathfrak{N} g(\boldsymbol{\xi})| \leq C\varepsilon \|g\|_{L_\infty(\partial F)}, \tag{2.155}$$

and

$$\max_{\overline{\Omega}_\varepsilon} |\mathfrak{N}_\varepsilon g_\varepsilon| \leq C\varepsilon \|g_\varepsilon\|_{L_\infty(\partial F)}. \tag{2.156}$$

It follows from (2.154) that $\frac{\partial}{\partial n} u_1(\mathbf{x}) = 0$ when $\mathbf{x} \in \partial\Omega$, and on the boundary ∂F_ε we have

$$\varphi_\varepsilon = g_\varepsilon + S_\varepsilon g_\varepsilon, \tag{2.157}$$

where

$$S_\varepsilon g_\varepsilon = -\frac{\partial}{\partial n} \Big(\mathfrak{N}_\Omega \big(\frac{\partial}{\partial n} (\mathfrak{N}_\varepsilon g_\varepsilon)_{\partial\Omega} \big) \Big) \quad \text{on } \partial F_\varepsilon. \tag{2.158}$$

Taking into account Lemma 2.1 and the definitions of $\mathfrak{N}_\Omega$ and $\mathfrak{N}_\varepsilon$, as in (2.114) and (2.59), (2.64), we deduce that

$$\max_{\partial\Omega} |\nabla(\mathfrak{N}_\varepsilon g_\varepsilon)| \leq \text{Const } \varepsilon^2 \|g_\varepsilon\|_{L_\infty(\partial F_\varepsilon)},$$

and

$$\|S_\varepsilon g_\varepsilon\|_{L_\infty(\partial F_\varepsilon)} \leq \text{Const } \varepsilon^2 \|g_\varepsilon\|_{L_\infty(\partial F_\varepsilon)}.$$

Owing to the smallness of the norm of the operator S_ε we can write

$$\|g_\varepsilon\|_{L_\infty(\partial F_\varepsilon)} \leq \text{Const } \|\varphi_\varepsilon\|_{L_\infty(\partial F_\varepsilon)}.$$

Following (2.118), (2.119), (2.154) and (2.156) we deduce (2.149) and

$$\max_{\overline{\Omega}_\varepsilon} |u_1| \leq \text{Const } \varepsilon \|\varphi_\varepsilon\|_{L_\infty(\partial F_\varepsilon)}. \tag{2.159}$$

(c) Next, we consider the problem (2.145)–(2.148) with the homogeneous data on $\partial\omega_\varepsilon$. The corresponding solution u_2 is written in the form

$$u_2 = \mathfrak{N}_\Omega \psi + v, \tag{2.160}$$

where the harmonic function v satisfies zero boundary condition on $\partial\Omega$, whereas the condition (2.153) is replaced by

$$\frac{\partial}{\partial n} v(\mathbf{x}) = -\frac{\partial}{\partial n}(\mathfrak{N}_\Omega \psi)(\mathbf{x}), \quad \mathbf{x} \in \partial F_\varepsilon,$$

and by part (b)

$$\max_{\overline{\Omega}_\varepsilon} |v| \le \text{Const } \|\psi\|_{C(\partial\Omega)}.$$

The function v and hence u_2 satisfy (2.149).
Following (2.118), (2.119) and (2.160) we deduce

$$\max_{\overline{\Omega}_\varepsilon} |u_2| \le \text{Const } \|\psi\|_{C(\partial\Omega)}. \tag{2.161}$$

Combining estimates (2.159) and (2.161) we complete the proof. □

2.3.3 *Asymptotic Approximation of* N_ε

Now we state the theorem, which gives a uniform asymptotic formula for the Neumann function N_ε.

Theorem 2.3. *The Neumann function* $N_\varepsilon(\mathbf{x}, \mathbf{y})$ *of the domain* Ω_ε *defined in* (2.141)–(2.144) *satisfies*

$$\begin{aligned} N_\varepsilon(\mathbf{x}, \mathbf{y}) = \; & N(\mathbf{x}, \mathbf{y}) - h_N(\varepsilon^{-1}\mathbf{x}, \varepsilon^{-1}\mathbf{y}) \\ & + \varepsilon \boldsymbol{\mathcal{D}}(\varepsilon^{-1}\mathbf{x}) \cdot \nabla_x R(0, \mathbf{y}) \\ & + \varepsilon \boldsymbol{\mathcal{D}}(\varepsilon^{-1}\mathbf{y}) \cdot \nabla_y R(\mathbf{x}, 0) + r_\varepsilon(\mathbf{x}, \mathbf{y}), \end{aligned} \tag{2.162}$$

where

$$|r_\varepsilon(\mathbf{x}, \mathbf{y})| \le \text{Const } \varepsilon^2 \tag{2.163}$$

uniformly with respect to $\mathbf{x}, \mathbf{y} \in \Omega_\varepsilon$.

Proof. We begin with a formal argument leading to the approximation (2.162). Consider the first three terms in the right-hand side of (2.162) and let

$$r_\varepsilon^{(1)}(\mathbf{x}, \mathbf{y}) = N_\varepsilon(\mathbf{x}, \mathbf{y}) - N(\mathbf{x}, \mathbf{y}) + h_N(\boldsymbol{\xi}, \boldsymbol{\eta}) - \varepsilon \boldsymbol{\mathcal{D}}(\boldsymbol{\xi}) \cdot \nabla_x R(0, \mathbf{y}). \tag{2.164}$$

The function $r_\varepsilon^{(1)}$ is harmonic in Ω_ε, and the direct substitution into the boundary conditions (2.142) and (2.143) gives

$$\begin{aligned}\frac{\partial r_\varepsilon^{(1)}}{\partial n_x}(\mathbf{x},\mathbf{y}) &= -\frac{\partial}{\partial n_x}\Big(\frac{1}{2\pi}\log|\mathbf{x}-\mathbf{y}|^{-1}\Big) + \frac{\partial}{\partial n_x}\big(h_N(\varepsilon^{-1}\mathbf{x},\varepsilon^{-1}\mathbf{y})\big)\\ &\quad+\mathbf{n}\cdot\nabla_x R(0,\mathbf{y}) - \varepsilon\frac{\partial}{\partial n_x}\mathcal{D}(\varepsilon^{-1}\mathbf{x})\cdot\nabla_x R(0,\mathbf{y}) + O(\varepsilon)\\ &= O(\varepsilon), \quad \text{for } \mathbf{x}\in\partial F_\varepsilon,\ \mathbf{y}\in\Omega_\varepsilon,\end{aligned} \tag{2.165}$$

and

$$\begin{aligned}\frac{\partial r_\varepsilon^{(1)}}{\partial n_x}(\mathbf{x},\mathbf{y}) &= \frac{\partial}{\partial n_x}\big(h_N(\varepsilon^{-1}\mathbf{x},\varepsilon^{-1}\mathbf{y})\big) + O(\varepsilon^2)\\ &= \varepsilon\mathcal{D}(\varepsilon^{-1}\mathbf{y})\cdot\frac{\partial}{\partial n_x}\nabla_y R(\mathbf{x},0) + O(\varepsilon^2),\\ &\quad \text{for } \mathbf{x}\in\partial\Omega,\ \mathbf{y}\in\Omega_\varepsilon.\end{aligned} \tag{2.166}$$

Thus, $r_\varepsilon^{(1)}$ can be approximated as

$$r_\varepsilon^{(1)}(\mathbf{x},\mathbf{y}) = \varepsilon\mathcal{D}(\varepsilon^{-1}\mathbf{y})\cdot\nabla_y R(\mathbf{x},0) + O(\varepsilon^2),$$

and together with the representation (2.164), this leads to the required formula (2.162).

Finally, the direct substitution of (2.162) into (2.141)–(2.143) yields that the remainder term $r_\varepsilon(\mathbf{x},\mathbf{y})$ satisfies the problem (2.145)–(2.148), with

$$\max_{\mathbf{x}\in\partial\Omega}|\psi(\mathbf{x},\mathbf{y})| \le \text{Const}\ \varepsilon^2$$

and

$$\max_{\mathbf{x}\in\partial F_\varepsilon}|\varphi_\varepsilon(\varepsilon^{-1}\mathbf{x},\varepsilon^{-1}\mathbf{y})| \le \text{Const}\ \varepsilon$$

for all $\mathbf{y}\in\Omega_\varepsilon$. Then the estimate (2.163) follows from Lemma 2.6. □

2.3.4 Simpler Asymptotic Representation of Neumann's Function N_ε

Two corollaries, formulated in this section, follow from Theorem 2.3. They include asymptotic formulae for the Neumann's function, which are efficient when either both $\mathbf{x}$ and $\mathbf{y}$ are distant from F_ε or both $\mathbf{x}$ and $\mathbf{y}$ are sufficiently close to F_ε.

Corollary 2.6. *Let* $\min\{|\mathbf{x}|, |\mathbf{y}|\} > 2\varepsilon$. *Then*

$$N_\varepsilon(\mathbf{x}, \mathbf{y}) = N(\mathbf{x}, \mathbf{y}) - \frac{\varepsilon^2}{4\pi^2} \frac{\mathbf{x}^T}{|\mathbf{x}|^2} \mathcal{P} \frac{\mathbf{y}^T}{|\mathbf{y}|^2}$$
$$+ \frac{\varepsilon^2}{2\pi} \Big\{ \frac{\mathbf{x}^T}{|\mathbf{x}|^2} \mathcal{P} \nabla_x R(0, \mathbf{y}) + \frac{\mathbf{y}^T}{|\mathbf{y}|^2} \mathcal{P} \nabla_y R(\mathbf{x}, 0) \Big\} \tag{2.167}$$
$$+ \varepsilon^2 O(|\mathbf{x}|^{-2} + |\mathbf{y}|^{-2}),$$

where R *is the regular part of Neumann's function* N *in* Ω, *and* $\mathcal{P}$ *is the dipole matrix for* F, *as defined in* (2.22).

Proof. The proof is similar to that of Corollary 2.2, and it uses formula (2.53) for the regular part h_N of the Neumann function in $\mathbb{R}^2 \setminus F$, together with the asymptotic representation (2.22) of the dipole fields $\mathcal{D}_j$ in $\mathbb{R}^2 \setminus F$. □

Next, we state a proposition similar to Corollaries 2.3 and 2.5 formulated earlier for Green's functions $G_\varepsilon^{(D)}$ and $G_\varepsilon^{(N)}$.

Corollary 2.7. *Neumann's function* N_ε, *defined by* (2.141)–(2.144), *satisfies the asymptotic formula*

$$N_\varepsilon(\mathbf{x}, \mathbf{y}) = (2\pi)^{-1} \log |\mathbf{x} - \mathbf{y}|^{-1} - R(0, 0) - h_N(\varepsilon^{-1}\mathbf{x}, \varepsilon^{-1}\mathbf{y}) \tag{2.168}$$
$$- \Big(\mathbf{x} - \varepsilon \mathcal{D}(\varepsilon^{-1}\mathbf{x})\Big) \cdot \nabla_x R(0, \mathbf{y}) - \Big(\mathbf{y} - \varepsilon \mathcal{D}(\varepsilon^{-1}\mathbf{y})\Big) \cdot \nabla_y R(\mathbf{x}, 0)$$
$$+ O(|\mathbf{x}|^2 + |\mathbf{y}|^2 + \varepsilon^2),$$

for $\mathbf{x}, \mathbf{y} \in \Omega_\varepsilon$. *(As in Corollaries* 2.3 *and* 2.5, ε^2 *in the remainder can be omitted if the interior of* F *is nonempty and contains the origin.)*

Proof. The proof is similar to that of Corollary 2.3, and it employs the linear approximation of the regular part R of Neumann's function in a neighbourhood of the origin. □

Although, the formulation of Corollary 2.7 is valid for all $\mathbf{x}, \mathbf{y} \in \Omega_\varepsilon$, the asymptotic formula (2.168) becomes effective when both $\mathbf{x}$ and $\mathbf{y}$ are sufficiently close to F_ε.

2.4 Asymptotic Approximations of Green's Kernels for Mixed and Neumann's Problems in Three Dimensions

This section includes asymptotic formulae for Green's kernels $G_\varepsilon^{(D)}$, $G_\varepsilon^{(N)}$ and N_ε in $\Omega_\varepsilon \subset \mathbb{R}^3$. The special solutions of model problems differ from the corresponding solutions used for the two-dimensional case. The uniform asymptotic formulae of

Green's kernels are accompanied by simpler representations, which are efficient when certain constraints are imposed on the independent variables.

2.4.1 Special Solutions of Model Problems in Limit Domains

Here, we describe the functions $G, \mathcal{G}, N, \mathcal{N}$, defined in the limit domains and used for the approximation of Green's kernels.

1. The notation G is used for Green's function of the Dirichlet problem in $\Omega \subset \mathbb{R}^3$:

$$G(\mathbf{x}, \mathbf{y}) = (4\pi|\mathbf{x} - \mathbf{y}|)^{-1} - H(\mathbf{x}, \mathbf{y}). \tag{2.169}$$

Here H is the regular part of G, and it is a unique solution of the Dirichlet problem

$$\Delta_x H(\mathbf{x}, \mathbf{y}) = 0, \ \ \mathbf{x}, \mathbf{y} \in \Omega, \tag{2.170}$$

$$H(\mathbf{x}, \mathbf{y}) = (4\pi|\mathbf{x} - \mathbf{y}|)^{-1}, \ \ \mathbf{x} \in \partial\Omega, \ \mathbf{y} \in \Omega. \tag{2.171}$$

2. Green's function $\mathcal{G}$ for the Dirichlet problem in $\mathbb{R}^3 \setminus F$ is defined as a unique solution of the problem

$$\Delta_{\xi} \mathcal{G}(\boldsymbol{\xi}, \boldsymbol{\eta}) + \delta(\boldsymbol{\xi} - \boldsymbol{\eta}) = 0, \ \ \boldsymbol{\xi}, \boldsymbol{\eta} \in \mathbb{R}^3 \setminus F, \tag{2.172}$$

$$\mathcal{G}(\boldsymbol{\xi}, \boldsymbol{\eta}) = 0, \ \ \boldsymbol{\xi} \in \partial F, \ \boldsymbol{\eta} \in \mathbb{R}^3 \setminus F, \tag{2.173}$$

$$\mathcal{G}(\boldsymbol{\xi}, \boldsymbol{\eta}) \to 0 \ \text{ as } |\boldsymbol{\xi}| \to \infty \text{ and } \boldsymbol{\eta} \in \mathbb{R}^3 \setminus F. \tag{2.174}$$

Here F is a contractible compact set of positive harmonic capacity.
The regular part h of Green's function $\mathcal{G}$ is

$$h(\boldsymbol{\xi}, \boldsymbol{\eta}) = (4\pi|\boldsymbol{\xi} - \boldsymbol{\eta}|)^{-1} - \mathcal{G}(\boldsymbol{\xi}, \boldsymbol{\eta}). \tag{2.175}$$

3. The components of the vector field $\mathbf{D}(\boldsymbol{\xi}) = (D_1(\boldsymbol{\xi}), D_2(\boldsymbol{\xi}), D_3(\boldsymbol{\xi}))^T$ (compare with (2.100)–(2.102)), for $\boldsymbol{\xi} \in \mathbb{R}^3 \setminus F$, satisfy the problem

$$\Delta D_j(\boldsymbol{\xi}) = 0, \ \ \boldsymbol{\xi} \in \mathbb{R}^3 \setminus F, \tag{2.176}$$

$$D_j(\boldsymbol{\xi}) = \xi_j, \ \ \boldsymbol{\xi} \in \partial F, \tag{2.177}$$

$$D_j(\boldsymbol{\xi}) \to 0, \ \ \text{as } |\boldsymbol{\xi}| \to \infty. \tag{2.178}$$

We shall use the matrix $\mathcal{T} = (\mathcal{T}_{jk})_{j,k=1}^3$ of coefficients in the asymptotic representation of D_j at infinity

$$D_j(\boldsymbol{\xi}) = \frac{1}{4\pi} \sum_{k=1}^{3} \frac{\mathcal{T}_{jk} \xi_k}{|\boldsymbol{\xi}|^3} + O(|\boldsymbol{\xi}|^{-3}). \tag{2.179}$$

The symmetry of $\mathcal{T}$ is verified by applying Green's formula in $B_R \setminus F$ to $\xi_j - D_j(\boldsymbol{\xi})$ and $D_k(\boldsymbol{\xi})$ and taking the limit $R \to \infty$. We have

$$\int_{\partial B_R} \Big\{(\xi_j - D_j(\boldsymbol{\xi}))\frac{\partial D_k(\boldsymbol{\xi})}{\partial|\boldsymbol{\xi}|} - D_k(\boldsymbol{\xi})\Big(\frac{\xi_j}{|\boldsymbol{\xi}|} - \frac{\partial D_j(\boldsymbol{\xi})}{\partial|\boldsymbol{\xi}|}\Big)\Big\}dS$$
$$+ \int_{\partial F} D_k(\boldsymbol{\xi})\Big(\frac{\partial D_j(\boldsymbol{\xi})}{\partial n} - n_j\Big)dS = 0, \tag{2.180}$$

where $\partial/\partial n$ is the normal derivative in the direction of the interior normal with respect to F. As $R \to \infty$, the first integral $\mathcal{I}(\partial B_R)$ in the left-hand side of (2.180) gives

$$\lim_{R\to\infty} \mathcal{I}(\partial B_R) = \lim_{R\to\infty} \int_{\partial B_R} \Big\{\xi_j \frac{\partial D_k(\boldsymbol{\xi})}{\partial|\boldsymbol{\xi}|} - D_k(\boldsymbol{\xi})\frac{\xi_j}{|\boldsymbol{\xi}|}\Big\}dS$$
$$= -\frac{3}{4\pi}\int_{\partial B_1} \sum_{q=1}^{3} \mathcal{T}_{kq}\xi_q\xi_j\, dS = -\mathcal{T}_{kj}. \tag{2.181}$$

The second integral $\mathcal{I}(\partial F)$ in the left-hand side of (2.180) becomes

$$\mathcal{I}(\partial F) = -\int_{\partial F} \xi_k n_j\, dS + \int_{\partial F} D_k(\boldsymbol{\xi})\frac{\partial D_j(\boldsymbol{\xi})}{\partial n}dS$$
$$= \delta_{jk}\ \mathrm{meas}_3(F) + \int_{\mathbb{R}^3\setminus F} \nabla D_k(\boldsymbol{\xi}) \cdot \nabla D_j(\boldsymbol{\xi})d\boldsymbol{\xi}, \tag{2.182}$$

where $\mathrm{meas}_3(F)$ is the three-dimensional Lebesgue measure of F. Using (2.181) and (2.182) we deduce

$$\mathcal{T}_{kj} = \delta_{jk}\ \mathrm{meas}_3(F) + \int_{\mathbb{R}^3\setminus F} \nabla D_k(\boldsymbol{\xi}) \cdot \nabla D_j(\boldsymbol{\xi})d\boldsymbol{\xi}, \tag{2.183}$$

which implies that $\mathcal{T}$ is *symmetric and positive definite.*

4. The Neumann function $N(\mathbf{x},\mathbf{y})$ in $\Omega \subset \mathbb{R}^3$ and its regular part are defined as follows

$$\Delta_x N(\mathbf{x},\mathbf{y}) + \delta(\mathbf{x}-\mathbf{y}) = 0,\ \ \mathbf{x},\mathbf{y} \in \Omega \subset \mathbb{R}^3, \tag{2.184}$$

$$\frac{\partial}{\partial n_x}\Big(N(\mathbf{x},\mathbf{y}) - (4\pi)^{-1}|\mathbf{x}|^{-1}\Big) = 0,\ \ \mathbf{x} \in \partial\Omega,\ \mathbf{y} \in \Omega, \tag{2.185}$$

and

$$\int_{\partial\Omega} N(\mathbf{x},\mathbf{y})\frac{\partial}{\partial n_x}|\mathbf{x}|^{-1}ds_x = 0, \tag{2.186}$$

where the last condition (2.186) implies the symmetry of $N(\mathbf{x}, \mathbf{y})$. The regular part of the Neumann function in three dimensions is defined by

$$R(\mathbf{x}, \mathbf{y}) = (4\pi)^{-1}|\mathbf{x} - \mathbf{y}|^{-1} - N(\mathbf{x}, \mathbf{y}). \tag{2.187}$$

5. In this section, the notation $\mathcal{N}(\boldsymbol{\xi}, \boldsymbol{\eta})$ will be used for the Neumann function in $\mathbb{R}^3 \setminus F$, where F is the compact closure of a domain with a smooth boundary, and $\mathcal{N}$ is defined by

$$\mathcal{N}(\boldsymbol{\xi}, \boldsymbol{\eta}) = (4\pi)^{-1}|\boldsymbol{\xi} - \boldsymbol{\eta}|^{-1} - h_N(\boldsymbol{\xi}, \boldsymbol{\eta}), \tag{2.188}$$

where h_N is the regular part of $\mathcal{N}$ subject to

$$\Delta_{\xi} h_N(\boldsymbol{\xi}, \boldsymbol{\eta}) = 0, \quad \boldsymbol{\xi}, \boldsymbol{\eta} \in \mathbb{R}^3 \setminus F, \tag{2.189}$$

$$\frac{\partial h_N}{\partial n_{\xi}}(\boldsymbol{\xi}, \boldsymbol{\eta}) = \frac{1}{4\pi} \frac{\partial}{\partial n_{\xi}}(|\boldsymbol{\xi} - \boldsymbol{\eta}|^{-1}), \quad \boldsymbol{\xi} \in \partial F, \ \boldsymbol{\eta} \in \mathbb{R}^3 \setminus F, \tag{2.190}$$

$$h_N(\boldsymbol{\xi}, \boldsymbol{\eta}) \to 0, \ \text{ as } |\boldsymbol{\xi}| \to \infty, \ \boldsymbol{\eta} \in \mathbb{R}^3 \setminus F. \tag{2.191}$$

The smoothness assumption on ∂F here and in the sequel is introduced for the simplicity of proofs and can be considerably weakened. In particular, the case of a piece-wise smooth planar crack can be included.

We note that the Neumann function $\mathcal{N}$ just defined is symmetric, i.e. $\mathcal{N}(\boldsymbol{\xi}, \boldsymbol{\eta}) = \mathcal{N}(\boldsymbol{\eta}, \boldsymbol{\xi})$.

6. The definition of the dipole vector field $\mathcal{D}(\boldsymbol{\xi}) = (\mathcal{D}_1(\boldsymbol{\xi}), \mathcal{D}_2(\boldsymbol{\xi}), \mathcal{D}_3(\boldsymbol{\xi}))^T$ is similar to (2.19)–(2.21), with $\boldsymbol{\xi} \in \mathbb{R}^3 \setminus F$. The components of the three-dimensional dipole matrix $\mathcal{P} = (\mathcal{P}_{jk})_{j,k=1}^3$ appear in the asymptotic representation of $\mathcal{D}_j(\boldsymbol{\xi})$ at infinity

$$\mathcal{D}_j(\boldsymbol{\xi}) = \frac{1}{4\pi} \sum_{k=1}^{3} \frac{\mathcal{P}_{jk} \xi_k}{|\boldsymbol{\xi}|^3} + O(|\boldsymbol{\xi}|^{-3}). \tag{2.192}$$

Similar to Sect. 2.1.2, it can be proved the *the dipole matrix $\mathcal{P}$ for the hole F is symmetric and negative definite.*

2.4.2 Approximations of Green's Kernels

The following assertions hold for uniform asymptotic approximations in three-dimensional domains with small holes (or cracks) or inclusions.

Theorem 2.4. *Green's function $G_{\varepsilon}^{(N)}(\mathbf{x}, \mathbf{y})$ for the mixed problem with the Neumann data on ∂F_{ε} and the Dirichlet data on $\partial \Omega$, has the asymptotic representation*

$$G_\varepsilon^{(N)}(\mathbf{x},\mathbf{y}) = G(\mathbf{x},\mathbf{y}) + \varepsilon^{-1}\mathcal{N}(\varepsilon^{-1}\mathbf{x},\varepsilon^{-1}\mathbf{y}) - (4\pi)^{-1}|\mathbf{x}-\mathbf{y}|^{-1}$$

$$+ \varepsilon\boldsymbol{\mathcal{D}}(\varepsilon^{-1}\mathbf{x})\cdot\nabla_x H(0,\mathbf{y}) + \varepsilon\boldsymbol{\mathcal{D}}(\varepsilon^{-1}\mathbf{y})\cdot\nabla_y H(\mathbf{x},0) + r_\varepsilon(\mathbf{x},\mathbf{y}), \tag{2.193}$$

where $\boldsymbol{\mathcal{D}}$ *is the three-dimensional dipole vector function in* $\mathbb{R}^3 \setminus F$, *and* $\mathcal{N}$ *is the Neumann function in* $\mathbb{R}^3 \setminus F$, *vanishing at infinity. Here*

$$|r_\varepsilon(\mathbf{x},\mathbf{y})| \le \text{Const}\ \varepsilon^2 \tag{2.194}$$

uniformly with respect to $\mathbf{x},\mathbf{y} \in \Omega_\varepsilon$.

The proof follows the same algorithm as in Theorem 2.1.

Now we give the analogues of Corollaries 2.2 and 2.3 formulated earlier in Sect. 2.1.7.

Corollary 2.8. *Let* $\min\{|\mathbf{x}|,|\mathbf{y}|\} > 2\varepsilon$. *Then the asymptotic formula* (2.193) *is simplified to the form*

$$\begin{aligned} G_\varepsilon^{(N)}(\mathbf{x},\mathbf{y}) &= G(\mathbf{x},\mathbf{y}) \\ &+ \frac{\varepsilon^3}{4\pi}\Big\{\frac{\mathbf{x}^T}{|\mathbf{x}|^3}\boldsymbol{\mathcal{P}}\nabla_x H(0,\mathbf{y}) + \frac{\mathbf{y}^T}{|\mathbf{y}|^3}\boldsymbol{\mathcal{P}}\nabla_y H(\mathbf{x},0)\Big\} \\ &- \frac{\varepsilon^3}{(4\pi)^2}\frac{\mathbf{x}^T}{|\mathbf{x}|^3}\boldsymbol{\mathcal{P}}\frac{\mathbf{y}}{|\mathbf{y}|^3} \\ &+ O(\varepsilon^2 + \varepsilon^4(|\mathbf{x}|+|\mathbf{y}|)|\mathbf{x}|^{-3}|\mathbf{y}|^{-3}), \end{aligned} \tag{2.195}$$

where H *is the regular part of Green's function* G *in* Ω, *and* $\boldsymbol{\mathcal{P}}$ *is the dipole matrix for* F, *as defined in* (2.192).

The next assertion is similar to Corollary 2.3 of Sect. 2.1.7.

Corollary 2.9. *The following asymptotic formula for Green's function* $G_\varepsilon^{(N)}$ *holds*

$$\begin{aligned} G_\varepsilon^{(N)}(\mathbf{x},\mathbf{y}) &= \varepsilon^{-1}\mathcal{N}(\varepsilon^{-1}\mathbf{x},\varepsilon^{-1}\mathbf{y}) - H(0,0) \\ &- (\mathbf{x} - \varepsilon\boldsymbol{\mathcal{D}}(\varepsilon^{-1}\mathbf{x}))\cdot\nabla_x H(0,\mathbf{y}) - (\mathbf{y} - \varepsilon\boldsymbol{\mathcal{D}}(\varepsilon^{-1}\mathbf{y}))\cdot\nabla_y H(\mathbf{x},0) \\ &+ O(\varepsilon^2 + |\mathbf{x}|^2 + |\mathbf{y}|^2), \end{aligned} \tag{2.196}$$

for $\mathbf{x},\mathbf{y} \in \Omega_\varepsilon$. *(As in Corollary* 2.3, ε^2 *in the remainder can be omitted if the interior of* F *is nonempty and contains the origin.)*

In turn, for the case when the Neumann and Dirichlet boundary conditions are set on $\partial\Omega$ and ∂F_ε, respectively, the modified version of formula (2.125) is given by

Theorem 2.5. *The Green's function* $G_\varepsilon^{(D)}(\mathbf{x},\mathbf{y})$ *for the mixed problem with the Dirichlet data on* ∂F_ε *and the Neumann data on* $\partial\Omega$, *admits the asymptotic representation*

$$G_\varepsilon^{(D)}(\mathbf{x}, \mathbf{y}) = \varepsilon^{-1}\mathcal{G}(\varepsilon^{-1}\mathbf{x}, \varepsilon^{-1}\mathbf{y}) + N(\mathbf{x}, \mathbf{y}) - (4\pi)^{-1}|\mathbf{x}-\mathbf{y}|^{-1} + R(0,0)$$
$$+ \varepsilon\mathbf{D}(\varepsilon^{-1}\mathbf{y}) \cdot \nabla_y R(\mathbf{x}, 0) + \varepsilon\mathbf{D}(\varepsilon^{-1}\mathbf{x}) \cdot \nabla_x R(0, \mathbf{y}) + r_\varepsilon(\mathbf{x}, \mathbf{y}), \qquad (2.197)$$

where

$$|r_\varepsilon(\mathbf{x}, \mathbf{y})| \le \text{Const}\, \varepsilon^2,$$

which is uniform with respect to $\mathbf{x}, \mathbf{y} \in \Omega_\varepsilon$.

The proof is similar to that of Theorem 2.2. We note that unlike the two-dimensional case, in three dimensions no orthogonality condition is required to ensure the decay of the solution of the exterior Dirichlet problem in $\mathbb{R}^3 \setminus F$.

The analogues of Corollaries 2.4 and 2.5 are formulated as follows.

Corollary 2.10. *Let* $\min\{|\mathbf{x}|, |\mathbf{y}|\} > 2\varepsilon$. *Then the asymptotic formula* (2.197) *is simplified to the form*

$$G_\varepsilon^{(D)}(\mathbf{x}, \mathbf{y}) = N(\mathbf{x}, \mathbf{y}) + R(0,0)$$
$$+ \frac{\varepsilon^3}{4\pi}\Big\{\frac{\mathbf{x}^T}{|\mathbf{x}|^3}\boldsymbol{T}\nabla_x R(0, \mathbf{y}) + \frac{\mathbf{y}^T}{|\mathbf{y}|^3}\boldsymbol{T}\nabla_y R(\mathbf{x}, 0)\Big\}$$
$$- \frac{\varepsilon^3}{(4\pi)^2}\frac{\mathbf{x}^T}{|\mathbf{x}|^3}\boldsymbol{T}\frac{\mathbf{y}}{|\mathbf{y}|^3}$$
$$+ O(\varepsilon^2 + \varepsilon^4(|\mathbf{x}| + |\mathbf{y}|)|\mathbf{x}|^{-3}|\mathbf{y}|^{-3}), \qquad (2.198)$$

where R *is the regular part of Neumann's function* N *in* Ω, *and* $\boldsymbol{T}$ *is the matrix of coefficients in* (2.179).

The next assertion is similar to Corollary 2.5 of Sect. 2.2.5.

Corollary 2.11. *The following asymptotic formula for Green's function* $G_\varepsilon^{(D)}$ *holds*

$$G_\varepsilon^{(D)}(\mathbf{x}, \mathbf{y}) = \varepsilon^{-1}\mathcal{G}(\varepsilon^{-1}\mathbf{x}, \varepsilon^{-1}\mathbf{y})$$
$$- (\mathbf{x} - \varepsilon\mathbf{D}(\varepsilon^{-1}\mathbf{x})) \cdot \nabla_x R(0, \mathbf{y}) - (\mathbf{y} - \varepsilon\mathbf{D}(\varepsilon^{-1}\mathbf{y})) \cdot \nabla_y R(\mathbf{x}, 0)$$
$$+ O(\varepsilon^2 + |\mathbf{x}|^2 + |\mathbf{y}|^2), \qquad (2.199)$$

for $\mathbf{x}, \mathbf{y} \in \Omega_\varepsilon$. *(The term* ε^2 *in the remainder can be omitted if the interior of* F *is nonempty and contains the origin.)*

Finally, we consider the *Neumann function* $N_\varepsilon(\mathbf{x}, \mathbf{y})$ for $\Omega_\varepsilon \subset \mathbb{R}^3$. Here, $\Omega_\varepsilon = \Omega \setminus F_\varepsilon$, and F_ε is the small hole with a smooth boundary. We define N_ε as a solution of the following boundary value problem

$$\Delta_x N_\varepsilon(\mathbf{x},\mathbf{y}) + \delta(\mathbf{x}-\mathbf{y}) = 0, \ \ \mathbf{x},\mathbf{y} \in \Omega_\varepsilon, \tag{2.200}$$

$$\frac{\partial}{\partial n_x}\Big(N_\varepsilon(\mathbf{x},\mathbf{y}) - (4\pi)^{-1}|\mathbf{x}|^{-1}\Big) = 0, \ \ \mathbf{x} \in \partial\Omega, \ \mathbf{y} \in \Omega_\varepsilon, \tag{2.201}$$

$$\frac{\partial N_\varepsilon}{\partial n_x}(\mathbf{x},\mathbf{y}) = 0, \ \ \mathbf{x} \in \partial F_\varepsilon, \ \mathbf{y} \in \Omega_\varepsilon. \tag{2.202}$$

In addition, we require the orthogonality condition, which provides the symmetry of $N_\varepsilon(\mathbf{x},\mathbf{y})$

$$\int_{\partial\Omega} N_\varepsilon(\mathbf{x},\mathbf{y})\frac{\partial}{\partial n}|\mathbf{x}|^{-1} dS_x = 0. \tag{2.203}$$

The asymptotic approximation of N_ε is given by

Theorem 2.6. *The Neumann function $N_\varepsilon(\mathbf{x},\mathbf{y})$ for the domain Ω_ε, defined in* (2.200)–(2.203) *satisfies the asymptotic formula*

$$\begin{aligned} N_\varepsilon(\mathbf{x},\mathbf{y}) = \ & N(\mathbf{x},\mathbf{y}) - \varepsilon^{-1} h_N(\varepsilon^{-1}\mathbf{x},\varepsilon^{-1}\mathbf{y}) + \varepsilon\boldsymbol{\mathcal{D}}(\varepsilon^{-1}\mathbf{x})\cdot\nabla_x R(0,\mathbf{y}) \\ & + \varepsilon\boldsymbol{\mathcal{D}}(\varepsilon^{-1}\mathbf{y})\cdot\nabla_y R(\mathbf{x},0) + r_\varepsilon(\mathbf{x},\mathbf{y}), \end{aligned} \tag{2.204}$$

where

$$|r_\varepsilon(\mathbf{x},\mathbf{y})| \le \text{Const}\ \varepsilon^2 \tag{2.205}$$

uniformly with respect to $\mathbf{x},\mathbf{y} \in \Omega_\varepsilon$. *Here* $\boldsymbol{\mathcal{D}}$ *is the three-dimensional dipole vector function in* $\mathbb{R}^3 \setminus F$*, and* h_N *is the regular part of the Neumann function* $\mathcal{N}$ *in* $\mathbb{R}^3 \setminus F$*, vanishing at infinity. The Neumann function* N *in* Ω *and its regular part* R *are the same as in* (2.184)–(2.187).

The proof follows the same algorithm as in Theorem 2.3.

At last, we formulate the analogues of Corollaries 2.6 and 2.7 for the Neumann problem in Ω_ε.

Corollary 2.12. *Let* $\min\{|\mathbf{x}|,|\mathbf{y}|\} > 2\varepsilon$. *Then* $N_\varepsilon(\mathbf{x},\mathbf{y})$ *is approximated in the form*

$$\begin{aligned} N_\varepsilon(\mathbf{x},\mathbf{y}) = N(\mathbf{x},\mathbf{y}) & - \frac{\varepsilon^3}{(4\pi)^2}\frac{\mathbf{x}^T}{|\mathbf{x}|^3}\boldsymbol{\mathcal{P}}\frac{\mathbf{y}^T}{|\mathbf{y}|^3} \\ & + \frac{\varepsilon^3}{4\pi}\Big\{\frac{\mathbf{x}^T}{|\mathbf{x}|^3}\boldsymbol{\mathcal{P}}\nabla_x R(0,\mathbf{y}) + \frac{\mathbf{y}^T}{|\mathbf{y}|^3}\boldsymbol{\mathcal{P}}\nabla_y R(\mathbf{x},0)\Big\} \\ & + O(\varepsilon^2 + \varepsilon^4(|\mathbf{x}|+|\mathbf{y}|)|\mathbf{x}|^{-3}|\mathbf{y}|^{-3}), \end{aligned} \tag{2.206}$$

where R *is the regular part of Neumann's function in* Ω*, and* $\boldsymbol{\mathcal{P}}$ *is the dipole matrix for* F*, as defined in* (2.192).

When both $\mathbf{x}$ and $\mathbf{y}$ are sufficiently close to F_ε the asymptotic approximation of N_ε is given in the next assertion.

Corollary 2.13. *Neumann's function N_ε satisfies the asymptotic formula*

$$\begin{aligned} N_\varepsilon(\mathbf{x},\mathbf{y}) &= \varepsilon^{-1}\mathcal{N}(\varepsilon^{-1}\mathbf{x},\varepsilon^{-1}\mathbf{y}) - R(0,0) \\ &\quad - (\mathbf{x}-\varepsilon\boldsymbol{\mathcal{D}}(\varepsilon^{-1}\mathbf{x}))\cdot\nabla_x R(0,\mathbf{y}) - (\mathbf{y}-\varepsilon\boldsymbol{\mathcal{D}}(\varepsilon^{-1}\mathbf{y}))\cdot\nabla_y R(\mathbf{x},0) \\ &\quad + O(\varepsilon^2+|\mathbf{x}|^2+|\mathbf{y}|^2), \end{aligned} \tag{2.207}$$

for $\mathbf{x},\mathbf{y}\in\Omega_\varepsilon$. The term ε^2 in the remainder can be omitted if the interior of F is nonempty and contains the origin.

Chapter 3
Green's Function for the Dirichlet Boundary Value Problem in a Domain with Several Inclusions

Here we focus on Green's kernels of the operator $-\Delta$ for the case of the domain containing multiple inclusions. The uniform asymptotic approximations, obtained here, can serve for the evaluation of Green's function for anti-plane shear in a domain with several inclusions. Formal asymptotic construction has been accompanied by the error estimates for the remainder term.

3.1 Domain of Definition and the Governing Equations for the Case of Multiple Inclusions

Let Ω defined as in Chap. 1. By $\omega^{(j)}$, $j = 1, \dots, N$, we denote domains in $\mathbb{R}^n$, $n = 2, 3$, with smooth boundary $\partial\omega^{(j)}$ and compact closure $\bar{\omega}^{(j)}$; its complement being $C\bar{\omega}^{(j)} = \mathbb{R}^n \setminus \bar{\omega}^{(j)}$. We shall assume that $\omega^{(j)}$, $j = 1, \dots, N$ contains the origin $\mathbf{O}$ as an interior point. We introduce the sets $\omega_\varepsilon^{(j)} = \{\mathbf{x} : \varepsilon^{-1}(\mathbf{x} - \mathbf{O}^{(j)}) \in \omega^{(j)}\}$, where ε is a small positive parameter and $\mathbf{O}^{(j)}$ is an interior point of $\omega_\varepsilon^{(j)}$. Also we have the open set $\Omega_\varepsilon = \Omega \setminus \bigcup_j \bar{\omega}_\varepsilon^{(j)}$. It is also assumed that the minimum distance between the $\mathbf{O}^{(j)}$ and the points of $\partial\Omega$ and $\partial\omega_\varepsilon^{(k)}$, $1 \le k \le N, k \ne j$, is equal to 1. In addition the maximum distance between $\mathbf{O}$ and the points of $\partial\omega^{(j)}$ will be taken as 1.

The main object of our study in Sects. 3.2 and 3.3 is Green's function for $-\Delta$ in $\Omega_\varepsilon \subset \mathbb{R}^2$, and we denote this function by G_ε. The function G_ε is a solution of

$$-\Delta_{\mathbf{x}} G_\varepsilon(\mathbf{x}, \mathbf{y}) = \delta(\mathbf{x} - \mathbf{y}), \quad \mathbf{x}, \mathbf{y} \in \Omega_\varepsilon, \tag{3.1}$$

$$G_\varepsilon(\mathbf{x}, \mathbf{y}) = 0, \quad \mathbf{x} \in \partial\Omega_\varepsilon, \mathbf{y} \in \Omega_\varepsilon. \tag{3.2}$$

In the sequel, along with $\mathbf{x}$ and $\mathbf{y}$, we shall use scaled variables $\boldsymbol{\xi}_j = \varepsilon^{-1}(\mathbf{x} - \mathbf{O}^{(j)})$ and $\boldsymbol{\eta}_j = \varepsilon^{-1}(\mathbf{y} - \mathbf{O}^{(j)})$, $j = 1, \dots, N$.

V. Maz'ya et al., *Green's Kernels and Meso-Scale Approximations in Perforated Domains*, Lecture Notes in Mathematics 2077, DOI 10.1007/978-3-319-00357-3_3,

By Const we always mean different positive constants which are independent of ε. The notation $f = O(g)$ for a scalar function f is equivalent to the inequality $|f| \le \text{Const}\, g$. Whenever we write $f = O(g)$ for a matrix (vector) function f, we mean a matrix (vector) f whose components are $O(g)$.

3.2 Green's Function for the Case of Anti-plane Shear for a Domain with Several Inclusions

Let $G(\mathbf{x},\mathbf{y})$ and $g^{(j)}(\boldsymbol{\xi}_j,\boldsymbol{\eta}_j)$ denote Green's function for the operator $-\Delta$ in the domains Ω and $C\bar{\omega}^{(j)}$, $j = 1,\dots,N$, respectively. The function G is a solution the following problem

$$-\Delta_{\mathbf{x}} G(\mathbf{x},\mathbf{y}) = \delta(\mathbf{x}-\mathbf{y}) , \quad \mathbf{x},\mathbf{y} \in \Omega , \tag{3.3}$$

$$G(\mathbf{x},\mathbf{y}) = 0 , \quad \mathbf{x} \in \partial\Omega, \mathbf{y} \in \Omega , \tag{3.4}$$

and the functions $g^{(j)}$ solve

$$-\Delta_{\boldsymbol{\xi}_j} g^{(j)}(\boldsymbol{\xi}_j,\boldsymbol{\eta}_j) = \delta(\boldsymbol{\xi}_j-\boldsymbol{\eta}_j) , \quad \boldsymbol{\xi}_j,\boldsymbol{\eta}_j \in C\bar{\omega}^{(j)} , \tag{3.5}$$

$$g^{(j)}(\boldsymbol{\xi}_j,\boldsymbol{\eta}_j) = 0 , \quad \boldsymbol{\xi}_j \in \partial C\bar{\omega}^{(j)}, \boldsymbol{\eta}_j \in C\bar{\omega}^{(j)} , \tag{3.6}$$

$$g^{(j)}(\boldsymbol{\xi}_j,\boldsymbol{\eta}_j) \text{ is bounded as } |\boldsymbol{\xi}_j| \to \infty , \boldsymbol{\eta}_j \in C\bar{\omega}^{(j)} . \tag{3.7}$$

We represent $G(\mathbf{x},\mathbf{y})$ as

$$G(\mathbf{x},\mathbf{y}) = -(2\pi)^{-1}\log|\mathbf{x}-\mathbf{y}| - H(\mathbf{x},\mathbf{y}) , \tag{3.8}$$

and $g^{(j)}(\boldsymbol{\xi}_j,\boldsymbol{\eta}_j)$ for $j = 1,\dots,N$, as

$$g^{(j)}(\boldsymbol{\xi}_j,\boldsymbol{\eta}_j) = -(2\pi)^{-1}\log|\boldsymbol{\xi}_j-\boldsymbol{\eta}_j| - h^{(j)}(\boldsymbol{\xi}_j,\boldsymbol{\eta}_j) , \tag{3.9}$$

where H and $h^{(j)}$ are the regular parts of G and $g^{(j)}$, respectively, and the first term in the right-hand sides of (3.8) and (3.9) is the fundamental solution of the operator $-\Delta$.

We introduce the function $\zeta^{(j)}$ as

$$\zeta^{(j)}(\boldsymbol{\eta}_j) = \lim_{|\boldsymbol{\xi}_j|\to\infty} g^{(j)}(\boldsymbol{\xi}_j,\boldsymbol{\eta}_j) , \tag{3.10}$$

and the constant

$$\zeta_\infty^{(j)} = \lim_{|\boldsymbol{\eta}_j| \to \infty} \{\zeta^{(j)}(\boldsymbol{\eta}_j) - (2\pi)^{-1} \log |\boldsymbol{\eta}_j|\} , \tag{3.11}$$

for $j = 1, \dots, N$.

3.2.1 Estimates for the Functions $h^{(j)}$ and $\zeta^{(j)}$ in the Unbounded Domain

In this subsection we state two results related to the functions $h^{(j)}$ and $\zeta^{(j)}$, $j = 1, \dots, N$, which will be used in the algorithm for the asymptotic expansion of the function G_ε.

The proof of the following lemma can be found in Maz'ya and Movchan [24].

Lemma 3.1. *For $|\boldsymbol{\xi}_j| > 2$ and $\boldsymbol{\eta}_j \in C\bar{\omega}^{(j)}$ the following estimate holds*

$$h^{(j)}(\boldsymbol{\xi}_j, \boldsymbol{\eta}_j) = -(2\pi)^{-1} \log |\boldsymbol{\xi}_j| - \zeta^{(j)}(\boldsymbol{\eta}_j) + O(|\boldsymbol{\xi}_j|^{-1}) , \tag{3.12}$$

for $j = 1, \dots, N$.

The proof of the next lemma follows from the asymptotic expansion of solutions to a general elliptic boundary value problem, by Kondratiev and Oleinik [18].

Lemma 3.2. *For $|\boldsymbol{\xi}_j| > 2$, the following representation for $\zeta^{(j)}$ holds*

$$\zeta^{(j)}(\boldsymbol{\xi}_j) = (2\pi)^{-1} \log |\boldsymbol{\xi}_j| + \zeta_\infty^{(j)} + O(|\boldsymbol{\xi}_j|^{-1}) , \tag{3.13}$$

for $j = 1, \dots, N$.

3.2.2 The Capacitary Potential

Let $P_\varepsilon^{(j)}(\mathbf{x})$ be the capacitary potential corresponding to the jth inclusion with centre $\mathbf{O}^{(j)}$. The function $P_\varepsilon^{(j)}(\mathbf{x})$ is defined as a solution of

$$\Delta P_\varepsilon^{(j)}(\mathbf{x}) = 0 , \quad \mathbf{x} \in \Omega_\varepsilon , \tag{3.14}$$

$$P_\varepsilon^{(j)}(\mathbf{x}) = 0 , \quad \mathbf{x} \in \partial\Omega , \tag{3.15}$$

$$P_\varepsilon^{(j)}(\mathbf{x}) = \delta_{ij} , \quad \mathbf{x} \in \partial\omega_\varepsilon^{(i)}, i = 1, \dots, N , \tag{3.16}$$

where δ_{ij} is the Kronecker delta.

We give a uniform approximation of the function $P_\varepsilon^{(j)}$, by considering the vector $P_\varepsilon(\mathbf{x}) = \{P_\varepsilon^{(j)}(\mathbf{x})\}_{j=1}^N$

Theorem 3.1. *The asymptotic approximation of $P_\varepsilon(\mathbf{x})$ is given by the formula,*

$$P_\varepsilon(\mathbf{x}) = \left(\operatorname*{diag}_{1\le j\le N} \{\alpha_\varepsilon^{(j)}\} - \mathfrak{M} \right)^{-1} \mathcal{S}(\mathbf{x}) + p_\varepsilon(\mathbf{x}) \tag{3.17}$$

where $\alpha_\varepsilon^{(j)} = (2\pi)^{-1}\log\varepsilon + H(\mathbf{O}^{(j)}, \mathbf{O}^{(j)}) - \zeta_\infty^{(j)}$, $\mathfrak{M} = \{(1-\delta_{kj})G(\mathbf{O}^{(k)}, \mathbf{O}^{(j)})\}_{k,j=1}^N$, $\mathcal{S}(\mathbf{x}) = \{-G(\mathbf{x}, \mathbf{O}^{(j)}) + \zeta^{(j)}(\boldsymbol{\xi}_j) - (2\pi)^{-1}\log|\boldsymbol{\xi}_j| - \zeta_\infty^{(j)}\}_{j=1}^N$, *and the vector* $p_\varepsilon(\mathbf{x})$ *is the remainder term such that*

$$|p_\varepsilon(\mathbf{x})| \le \text{Const}\,\varepsilon|\log\varepsilon|^{-1}, \tag{3.18}$$

uniformly with respect to $\mathbf{x} \in \Omega_\varepsilon$.

Prior to the proof of Theorem 3.1 we shall show that the leading order term of the functions $P_\varepsilon^{(j)}$ are solutions of a certain algebraic system.

Lemma 3.3. *The leading order part* $\mathcal{P}_\varepsilon^{(j)}$ *of the functions* $P_\varepsilon^{(j)}$ *are solutions of*

$$\left(\operatorname*{diag}_{1\le j\le N} \{\alpha_\varepsilon^{(j)}\} - \mathfrak{M} \right) \mathcal{P}_\varepsilon(\mathbf{x}) = \mathcal{S}(\mathbf{x}) \,, \tag{3.19}$$

where $\mathcal{P}_\varepsilon = \{\mathcal{P}_\varepsilon^{(j)}\}_{j=1}^N$.

Proof. We represent $P_\varepsilon^{(j)}(\mathbf{x})$ in the form

$$P_\varepsilon^{(j)}(\mathbf{x}) = \frac{-G(\mathbf{x}, \mathbf{O}^{(j)}) + \zeta^{(j)}(\boldsymbol{\xi}_j) - (2\pi)^{-1}\log|\boldsymbol{\xi}_j| - \zeta_\infty^{(j)}}{(2\pi)^{-1}\log\varepsilon + H(\mathbf{O}^{(j)}, \mathbf{O}^{(j)}) - \zeta_\infty^{(j)}} + R_\varepsilon^{(j)}(\mathbf{x}), \tag{3.20}$$

for $1 \le j \le N$, where the function $R_\varepsilon^{(j)}(\mathbf{x})$ satisfies

$$\Delta R_\varepsilon^{(j)}(\mathbf{x}) = 0, \quad \mathbf{x} \in \Omega_\varepsilon \,, \tag{3.21}$$

$$R_\varepsilon^{(j)}(\mathbf{x}) = -\frac{\zeta^{(j)}(\boldsymbol{\xi}_j) - (2\pi)^{-1}\log|\boldsymbol{\xi}_j| - \zeta_\infty^{(j)}}{(2\pi)^{-1}\log\varepsilon + H(\mathbf{O}^{(j)}, \mathbf{O}^{(j)}) - \zeta_\infty^{(j)}}, \quad \mathbf{x} \in \partial\Omega \,, \tag{3.22}$$

$$R_\varepsilon^{(j)}(\mathbf{x}) = 1 - \frac{(2\pi)^{-1}\log\varepsilon + H(\mathbf{x}, \mathbf{O}^{(j)}) - \zeta_\infty^{(j)}}{(2\pi)^{-1}\log\varepsilon + H(\mathbf{O}^{(j)}, \mathbf{O}^{(j)}) - \zeta_\infty^{(j)}}, \quad \mathbf{x} \in \partial\omega_\varepsilon^{(j)} \,, \tag{3.23}$$

$$R_\varepsilon^{(j)}(\mathbf{x}) = \frac{G(\mathbf{x}, \mathbf{O}^{(j)}) - \zeta^{(j)}(\boldsymbol{\xi}_j) + (2\pi)^{-1}\log|\boldsymbol{\xi}_j| + \zeta_\infty^{(j)}}{(2\pi)^{-1}\log\varepsilon + H(\mathbf{O}^{(j)}, \mathbf{O}^{(j)}) - \zeta_\infty^{(j)}} \,,$$
$$\mathbf{x} \in \partial\omega_\varepsilon^{(k)} \,, 1 \le k \le N \,, k \ne j \,. \tag{3.24}$$

The boundary condition (3.23) is equivalent to

$$R_\varepsilon^{(j)}(\mathbf{x}) = -\frac{H(\mathbf{x}, \mathbf{O}^{(j)}) - H(\mathbf{O}^{(j)}, \mathbf{O}^{(j)})}{(2\pi)^{-1}\log\varepsilon + H(\mathbf{O}^{(j)}, \mathbf{O}^{(j)}) - \zeta_\infty^{(j)}}\,, \quad \mathbf{x} \in \partial\omega_\varepsilon^{(j)}\,, \tag{3.25}$$

so $R_\varepsilon^{(j)}(\mathbf{x}) = O(\varepsilon|\log\varepsilon|^{-1})$ for $\mathbf{x} \in \partial\omega_\varepsilon^{(j)}$. Using the asymptotic approximation of $\zeta^{(j)}(\boldsymbol{\xi}_j)$ given in Lemma 3.2, we have from (3.22) that $R_\varepsilon^{(j)}(\mathbf{x}) = O(\varepsilon|\log\varepsilon|^{-1})$ for $\mathbf{x} \in \partial\Omega$. Then from (3.24), also using Lemma 3.2 and the fact $G(\mathbf{x}, \mathbf{O}^{(j)})$ is smooth in Ω_ε, we have

$$R_\varepsilon^{(j)}(\mathbf{x}) = \frac{G(\mathbf{O}^{(k)}, \mathbf{O}^{(j)})}{(2\pi)^{-1}\log\varepsilon + H(\mathbf{O}^{(j)}, \mathbf{O}^{(j)}) - \zeta_\infty^{(j)}} + O(\varepsilon|\log\varepsilon|^{-1})\,, \tag{3.26}$$

for $\mathbf{x} \in \partial\omega_\varepsilon^{(k)}$, $1 \le k \le N, k \ne j$.

Then we may write $R_\varepsilon^{(j)}(\mathbf{x})$, using the capacitary potential $P_\varepsilon^{(k)}$, $k \ne j$, as

$$R_\varepsilon^{(j)}(\mathbf{x}) = \frac{\sum\limits_{\substack{k\ne j\\ 1\le k\le N}} G(\mathbf{O}^{(k)}, \mathbf{O}^{(j)}) P_\varepsilon^{(k)}(\mathbf{x})}{(2\pi)^{-1}\log\varepsilon + H(\mathbf{O}^{(j)}, \mathbf{O}^{(j)}) - \zeta_\infty^{(j)}} + \mathfrak{p}_\varepsilon^{(j)}(\mathbf{x})\,, \tag{3.27}$$

where $\mathfrak{p}_\varepsilon^{(j)}(\mathbf{x})$ is the remainder term.

Now combining (3.27) with (3.20), we obtain the following

$$\begin{aligned} P_\varepsilon^{(j)}(\mathbf{x}) = \Big(&- G(\mathbf{x}, \mathbf{O}^{(j)}) + \zeta^{(j)}(\boldsymbol{\xi}_j) - (2\pi)^{-1}\log|\boldsymbol{\xi}_j| - \zeta_\infty^{(j)} \\ &+ \sum_{\substack{k\ne j\\ 1\le k\le N}} G(\mathbf{O}^{(k)}, \mathbf{O}^{(j)}) P_\varepsilon^{(k)}(\mathbf{x}) \Big) (\alpha_\varepsilon^{(j)})^{-1} + \mathfrak{p}_\varepsilon^{(j)}(\mathbf{x})\,, \end{aligned} \tag{3.28}$$

where $\alpha_\varepsilon^{(j)}$ is as in the formulation of Theorem 3.1, and $\mathfrak{p}_\varepsilon^{(j)}(\mathbf{x})$ is a function which is harmonic in Ω_ε and is $O(\varepsilon|\log\varepsilon|^{-1})$ for $\mathbf{x} \in \partial\Omega$ and $\mathbf{x} \in \partial\omega_\varepsilon^{(j)}$, $1 \le j \le N$. Therefore by the maximum principle $\mathfrak{p}_\varepsilon^{(j)}(\mathbf{x}) = O(\varepsilon|\log\varepsilon|^{-1})$ for $\mathbf{x} \in \Omega_\varepsilon$.

Then, (3.28) gives us the following system of algebraic equations in terms of the functions $P_\varepsilon^{(j)}$, whose solution will give us the approximation of the functions $P_\varepsilon^{(j)}$,

$$\left(\operatorname*{diag}_{1\le j\le N} \{\alpha_\varepsilon^{(j)}\} - \mathfrak{M} \right) P_\varepsilon(\mathbf{x}) = \mathcal{S}(\mathbf{x}) + \mathfrak{R}_\varepsilon\,, \tag{3.29}$$

where $P_\varepsilon(\mathbf{x}) = \{P_\varepsilon^{(j)}(\mathbf{x})\}_{j=1}^N$, $\mathcal{S}$ and $\mathfrak{M}$ are as in the formulation of Theorem 3.1, and $\mathfrak{R}_\varepsilon = \{\alpha_\varepsilon^{(j)}\mathfrak{p}_\varepsilon^{(j)}\}_{j=1}^N$. The leading order part of (3.29) is equivalent to (3.19). □

Let

$$\Xi = \left(\operatorname*{diag}_{1\le j\le N} \{\alpha_\varepsilon^{(j)}\} - \mathfrak{M} \right)^{-1} , \tag{3.30}$$

and Ξ_{ij}, $i, j = 1, \dots, N$ denote the components of this matrix. Multiplying both sides of (3.29) by Ξ, we have

$$P_\varepsilon(\mathbf{x}) = \Xi \mathcal{S}(\mathbf{x}) + p_\varepsilon , \tag{3.31}$$

where $p_\varepsilon = \Xi \mathfrak{R}_\varepsilon$ is the remainder. We shall now estimate the remainder in (3.31).

The proof of Theorem 3.1 is given via estimation of the remainder term p_ε. For the estimate of the norm of the vector $p_\varepsilon(\mathbf{x})$ in (3.17), we shall need an estimate for the entries Ξ_{ij} of the matrix Ξ, which is contained in the following Lemma.

Lemma 3.4. *For the matrix* $\Xi = [\Xi_{ij}]_{i,j=1}^N$, *we have*

$$\Xi_{ij} = \begin{cases} O(|\log\varepsilon|^{-1}) & \textit{for } i = j , \\ O((\log\varepsilon)^{-2}) & \textit{for } i \ne j . \end{cases}$$

Proof. Since $\mathfrak{M}$ is a symmetric matrix, it follows from (3.30) that Ξ is also symmetric. We have

$$\Xi = (\det(\Xi^{-1}))^{-1} \operatorname{adj}(\Xi^{-1}) , \tag{3.32}$$

where $\det(\Xi^{-1})$ is the determinant of the $N \times N$ matrix Ξ^{-1} and $\operatorname{adj}(\Xi^{-1})$ is the adjoint of the matrix Ξ^{-1}. Let the matrix of cofactors for Ξ^{-1} be denoted by C with entries

$$C_{ij} = (-1)^{i+j} T_{ij}, \quad i, j = 1, \dots, N ,$$

where T_{ij} are the corresponding minors of Ξ^{-1}.

First, we consider T_{ij} when $i = j$. In this case we shall need to compute the determinant of an $(N-1) \times (N-1)$ matrix, with $N-1$ terms each of $O(|\log\varepsilon|)$ along the diagonal, and with off-diagonal components of $O(1)$. Thus T_{ij} for $i = j$ is then is $O(|\log\varepsilon|^{N-1})$.

Next consider T_{ij}, when $i \ne j$, so that we compute the determinant of an $(N-1) \times (N-1)$ matrix, containing $N-2$ components of $O(|\log\varepsilon|)$ and all other components of $O(1)$. Then T_{ij}, for $i \ne j$ is $O(|\log\varepsilon|^{N-2})$. Therefore

$$C_{ij} = \begin{cases} O(|\log\varepsilon|^{N-1}) & \text{for } i = j , \\ O(|\log\varepsilon|^{N-2}) & \text{for } i \ne j . \end{cases}$$

Since $\det(\Xi^{-1})$ is $O(|\log\varepsilon|^N)$ we complete the proof of the Lemma. □

Now, we finalize the proof of Theorem 3.1

Proof of Theorem 3.1. The asymptotic approximation of the vector P_ε admits the representation given in (3.31) as a consequence of Lemma 3.3, with the remainder term given by $p_\varepsilon = \Xi\mathfrak{R}_\varepsilon$, where $\mathfrak{R}_\varepsilon = \{\alpha_\varepsilon^{(j)}\mathfrak{p}_\varepsilon^{(j)}\}_{j=1}^N$. In the proof of Lemma 3.3, it was shown that $\mathfrak{p}_\varepsilon^{(j)} = O(\varepsilon|\log\varepsilon|^{-1})$ and noting $\alpha_\varepsilon^{(j)} = O(|\log\varepsilon|)$, we have by the preceding Lemma, the remainder term p_ε has the vector norm $|p_\varepsilon| = O(\varepsilon|\log\varepsilon|^{-1})$. The proof of Theorem 3.1 is complete. □

3.2.3 A Uniform Asymptotic Approximation of Green's Function for $-\Delta$ in a Two-Dimensional Domain with Several Small Inclusions

Here we consider the approximation of Green's function G_ε for the Laplacian in a planar domain with several inclusions.

Theorem 3.2. *Green's function for the operator $-\Delta$ in $\Omega_\varepsilon \subset \mathbb{R}^2$ admits the representation*

$$\begin{aligned}G_\varepsilon(\mathbf{x},\mathbf{y}) = G(\mathbf{x},\mathbf{y}) &+ \sum_{j=1}^N g^{(j)}(\boldsymbol{\xi}_j,\boldsymbol{\eta}_j) + N(2\pi)^{-1}\log(\varepsilon^{-1}|\mathbf{x}-\mathbf{y}|)\\ &+\sum_{j=1}^N \{\alpha_\varepsilon^{(j)}P_\varepsilon^{(j)}(\mathbf{y})P_\varepsilon^{(j)}(\mathbf{x}) - \zeta^{(j)}(\boldsymbol{\xi}_j) - \zeta^{(j)}(\boldsymbol{\eta}_j) + \zeta_\infty^{(j)}\}\\ &-\sum_{j=1}^N \sum_{\substack{k\neq j\\1\le k\le N}} G(\mathbf{O}^{(k)},\mathbf{O}^{(j)})P_\varepsilon^{(k)}(\mathbf{y})P_\varepsilon^{(j)}(\mathbf{x}) + O(\varepsilon)\,,\end{aligned} \tag{3.33}$$

uniformly with respect to $(\mathbf{x},\mathbf{y}) \in \Omega_\varepsilon \times \Omega_\varepsilon$.

Proof. For this we propose that G_ε may be given as

$$G_\varepsilon(\mathbf{x},\mathbf{y}) = -(2\pi)^{-1}\log|\mathbf{x}-\mathbf{y}| - H_\varepsilon(\mathbf{x},\mathbf{y}) - \sum_{j=1}^N h_\varepsilon^{(j)}(\mathbf{x},\mathbf{y})\,, \tag{3.34}$$

where it suffices to seek the approximation of the functions $H_\varepsilon(\mathbf{x},\mathbf{y})$ and $h_\varepsilon^{(j)}(\mathbf{x},\mathbf{y})$, which are solutions of the problems

$$\Delta_{\mathbf{x}} H_\varepsilon(\mathbf{x},\mathbf{y}) = 0\,, \quad \mathbf{x},\mathbf{y}\in\Omega_\varepsilon\,, \tag{3.35}$$

$$H_\varepsilon(\mathbf{x},\mathbf{y}) = -(2\pi)^{-1}\log|\mathbf{x}-\mathbf{y}|\,, \quad \mathbf{x}\in\partial\Omega, \mathbf{y}\in\Omega_\varepsilon\,, \tag{3.36}$$

$$H_\varepsilon(\mathbf{x},\mathbf{y}) = 0\,, \quad \mathbf{x}\in\partial\omega_\varepsilon^{(j)}, \mathbf{y}\in\Omega_\varepsilon, 1\le j\le N\,, \tag{3.37}$$

and

$$\Delta_{\mathbf{x}} h_\varepsilon^{(j)}(\mathbf{x},\mathbf{y}) = 0\,, \quad \mathbf{x},\mathbf{y} \in \Omega_\varepsilon\,, \tag{3.38}$$

$$h_\varepsilon^{(j)}(\mathbf{x},\mathbf{y}) = 0\,, \quad \mathbf{x} \in \partial\Omega, \mathbf{y} \in \Omega_\varepsilon\,, \tag{3.39}$$

$$h_\varepsilon^{(j)}(\mathbf{x},\mathbf{y}) = -(2\pi)^{-1}\log|\mathbf{x}-\mathbf{y}|\,, \quad \mathbf{x} \in \partial\omega_\varepsilon^{(j)}, \mathbf{y} \in \Omega_\varepsilon\,, \tag{3.40}$$

$$h_\varepsilon^{(j)}(\mathbf{x},\mathbf{y}) = 0\,, \quad \mathbf{x} \in \partial\omega_\varepsilon^{(k)}, \mathbf{y} \in \Omega_\varepsilon\,, 1 \le k \le N\,, k \ne j\,. \tag{3.41}$$

The Approximation of $H_\varepsilon(\mathbf{x},\mathbf{y})$

Let $H_\varepsilon(\mathbf{x},\mathbf{y})$ be given by

$$H_\varepsilon(\mathbf{x},\mathbf{y}) = -H(\mathbf{O}^{(j)},\mathbf{y})P_\varepsilon^{(j)}(\mathbf{x}) + H(\mathbf{x},\mathbf{y}) + V(\mathbf{x},\mathbf{y})\,, \tag{3.42}$$

where the index j is fixed (it is not the index of summation) and $V(\mathbf{x},\mathbf{y})$ satisfies

$$\Delta_{\mathbf{x}} V(\mathbf{x},\mathbf{y}) = 0\,, \quad \mathbf{x},\mathbf{y} \in \Omega_\varepsilon\,, \tag{3.43}$$

$$V(\mathbf{x},\mathbf{y}) = 0\,, \quad \mathbf{x} \in \partial\Omega, \mathbf{y} \in \Omega_\varepsilon\,, \tag{3.44}$$

$$V(\mathbf{x},\mathbf{y}) = H(\mathbf{O}^{(j)},\mathbf{y}) - H(\mathbf{x},\mathbf{y})\,, \quad \mathbf{x} \in \partial\omega_\varepsilon^{(j)}, \mathbf{y} \in \Omega_\varepsilon\,, \tag{3.45}$$

$$V(\mathbf{x},\mathbf{y}) = -H(\mathbf{x},\mathbf{y})\,, \quad \mathbf{x} \in \partial\omega_\varepsilon^{(k)}, \mathbf{y} \in \Omega_\varepsilon\,, k \ne j\,, 1 \le k \le N\,. \tag{3.46}$$

Since $\omega_\varepsilon^{(j)}$, $1 \le j \le N$, are small inclusions and H is a smooth function in Ω, we may expand H about the centres of the inclusions. Namely, for the boundary condition (3.45) we have

$$V(\mathbf{x},\mathbf{y}) = H(\mathbf{O}^{(j)},\mathbf{y}) - H(\mathbf{x},\mathbf{y}) = O(\varepsilon)\,, \quad \mathbf{x} \in \partial\omega_\varepsilon^{(j)}, \mathbf{y} \in \Omega_\varepsilon\,, \tag{3.47}$$

and from (3.46)

$$\begin{aligned} V(\mathbf{x},\mathbf{y}) = -H(\mathbf{x},\mathbf{y}) &= -H(\mathbf{O}^{(k)},\mathbf{y}) + O(\varepsilon)\,, \\ &\mathbf{x} \in \partial\omega_\varepsilon^{(k)}, \mathbf{y} \in \Omega_\varepsilon\,, k \ne j\,, 1 \le k \le N\,. \end{aligned} \tag{3.48}$$

We therefore write the function $V(\mathbf{x},\mathbf{y})$ as

$$V(\mathbf{x},\mathbf{y}) = -\sum_{\substack{k \ne j \\ 1 \le k \le N}} H(\mathbf{O}^{(k)},\mathbf{y})P_\varepsilon^{(k)}(\mathbf{x}) + \mathfrak{H}_\varepsilon(\mathbf{x},\mathbf{y})\,, \tag{3.49}$$

where $\mathfrak{H}_\varepsilon$ is the remainder term. Substituting (3.49) into (3.42) we have

$$H_\varepsilon(\mathbf{x},\mathbf{y}) = -\sum_{j=1}^{N} H(\mathbf{O}^{(j)},\mathbf{y})P_\varepsilon^{(j)}(\mathbf{x}) + H(\mathbf{x},\mathbf{y}) + \mathfrak{H}_\varepsilon(\mathbf{x},\mathbf{y})\,, \tag{3.50}$$

where $\mathfrak{H}_\varepsilon(\mathbf{x}, \mathbf{y})$ satisfies

$$\Delta_{\mathbf{x}} \mathfrak{H}_\varepsilon(\mathbf{x}, \mathbf{y}) = 0 \,, \quad \mathbf{x}, \mathbf{y} \in \Omega_\varepsilon \,, \tag{3.51}$$

$$\mathfrak{H}_\varepsilon(\mathbf{x}, \mathbf{y}) = 0 \,, \quad \mathbf{x} \in \partial\Omega, \mathbf{y} \in \Omega_\varepsilon \,, \tag{3.52}$$

$$\begin{aligned} \mathfrak{H}_\varepsilon(\mathbf{x}, \mathbf{y}) &= H(\mathbf{O}^{(j)}, \mathbf{y}) - H(\mathbf{x}, \mathbf{y}) \\ &= O(\varepsilon) \,, \quad \mathbf{x} \in \partial\omega_\varepsilon^{(j)}, \mathbf{y} \in \Omega_\varepsilon, 1 \le j \le N \,, \end{aligned} \tag{3.53}$$

and therefore by the maximum principle $\mathfrak{H}_\varepsilon(\mathbf{x}, \mathbf{y}) = O(\varepsilon)$, uniformly with respect to $\mathbf{x}, \mathbf{y} \in \Omega_\varepsilon$.

The Approximation of $h_\varepsilon^{(j)}(\mathbf{x}, \mathbf{y})$

We begin by writing the boundary condition (3.40) on $\partial\omega_\varepsilon^{(j)}$ as

$$h_\varepsilon^{(j)}(\mathbf{x}, \mathbf{y}) = -(2\pi)^{-1} \log \varepsilon - (2\pi)^{-1} \log(\varepsilon^{-1}|\mathbf{x} - \mathbf{y}|) \,, \quad \mathbf{x} \in \partial\omega_\varepsilon^{(j)} \,, \mathbf{y} \in \Omega_\varepsilon \,. \tag{3.54}$$

We seek $h_\varepsilon^{(j)}(\mathbf{x}, \mathbf{y})$ in the form

$$h_\varepsilon^{(j)}(\mathbf{x}, \mathbf{y}) = -(2\pi)^{-1} \log \varepsilon + h^{(j)}(\boldsymbol{\xi}_j, \boldsymbol{\eta}_j) + \chi_\varepsilon^{(j)}(\mathbf{x}, \mathbf{y}) \,, \tag{3.55}$$

where the remainder $\chi_\varepsilon^{(j)}$ satisfies

$$\Delta_{\mathbf{x}} \chi_\varepsilon^{(j)}(\mathbf{x}, \mathbf{y}) = 0 \,, \quad \mathbf{x}, \mathbf{y} \in \Omega_\varepsilon \,, \tag{3.56}$$

$$\chi_\varepsilon^{(j)}(\mathbf{x}, \mathbf{y}) = (2\pi)^{-1} \log \varepsilon - h^{(j)}(\boldsymbol{\xi}_j, \boldsymbol{\eta}_j) \,, \quad \mathbf{x} \in \partial\Omega, \mathbf{y} \in \Omega_\varepsilon \,, \tag{3.57}$$

$$\chi_\varepsilon^{(j)}(\mathbf{x}, \mathbf{y}) = 0 \,, \quad \mathbf{x} \in \partial\omega_\varepsilon^{(j)}, \mathbf{y} \in \Omega_\varepsilon \,, \tag{3.58}$$

and

$$\chi_\varepsilon^{(j)}(\mathbf{x}, \mathbf{y}) = (2\pi)^{-1} \log \varepsilon - h^{(j)}(\boldsymbol{\xi}_j, \boldsymbol{\eta}_j) \,, \quad \mathbf{x} \in \partial\omega_\varepsilon^{(k)}, \mathbf{y} \subset \Omega_\varepsilon \,, \tag{3.59}$$

for $1 \le k \le N, k \ne j$.

From Lemma 3.1, we may write boundary conditions (3.57) and (3.59) as

$$\chi_\varepsilon^{(j)}(\mathbf{x}, \mathbf{y}) = (2\pi)^{-1} \log |\mathbf{x} - \mathbf{O}^{(j)}| + \zeta^{(j)}(\boldsymbol{\eta}_j) + O(\varepsilon) \,, \quad \mathbf{x} \in \partial\Omega, \mathbf{y} \in \Omega_\varepsilon \,, \tag{3.60}$$

$$\begin{aligned} \chi_\varepsilon^{(j)}(\mathbf{x}, \mathbf{y}) &= (2\pi)^{-1} \log |\mathbf{x} - \mathbf{O}^{(j)}| + \zeta^{(j)}(\boldsymbol{\eta}_j) + O(\varepsilon) \,, \\ &\text{for } \mathbf{x} \in \partial\omega_\varepsilon^{(k)}, \mathbf{y} \in \Omega_\varepsilon \,, 1 \le k \le N \,, k \ne j \,. \end{aligned} \tag{3.61}$$

Then we represent $\chi_\varepsilon^{(j)}$ as

$$\chi_\varepsilon^{(j)}(\mathbf{x},\mathbf{y}) = -H(\mathbf{x},\mathbf{O}^{(j)}) + (1 - P_\varepsilon^{(j)}(\mathbf{x}))\zeta^{(j)}(\boldsymbol{\eta}_j) + \mathfrak{h}_\varepsilon^{(j)}(\mathbf{x},\mathbf{y}) \,, \tag{3.62}$$

where $\mathfrak{h}_\varepsilon^{(j)}(\mathbf{x},\mathbf{y})$ satisfies

$$\Delta_{\mathbf{x}}\mathfrak{h}_\varepsilon^{(j)}(\mathbf{x},\mathbf{y}) = 0 \,, \quad \mathbf{x},\mathbf{y} \in \Omega_\varepsilon \,, \tag{3.63}$$

$$\mathfrak{h}_\varepsilon^{(j)}(\mathbf{x},\mathbf{y}) = O(\varepsilon) \,, \quad \mathbf{x} \in \partial\Omega, \mathbf{y} \in \Omega_\varepsilon \,, \tag{3.64}$$

$$\mathfrak{h}_\varepsilon^{(j)}(\mathbf{x},\mathbf{y}) = H(\mathbf{x},\mathbf{O}^{(j)}) \,, \quad \mathbf{x} \in \partial\omega_\varepsilon^{(j)}, \mathbf{y} \in \Omega_\varepsilon \,, \tag{3.65}$$

$$\mathfrak{h}_\varepsilon^{(j)}(\mathbf{x},\mathbf{y}) = -G(\mathbf{x},\mathbf{O}^{(j)}) + O(\varepsilon) \,, \quad \mathbf{x} \in \partial\omega_\varepsilon^{(k)}, \mathbf{y} \in \Omega_\varepsilon \,, 1 \le k \le N \,, k \neq j \,. \tag{3.66}$$

From the fact that $G(\mathbf{x},\mathbf{O}^{(j)})$ and its regular part are smooth functions in Ω_ε, we expand these functions about the centres of the small inclusions in such a way that boundary conditions (3.65) and (3.66) become

$$\mathfrak{h}_\varepsilon^{(j)}(\mathbf{x},\mathbf{y}) = H(\mathbf{O}^{(j)},\mathbf{O}^{(j)}) + O(\varepsilon) \,, \quad \mathbf{x} \in \partial\omega_\varepsilon^{(j)}, \mathbf{y} \in \Omega_\varepsilon \,, \tag{3.67}$$

$$\mathfrak{h}_\varepsilon^{(j)}(\mathbf{x},\mathbf{y}) = -G(\mathbf{O}^{(k)},\mathbf{O}^{(j)}) + O(\varepsilon) \,, \quad \mathbf{x} \in \partial\omega_\varepsilon^{(k)}, \mathbf{y} \in \Omega_\varepsilon \,, 1 \le k \le N \,, k \neq j \,. \tag{3.68}$$

Then $\mathfrak{h}_\varepsilon^{(j)}(\mathbf{x},\mathbf{y})$ is given by

$$\mathfrak{h}_\varepsilon^{(j)}(\mathbf{x},\mathbf{y}) = H(\mathbf{O}^{(j)},\mathbf{O}^{(j)})P_\varepsilon^{(j)}(\mathbf{x}) - \sum_{\substack{k\neq j\\ 1\le k\le N}} G(\mathbf{O}^{(k)},\mathbf{O}^{(j)})P_\varepsilon^{(k)}(\mathbf{x}) + O(\varepsilon) \,. \tag{3.69}$$

Placing (3.62) and (3.69) into (3.55), we obtain the following approximation of $h_\varepsilon^{(j)}(\mathbf{x},\mathbf{y})$

$$\begin{aligned} h_\varepsilon^{(j)}(\mathbf{x},\mathbf{y}) = {} & -(2\pi)^{-1}\log\varepsilon + h^{(j)}(\boldsymbol{\xi}_j,\boldsymbol{\eta}_j) - H(\mathbf{x},\mathbf{O}^{(j)}) \\ & +(1 - P_\varepsilon^{(j)}(\mathbf{x}))\zeta^{(j)}(\boldsymbol{\eta}_j) + H(\mathbf{O}^{(j)},\mathbf{O}^{(j)})P_\varepsilon^{(j)}(\mathbf{x}) \\ & - \sum_{\substack{k\neq j\\ 1\le k\le N}} G(\mathbf{O}^{(k)},\mathbf{O}^{(j)})P_\varepsilon^{(k)}(\mathbf{x}) + O(\varepsilon) \,, \end{aligned} \tag{3.70}$$

which is uniform with respect to $\mathbf{x},\mathbf{y} \in \Omega_\varepsilon$.

Combined Formula

Now substituting (3.50), (3.70) into (3.34) we obtain

$$\begin{aligned}G_\varepsilon(\mathbf{x},\mathbf{y}) = G(\mathbf{x},\mathbf{y}) &+ \sum_{j=1}^{N} g^{(j)}(\boldsymbol{\xi}_j,\boldsymbol{\eta}_j) + N(2\pi)^{-1}\log(|\mathbf{x}-\mathbf{y}|)\\
&+\sum_{j=1}^{N}(1-P_\varepsilon^{(j)}(\mathbf{x}))(H(\mathbf{O}^{(j)},\mathbf{O}^{(j)})-\zeta^{(j)}(\boldsymbol{\eta}_j)-H(\mathbf{O}^{(j)},\mathbf{y}))\\
&+\sum_{j=1}^{N}(H(\mathbf{x},\mathbf{O}^{(j)})+H(\mathbf{O}^{(j)},\mathbf{y})-H(\mathbf{O}^{(j)},\mathbf{O}^{(j)}))\\
&+\sum_{j=1}^{N}\sum_{\substack{k\neq j\\ 1\le k\le N}} G(\mathbf{O}^{(k)},\mathbf{O}^{(j)})P_\varepsilon^{(k)}(\mathbf{x}) + O(\varepsilon)\,. \qquad (3.71)\end{aligned}$$

Using the following relation obtained from the approximation of $P_\varepsilon^{(j)}(\mathbf{x})$ (see (3.28)),

$$\begin{aligned}&(H(\mathbf{O}^{(j)},\mathbf{O}^{(j)})-\zeta^{(j)}(\boldsymbol{\eta}_j)-H(\mathbf{O}^{(j)},\mathbf{y}))(\alpha_\varepsilon^{(j)})^{-1}\\
&=1-P_\varepsilon^{(j)}(\mathbf{y})+(\alpha_\varepsilon^{(j)})^{-1}\sum_{\substack{k\neq j\\ 1\le k\le N}} G(\mathbf{O}^{(k)},\mathbf{O}^{(j)})P_\varepsilon^{(k)}(\mathbf{y})+O(\varepsilon|\log\varepsilon|^{-1})\,,\end{aligned} \qquad (3.72)$$

and substituting into (3.71), we have

$$\begin{aligned}G_\varepsilon(\mathbf{x},\mathbf{y}) = G(\mathbf{x},\mathbf{y}) &+ \sum_{j=1}^{N} g^{(j)}(\boldsymbol{\xi}_j,\boldsymbol{\eta}_j) + N(2\pi)^{-1}\log|\mathbf{x}-\mathbf{y}|\\
&+\sum_{j=1}^{N}\alpha_\varepsilon^{(j)}(1-P_\varepsilon^{(j)}(\mathbf{x}))(1-P_\varepsilon^{(j)}(\mathbf{y}))\\
&+\sum_{j=1}^{N}(H(\mathbf{x},\mathbf{O}^{(j)})+H(\mathbf{O}^{(j)},\mathbf{y})-H(\mathbf{O}^{(j)},\mathbf{O}^{(j)}))\\
&+\sum_{j=1}^{N}\sum_{\substack{k\neq j\\ 1\le k\le N}} G(\mathbf{O}^{(k)},\mathbf{O}^{(j)})\{P_\varepsilon^{(k)}(\mathbf{y})+P_\varepsilon^{(k)}(\mathbf{x})\\
&-P_\varepsilon^{(k)}(\mathbf{y})P_\varepsilon^{(j)}(\mathbf{x})\}+O(\varepsilon)\,. \qquad (3.73)\end{aligned}$$

Then, expanding the fourth term on the right-hand side of (3.73) and using (3.72), we have

$$\sum_{j=1}^{N} \alpha_\varepsilon^{(j)}(1 - P_\varepsilon^{(j)}(\mathbf{x}))(1 - P_\varepsilon^{(j)}(\mathbf{y}))$$

$$= -\sum_{j=1}^{N} (H(\mathbf{x}, \mathbf{O}^{(j)}) + H(\mathbf{O}^{(j)}, \mathbf{y}) - H(\mathbf{O}^{(j)}, \mathbf{O}^{(j)}))$$

$$-\sum_{j=1}^{N} (\zeta^{(j)}(\boldsymbol{\xi}_j) + \zeta^{(j)}(\boldsymbol{\eta}_j) - \zeta_\infty^{(j)}) - N(2\pi)^{-1} \log \varepsilon$$

$$-\sum_{j=1}^{N} \sum_{\substack{k \neq j \\ 1 \le k \le N}} G(\mathbf{O}^{(k)}, \mathbf{O}^{(j)})\{P_\varepsilon^{(k)}(\mathbf{y}) + P_\varepsilon^{(k)}(\mathbf{x})\}$$

$$+\sum_{j=1}^{N} \alpha_\varepsilon^{(j)} P_\varepsilon^{(j)}(\mathbf{y}) P_\varepsilon^{(j)}(\mathbf{x}) + O(\varepsilon)\,. \tag{3.74}$$

Substitution of (3.74) in (3.73) leads to the formula (3.33). The proof is complete. □

3.3 Simplified Asymptotic Formulae for Green's Function Subject to Constraints on the Independent Variables

Analogous to Sect. 1.3, we now show how the asymptotic formula for G_ε (see (3.33)), simplifies under restrictions on the points $\mathbf{x}$ and $\mathbf{y}$. We again consider two cases, the first being the situation when the points $\mathbf{x}, \mathbf{y}$ are sufficiently far away from each of the inclusions, the second is when the points are within a small neighborhood of a particular inclusion.

Corollary 3.1. *a) Let* $\mathbf{x}, \mathbf{y} \in \Omega_\varepsilon \subset \mathbb{R}^2$ *such that*

$$\min\{|\mathbf{x} - \mathbf{O}^{(j)}|, |\mathbf{y} - \mathbf{O}^{(j)}|\} > 2\,\varepsilon \text{ for all } j = 1, \dots, N. \tag{3.75}$$

Then

$$G_\varepsilon(\mathbf{x}, \mathbf{y}) = G(\mathbf{x}, \mathbf{y}) + \sum_{i,m=1}^{N} \Xi_{im} G(\mathbf{y}, \mathbf{O}^{(m)}) G(\mathbf{x}, \mathbf{O}^{(i)})$$
$$+O\left(\sum_{i=1}^{N} \varepsilon(\min\{|\mathbf{x} - \mathbf{O}^{(i)}|, |\mathbf{y} - \mathbf{O}^{(i)}|\})^{-1}\right), \tag{3.76}$$

where $\Xi = [\Xi_{ij}]_{i,j=1}^{2}$*, is given by (3.30).*

b) If $\max\{|\mathbf{x}-\mathbf{O}^{(m)}|, |\mathbf{y}-\mathbf{O}^{(m)}|\} < 1/2$*, then*

$$G_\varepsilon(\mathbf{x},\mathbf{y}) = g^{(m)}(\boldsymbol{\xi}_m, \boldsymbol{\eta}_m) - \alpha_\varepsilon^{(m)} - \zeta^{(m)}(\boldsymbol{\xi}_m) - \zeta^{(m)}(\boldsymbol{\eta}_m)$$
$$+ \sum_{i,j=1}^{N} S_m^{(i)}(\mathbf{x}) \Xi_{ij} S_m^{(j)}(\mathbf{y})$$
$$+ O(\max\{|\mathbf{x}-\mathbf{O}^{(m)}|, |\mathbf{y}-\mathbf{O}^{(m)}|\}) , \quad (3.77)$$

where $\alpha_\varepsilon^{(j)} = (2\pi)^{-1}\log\varepsilon + H(\mathbf{O}^{(j)}, \mathbf{O}^{(j)}) - \zeta_\infty^{(j)}$ *and* $S_m^{(i)}(\mathbf{x}) = \delta_{im}(\alpha_\varepsilon^{(m)} + \zeta^{(m)}(\boldsymbol{\xi}_m)) - (1-\delta_{im})G(\mathbf{O}^{(m)}, \mathbf{O}^{(i)})$.

Both (3.76) *and* (3.77) *are uniform with respect to* $(\mathbf{x},\mathbf{y}) \in \Omega_\varepsilon \times \Omega_\varepsilon$.

Proof. (a) From (3.33), G_ε may be written as

$$G_\varepsilon(\mathbf{x},\mathbf{y}) = G(\mathbf{x},\mathbf{y}) - \sum_{j=1}^{N} h^{(j)}(\boldsymbol{\xi}_j, \boldsymbol{\eta}_j)$$
$$+ \sum_{j=1}^{N} \{\alpha_\varepsilon^{(j)} P_\varepsilon^{(j)}(\mathbf{y}) P_\varepsilon^{(j)}(\mathbf{x}) - \zeta^{(j)}(\boldsymbol{\xi}_j) - \zeta^{(j)}(\boldsymbol{\eta}_j) + \zeta_\infty^{(j)}\}$$
$$- \sum_{j=1}^{N} \sum_{\substack{k\neq j \\ 1\le k\le N}} G(\mathbf{O}^{(k)}, \mathbf{O}^{(j)}) P_\varepsilon^{(k)}(\mathbf{y}) P_\varepsilon^{(j)}(\mathbf{x}) + O(\varepsilon) . \quad (3.78)$$

Owing to Lemma 3.2, we have the estimate for the function $\zeta^{(j)}$

$$\zeta^{(j)}(\boldsymbol{\xi}_j) = (2\pi)^{-1}\log|\boldsymbol{\xi}_j| + \zeta_\infty^{(j)} + O(|\boldsymbol{\xi}_j|^{-1}) , \quad (3.79)$$

and, as a result of condition (3.75), along with the estimate for $h^{(j)}$ given in Lemma 3.1 we obtain

$$h^{(j)}(\boldsymbol{\xi}_j, \boldsymbol{\eta}_j) = -(2\pi)^{-1}\log|\boldsymbol{\xi}_j| - \zeta^{(j)}(\boldsymbol{\eta}_j) + O(|\boldsymbol{\xi}_j|^{-1})$$
$$= -(2\pi)^{-1}\log|\boldsymbol{\xi}_j| - (2\pi)^{-1}\log|\boldsymbol{\eta}_j| - \zeta_\infty^{(j)}$$
$$+ O(\varepsilon(\min\{|\mathbf{x}-\mathbf{O}^{(j)}|, |\mathbf{y}-\mathbf{O}^{(j)}|\})^{-1}) . \quad (3.80)$$

Using the latter estimates in (3.78), yields

$$G_\varepsilon(\mathbf{x},\mathbf{y}) = G(\mathbf{x},\mathbf{y}) + \sum_{j=1}^{N} \alpha_\varepsilon^{(j)} P_\varepsilon^{(j)}(\mathbf{y}) P_\varepsilon^{(j)}(\mathbf{x})$$

$$-\sum_{j=1}^{N} \sum_{\substack{k\neq j\\ 1\le k\le N}} G(\mathbf{O}^{(k)},\mathbf{O}^{(j)}) P_\varepsilon^{(k)}(\mathbf{y}) P_\varepsilon^{(j)}(\mathbf{x})$$

$$+O\Big(\sum_{j=1}^{N} \varepsilon(\min\{|\mathbf{x}-\mathbf{O}^{(j)}|, |\mathbf{y}-\mathbf{O}^{(j)}|\})^{-1}\Big)\,. \tag{3.81}$$

The second and third terms in the right-hand side of (3.81) may be written as

$$\sum_{j=1}^{N} \alpha_\varepsilon^{(j)} P_\varepsilon^{(j)}(\mathbf{y}) P_\varepsilon^{(j)}(\mathbf{x}) - \sum_{j=1}^{N} \sum_{\substack{k\neq j\\ 1\le k\le N}} G(\mathbf{O}^{(k)},\mathbf{O}^{(j)}) P_\varepsilon^{(k)}(\mathbf{y}) P_\varepsilon^{(j)}(\mathbf{x})$$

$$= P_\varepsilon^T(\mathbf{x}) \operatorname*{diag}_{1\le j\le N} \{\alpha_\varepsilon^{(j)}\} P_\varepsilon(\mathbf{y}) - P_\varepsilon^T(\mathbf{x}) \mathfrak{M} P_\varepsilon(\mathbf{y})$$

$$= P_\varepsilon^T(\mathbf{x}) \Xi^{-1} P_\varepsilon(\mathbf{y})\,, \tag{3.82}$$

where $P_\varepsilon = \{P_\varepsilon^{(j)}\}_{j=1}^N$, $\mathfrak{M} = \{(1-\delta_{jk}) G(\mathbf{O}^{(k)},\mathbf{O}^{(j)})\}_{k,j=1}^N$, and Ξ is given by (3.30).

From Theorem 3.1,

$$P_\varepsilon(\mathbf{x}) = \Xi \mathcal{S}(\mathbf{x}) + O(\varepsilon|\log\varepsilon|^{-1})\,, \tag{3.83}$$

where $\mathcal{S}(\mathbf{x}) = \{-G(\mathbf{x},\mathbf{O}^{(j)}) + \zeta^{(j)}(\boldsymbol{\xi}_j) - (2\pi)^{-1}\log|\boldsymbol{\xi}_j| - \zeta_\infty^{(j)}\}_{j=1}^N$, which by Lemma 3.2, $\mathcal{S}(\mathbf{x}) = \{-G(\mathbf{x},\mathbf{O}^{(j)}) + O(|\boldsymbol{\xi}_j|^{-1})\}_{j=1}^N$. Then, combining this with (3.83) in (3.82), we may write (3.82) as

$$\sum_{j=1}^{N} \alpha_\varepsilon^{(j)} P_\varepsilon^{(j)}(\mathbf{x}) P_\varepsilon^{(j)}(\mathbf{y}) - \sum_{j=1}^{N} \sum_{\substack{k\neq j\\ 1\le k\le N}} G(\mathbf{O}^{(k)},\mathbf{O}^{(j)}) P_\varepsilon^{(j)}(\mathbf{x}) P_\varepsilon^{(k)}(\mathbf{y})$$

$$= \sum_{i,m=1}^{N} \Xi_{im} G(\mathbf{y},\mathbf{O}^{(m)}) G(\mathbf{x},\mathbf{O}^{(i)})$$

$$+O\left(\sum_{i=1}^{N} \varepsilon(\min\{|\mathbf{x}-\mathbf{O}^{(i)}|, |\mathbf{y}-\mathbf{O}^{(i)}|\})^{-1}\right)\,, \tag{3.84}$$

where $\Xi_{im}, i, m = 1, \dots, N$ are the entries of Ξ. Next, substituting (3.84) into (3.81), we arrive at (3.76).

(b) Now we assume $\max\{|\mathbf{x} - \mathbf{O}^{(m)}|, |\mathbf{y} - \mathbf{O}^{(m)}|\} < 1/2$. In (3.33), the definitions of G and $g^{(j)}$ in (3.8) and (3.9), respectively, and estimates (3.79) and (3.80), for $j \neq m$, lead to

$$G_\varepsilon(\mathbf{x}, \mathbf{y}) = -H(\mathbf{x}, \mathbf{y}) + g^{(m)}(\boldsymbol{\xi}_m, \boldsymbol{\eta}_m) - \frac{1}{2\pi} \log \varepsilon - \zeta^{(m)}(\boldsymbol{\xi}_m)$$
$$-\zeta^{(m)}(\boldsymbol{\eta}_m) + \zeta_\infty^{(m)} + P_\varepsilon^T(\mathbf{x}) \Xi^{-1} P_\varepsilon(\mathbf{y}) + O(\varepsilon), \qquad (3.85)$$

where (3.82) has also been used. Formula (3.83) then yields

$$P_\varepsilon^T(\mathbf{x}) \Xi^{-1} P_\varepsilon(\mathbf{y}) = \mathcal{S}^T(\mathbf{x}) \Xi \mathcal{S}(\mathbf{y}) + O(\varepsilon |\log \varepsilon|^{-1}) \,. \qquad (3.86)$$

Here, the vector function $\mathcal{S}(\mathbf{x})$, owing to the estimate (3.79) and the Taylor expansion of the functions $H(\mathbf{x}, \mathbf{O}^{(m)})$ and $G(\mathbf{x}, \mathbf{O}^{(k)})$, $k \neq m$, about $\mathbf{x} = \mathbf{O}^{(m)}$, admit the representation

$$\mathcal{S}(\mathbf{x}) = \mathcal{S}_m(\mathbf{x}) + O(|\mathbf{x} - \mathbf{O}^{(m)}|) \,.$$

where the components of $\mathcal{S}_m(\mathbf{x}) = \{S_m^{(j)}(\mathbf{x})\}_{j=1}^N$ are defined in the statement of the present corollary. A similar representation for $\mathcal{S}(\mathbf{y})$ also holds in the vicinity of $\mathbf{y} = \mathbf{O}^{(m)}$ and using these expressions in (3.86) gives

$$P_\varepsilon^T(\mathbf{x}) \Xi^{-1} P_\varepsilon(\mathbf{y}) = \mathcal{S}_m^T(\mathbf{x}) \Xi \mathcal{S}_m(\mathbf{y}) + O(\max\{|\mathbf{x} - \mathbf{O}^{(m)}|, |\mathbf{y} - \mathbf{O}^{(m)}|\}) \,.$$

Finally, substituting this into (3.85) and expanding $H(\mathbf{x}, \mathbf{y})$ about $\mathbf{x} = \mathbf{O}^{(m)}$ and $\mathbf{y} = \mathbf{O}^{(m)}$, we arrive at (3.77).

Chapter 4
Numerical Simulations Based on the Asymptotic Approximations

Throughout this chapter, we shall implement the asymptotic formulae in numerical simulations. The objective here is to investigate the accuracy of the asymptotic formulae obtained in Chap. 3 for the two-dimensional Green's kernels. We will compare the formulae, by considering the regular part $\mathcal{H}_\varepsilon$ of the function G_ε for the operator $-\Delta$ with a solution produced by the method of finite elements in FEMLAB/COMSOL.

We begin with the Green's function for the Laplacian in a domain with multiple inclusions in Sect. 4.1. Section 4.1.1, describes the numerical settings for the case of a planar circular domain with circular inclusions, where we will be concerned with two particular configurations for the numerical experiments. The first is that of a disk with a relatively large number of inclusions in Sect. 4.1.2, the second situation is when the inclusions are allowed to become relatively large and we consider this in Sect. 4.1.3. We then analyse the error between the asymptotic formula and the solution given in FEMLAB/COMSOL for both these cases.

4.1 Asymptotic Formulae Versus Numerical Solution for the Operator $-\Delta$

In the present section, for the case of when Ω_ε is a planar circular domain with several circular inclusions, we shall compare the asymptotic formula for the regular part $\mathcal{H}_\varepsilon$ of the function G_ε for the operator $-\Delta$, with a solution produced by the method of finite elements in FEMLAB/COMSOL.

The aim of this section is to illustrate through two examples:

(i) That the asymptotic formulae can produce a solution to the problem, even when the finite element package cannot.
(ii) That we are able to take the inclusions in our example configurations to be rather large (by increasing ε) and still obtain a good accuracy by the asymptotic formulae.

V. Maz'ya et al., *Green's Kernels and Meso-Scale Approximations in Perforated Domains*, Lecture Notes in Mathematics 2077, DOI 10.1007/978-3-319-00357-3_4,

4.1.1 Domain and the Asymptotic Approximation

Let $\Omega \subset \mathbb{R}^2$ be a disk of radius R and let $\mathbf{O}^{(1)}, \dots, \mathbf{O}^{(N)}$ be interior points of Ω. We introduce the sets $\omega_\varepsilon^{(j)}$ as disks in $\mathbb{R}^2$ each with centres $\mathbf{O}^{(j)}$ and small radii $\rho^{(j)}$ for $j = 1, \dots, N$, and we have the set $\Omega_\varepsilon = \Omega \setminus \bigcup_j \bar{\omega}_\varepsilon^{(j)}$. The function $\mathcal{H}_\varepsilon$ is a solution of the problem

$$\Delta_{\mathbf{x}} \mathcal{H}_\varepsilon(\mathbf{x}, \mathbf{y}) = 0\,, \quad \mathbf{x}, \mathbf{y} \in \Omega_\varepsilon\,, \tag{4.1}$$

$$\mathcal{H}_\varepsilon(\mathbf{x}, \mathbf{y}) = -(2\pi)^{-1} \log|\mathbf{x} - \mathbf{y}|\,, \quad \mathbf{x} \in \partial\Omega_\varepsilon, \mathbf{y} \in \Omega_\varepsilon\,. \tag{4.2}$$

The regular part $\mathcal{H}_\varepsilon$ of Green's function G_ε for $-\Delta$ in the domain Ω_ε is given by

$$\begin{aligned}
\mathcal{H}_\varepsilon(\mathbf{x}, \mathbf{y}) = H(\mathbf{x}, \mathbf{y}) &- \sum_{j=1}^{N} g^{(j)}(\mathbf{x} - \mathbf{O}^{(j)}, \mathbf{y} - \mathbf{O}^{(j)}) \\
&-(2\pi)^{-1} N \log|\mathbf{x} - \mathbf{y}| - \sum_{j=1}^{N} \{\alpha^{(j)} \mathcal{P}_\varepsilon^{(j)}(\mathbf{y}) \mathcal{P}_\varepsilon^{(j)}(\mathbf{x}) \\
&+(2\pi)^{-1} \log(\rho^{(j)}(|\mathbf{x} - \mathbf{O}^{(j)}||\mathbf{y} - \mathbf{O}^{(j)}|)^{-1})\} \\
&+\sum_{j=1}^{N} \sum_{\substack{k \neq j \\ 1 \leq k \leq N}} G(\mathbf{O}^{(k)}, \mathbf{O}^{(j)}) \mathcal{P}_\varepsilon^{(k)}(\mathbf{y}) \mathcal{P}_\varepsilon^{(j)}(\mathbf{x}) + O(\varepsilon)\,,
\end{aligned} \tag{4.3}$$

which is uniform with respect to $(\mathbf{x}, \mathbf{y}) \in \Omega_\varepsilon \times \Omega_\varepsilon$. We will use the leading order part of this approximation for our calculations.

Here $\varepsilon = m/d$ is the small parameter, with m being the maximum radius of all the disks $\omega_\varepsilon^{(j)}$ and

$$d = \min\{\min_{1 \leq j \leq N} \{\text{dist}(\mathbf{O}^{(j)}, \partial\Omega)\}, \min_{\substack{1 \leq i,k \leq N \\ i \neq k}} \{\text{dist}(\mathbf{O}^{(i)}, \mathbf{O}^{(k)})\}\}\,, \tag{4.4}$$

the function H is the regular part of Green's function G for the domain Ω

$$H(\mathbf{x}, \mathbf{y}) = \frac{1}{2\pi} \log\left(\frac{R}{|\mathbf{y}||\mathbf{x} - \bar{\mathbf{y}}|}\right), \quad \bar{\mathbf{y}} = \frac{R^2}{|\mathbf{y}|^2}\mathbf{y}\,,$$

$g^{(j)}$ is the Green's function for the exterior of the set $\omega_\varepsilon^{(j)}$, $j = 1, \dots, N$, given by

$$g^{(j)}(\mathbf{x} - \mathbf{O}^{(j)}, \mathbf{y} - \mathbf{O}^{(j)}) = \frac{1}{2\pi} \log\left(\frac{|\mathbf{y} - \mathbf{O}^{(j)}||\mathbf{x} - \mathbf{O}^{(j)} - \frac{(\rho^{(j)})^2}{|\mathbf{y} - \mathbf{O}^{(j)}|^2}(\mathbf{y} - \mathbf{O}^{(j)})|}{\rho^{(j)}|\mathbf{x} - \mathbf{y}|}\right).$$

The function $\mathcal{P}_\varepsilon^{(j)}$ is the leading part of the approximation of the function $P_\varepsilon^{(j)}$, $j = 1, \dots, N$ which is a solution of

$$\Delta_{\mathbf{x}} P_\varepsilon^{(j)}(\mathbf{x}) = 0 , \quad \mathbf{x} \in \Omega_\varepsilon , \tag{4.5}$$

$$P_\varepsilon^{(j)}(\mathbf{x}) = 0 , \quad \mathbf{x} \in \partial\Omega , \tag{4.6}$$

$$P_\varepsilon^{(j)}(\mathbf{x}) = \delta_{kj} , \quad \mathbf{x} \in \partial\omega_\varepsilon^{(k)} , k = 1, \dots, N . \tag{4.7}$$

Let $\mathcal{P}_\varepsilon = \{\mathcal{P}_\varepsilon^{(j)}\}_{j=1}^N$, then the entries $\mathcal{P}_\varepsilon^{(j)}$ are obtained from

$$\mathcal{P}_\varepsilon(\mathbf{x}) = \left(\operatorname*{diag}_{1 \le j \le N} \{\alpha^{(j)}\} - \mathfrak{M} \right)^{-1} \mathcal{S}(\mathbf{x}) , \tag{4.8}$$

where $\alpha^{(j)} = (2\pi)^{-1} \log \rho^{(j)} + H(\mathbf{O}^{(j)}, \mathbf{O}^{(j)})$, $\mathfrak{M} = ((1 - \delta_{kj}) G(\mathbf{O}^{(k)}, \mathbf{O}^{(j)}))_{j,k=1}^N$, with

$$G(\mathbf{x}, \mathbf{y}) = -\frac{1}{2\pi} \log |\mathbf{x} - \mathbf{y}| - H(\mathbf{x}, \mathbf{y}) ,$$

and $\mathcal{S} = \{\mathcal{S}^{(j)}\}_{j=1}^N$ with entries being given by $\mathcal{S}^{(j)}(\mathbf{x}) = -G(\mathbf{x}, \mathbf{O}^{(j)})$.

The formula (4.3) can be written via solutions of model problems in domains independent of the small parameter.

Let the sets $\omega^{(j)} = \{\varepsilon^{-1}(\mathbf{x} - \mathbf{O}^{(j)}) : \mathbf{x} \in \omega_\varepsilon^{(j)}\}$, $j = 1, \dots, N$ with radii $r^{(j)} = \varepsilon^{-1} \rho^{(j)}$, and denote their complements by $C\bar{\omega}^{(j)} = \mathbb{R}^2 \backslash \omega^{(j)}$, $j = 1, \dots, N$.

We will assume that all of $\omega^{(j)}$ contain the origin and that the maximum distance between the $\mathbf{O}$ and $\partial\omega^{(j)}$ is equal to d.

In the following we use the scaled variables $\boldsymbol{\xi}_j = \varepsilon^{-1}(\mathbf{x} - \mathbf{O}^{(j)})$ and $\boldsymbol{\eta}_j = \varepsilon^{-1}(\mathbf{y} - \mathbf{O}^{(j)})$. The Green's functions for the sets $\omega^{(j)}$, $j = 1, \dots, N$ are given by

$$g^{(j)}(\boldsymbol{\xi}_j, \boldsymbol{\eta}_j) = \frac{1}{2\pi} \log \left(\frac{|\boldsymbol{\eta}_j| |\boldsymbol{\xi}_j - \bar{\boldsymbol{\eta}}_j|}{r^{(j)} |\boldsymbol{\xi}_j - \boldsymbol{\eta}_j|} \right) , \quad \bar{\boldsymbol{\eta}}_j = \frac{(r^{(j)})^2}{|\boldsymbol{\eta}_j|^2} \boldsymbol{\eta}_j . \tag{4.9}$$

We introduce the functions $\zeta^{(j)}$ by

$$\zeta^{(j)}(\boldsymbol{\eta}_j) = \lim_{|\boldsymbol{\xi}_j| \to \infty} g^{(j)}(\boldsymbol{\xi}_j, \boldsymbol{\eta}_j) , \tag{4.10}$$

and the constants

$$\zeta_\infty^{(j)} = \lim_{|\boldsymbol{\eta}_j| \to \infty} \{\zeta^{(j)}(\boldsymbol{\eta}_j) - (2\pi)^{-1} \log |\boldsymbol{\eta}_j|\} , \tag{4.11}$$

for $j = 1, \dots, N$. For the domain Ω_ε described above

$$\zeta^{(j)}(\boldsymbol{\eta}_j) = \frac{1}{2\pi} \log\left(\frac{|\boldsymbol{\eta}_j|}{r^{(j)}}\right), \quad \zeta^{(j)}_\infty = -\frac{1}{2\pi} \log r^{(j)} . \tag{4.12}$$

We may then rewrite (4.3), incorporating the small parameter ε with the use of (4.9), (4.10) and (4.12) as follows

$$\begin{aligned}\mathcal{H}_\varepsilon(\mathbf{x}, \mathbf{y}) &= H(\mathbf{x}, \mathbf{y}) - \sum_{j=1}^{N} g^{(j)}(\boldsymbol{\xi}_j, \boldsymbol{\eta}_j) - (2\pi)^{-1} N \log(\varepsilon^{-1}|\mathbf{x} - \mathbf{y}|) \\ &- \sum_{j=1}^{N} \{\alpha^{(j)}_\varepsilon \mathcal{P}^{(j)}_\varepsilon(\mathbf{y}) \mathcal{P}^{(j)}_\varepsilon(\mathbf{x}) - \zeta^{(j)}(\boldsymbol{\xi}_j) - \zeta^{(j)}(\boldsymbol{\eta}_j) + \zeta^{(j)}_\infty\} \\ &+ \sum_{j=1}^{N} \sum_{\substack{k \neq j \\ 1 \leq k \leq N}} G(\mathbf{O}^{(k)}, \mathbf{O}^{(j)}) \mathcal{P}^{(k)}_\varepsilon(\mathbf{y}) \mathcal{P}^{(j)}_\varepsilon(\mathbf{x}) + O(\varepsilon) ,\end{aligned} \tag{4.13}$$

where $\alpha^{(j)}_\varepsilon = (2\pi)^{-1} \log \varepsilon + (2\pi)^{-1} \log r^{(j)} + H(\mathbf{O}^{(j)}, \mathbf{O}^{(j)})$.

4.1.2 Example: A Configuration with a Large Number of Small Inclusions

For our first illustrative example, we shall plot the regular part $\mathcal{H}_\varepsilon$ of Green's function G_ε.

We produced the surface plot of the asymptotic solution for $\mathcal{H}_\varepsilon$, on a mesh consisting of 752,448 elements, (see Fig. 4.3). On this mesh, FEMLAB was unable to produce an accurate numerical solution, but the asymptotic formula is still efficient for this case.

The numerical settings are as follows. Let Ω be the disk of radius $R = 70$, centered at the origin. We consider the situation when we have $N = 50$ small disks, whose radii in scaled coordinates do not exceed 10.0449, and the small parameter $\varepsilon = 0.0498$. The location of the point force is given by $\mathbf{y} = (-20, 15)$.

For a mesh containing 188,112 elements, we produced a surface plot of the asymptotic formula for $\mathcal{H}_\varepsilon$ given in (4.13) and the numerical solution given in FEMLAB by the method of finite elements, and the corresponding diagrams are shown in Fig 4.1 a, b.

We compared both the asymptotic representation for the regular part of G_ε and the numerical solution produced in FEMLAB on this mesh, by taking the absolute difference between the two (see Fig. 4.2a) and then the relative error (see Fig. 4.2b). From both of these figures it can be seen that the asymptotic formula gives a good approximation to the numerical solution produced in FEMLAB.

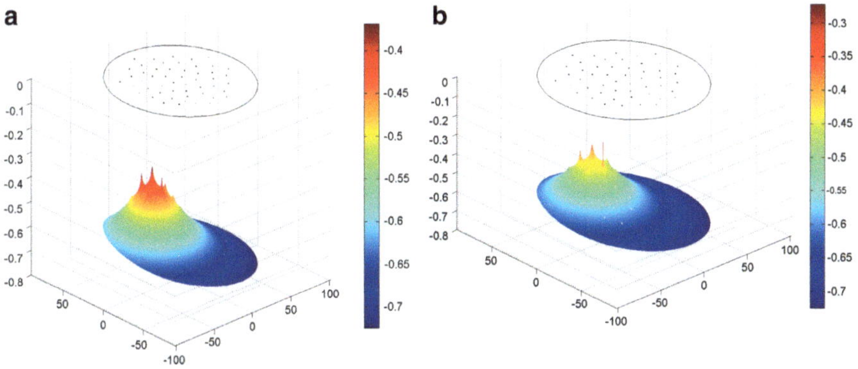

Fig. 4.1 (**a**) Numerical solution produced in FEMLAB on a mesh containing 188,112 elements. (**b**) Computation based on the asymptotic formula for $\mathcal{H}_\varepsilon$, when $\mathbf{y} = (-20, 15)$ and $\varepsilon = 0.0498$

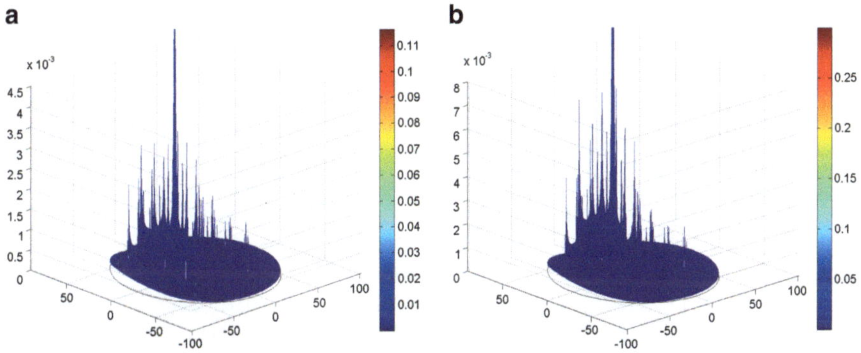

Fig. 4.2 (**a**) Absolute error and (**b**) relative error between numerical solution and the computations produced by the asymptotic formula for $\mathcal{H}_\varepsilon$, when $\varepsilon = 0.0498$ and the mesh contains 188,112 elements. All the spikes occur on the boundaries of the inclusions. Maximum absolute error is 0.1162, maximum relative error is 0.2995, which is attained on the boundary of the inclusion with centre $(-20, 4)$, near the point $(-20, 15)$ where the force is applied

The critical case when FEMLAB failed but the asymptotic formula still produced an accurate solution is shown in Fig. 4.3.

4.1.3 Example: A Configuration with Inclusions of Relatively Large Size

In this example, we shall once again take the asymptotic formula for the regular part $\mathcal{H}_\varepsilon$ of the function G_ε and compare this with numerical solutions produced in FEMLAB, for a configuration with few inclusions, and we shall experiment with

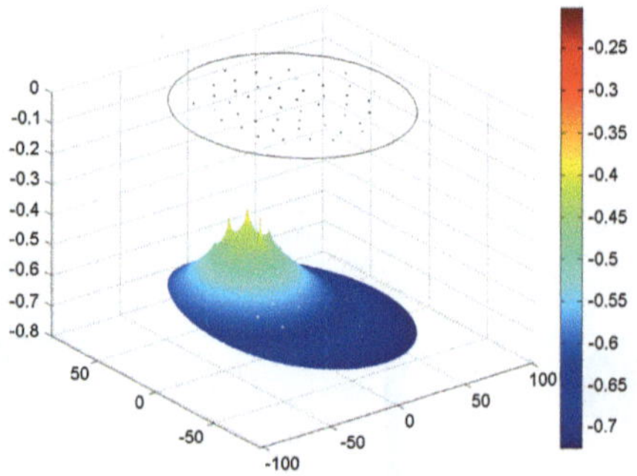

Fig. 4.3 The computation based on the asymptotic formula for the regular part $\mathcal{H}_\varepsilon$ of Green's function on the refined mesh, when $\mathbf{y} = (-20, 15)$, $\varepsilon = 0.0498$ and the mesh contains 752,448 elements

Table 4.1 Maximum absolute and relative error corresponding to various values of ε and when $\mathbf{y} = (-25, 70)$

m	ε	A_{max}	R_{max}
40	0.7436	0.1219	0.1991
36	0.6692	0.09741	0.157
32	0.5949	0.07637	0.1216
28	0.5205	0.05845	0.09204
24	0.4462	0.04335	0.06752
20	0.3718	0.0308	0.04749
16	0.2974	0.0206	0.03156
12	0.2231	0.01298	0.02
8	0.1487	0.007266	0.0111
4	0.0744	0.002993	0.004503
2	0.0372	0.001395	0.001991
1	0.0186	0.0006608	0.0009269
0.5	0.0093	0.0003156	0.0004448
0.25	0.0046	0.0001515	0.0002171

the parameter ε. We show that we are able to consider a configuration where the inclusions are rather large (by increasing ε) and the asymptotic formula for $\mathcal{H}_\varepsilon$ still gives a good approximation to the numerical solution.

Let Ω now be a disk of radius 150, and we consider the case when we have 5 inclusions $\omega_\varepsilon^{(j)}$, $j = 1, \dots, 5$, with centres $\mathbf{O}^{(1)} = (44, 66)$, $\mathbf{O}^{(2)} = (-90, 34)$, $\mathbf{O}^{(3)} = (-36, -68)$, $\mathbf{O}^{(4)} = (68, -26)$, $\mathbf{O}^{(5)} = (-14, 0)$, and whose radii in scaled coordinates do not exceed 53.7919. The position of the point force is $\mathbf{y} = (-25, 70)$.

In Table 4.1, we present data showing how the error between the numerical solution given in FEMLAB and the asymptotic formula for the regular part of Green's function $\mathcal{H}_\varepsilon$ changes as we decrease ε. Here m denotes the maximum radius of the inclusions and A_{max} and R_{max} are absolute and relative error, respectively.

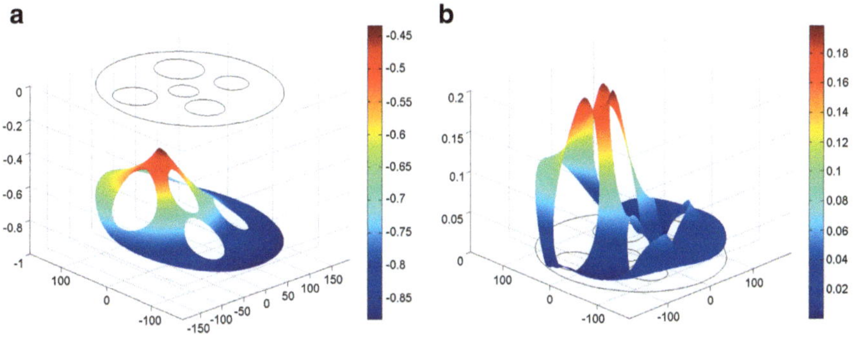

Fig. 4.4 (**a**) Computations produced by the asymptotic formula for $\mathcal{H}_\varepsilon$, (**b**) The relative error between the numerical solution and the asymptotic formula for the case when $\mathbf{y} = (-25, 70)$ and $\varepsilon = 0.7436$

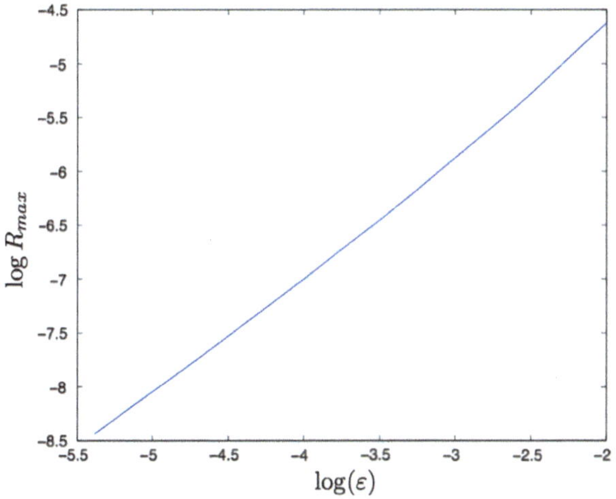

Fig. 4.5 Plot of $\log(\varepsilon)$ against $\log R_{max}$ when $\mathbf{y} = (-25, 70)$

We also have for the situation when $\varepsilon = 0.7436$ the surface plot of the asymptotic formula for the regular part of Green's function and the relative error between the numerical solution and the asymptotic formula; we note that inclusions are rather large in this case (see Fig. 4.4a and b). It can be seen from Fig. 4.4b that although the maximum relative error is larger near where the point force is applied ($R_{max} = 0.1991$), the asymptotic formula still gives a good match with the numerical solution everywhere else.

The plot of ε against R_{max} on a logarithmic scale is shown in Fig. 4.5. It can be seen from this that for small ε the graph appears to be linear and from this we can conclude the numerical evaluation of the relative error R_{max} is consistent with the theoretical prediction of formula (4.13).

Chapter 5
Other Examples of Asymptotic Approximations of Green's Functions in Singularly Perturbed Domains

The structure of this chapter can be described as follows. Sections 5.1 and 5.2 give several results for asymptotic approximations of Green's kernels in domains with singularly perturbed smooth or conical exterior boundaries. Section 5.3 presents a detailed analysis of Green's function of the Dirichlet–Neumann problem in a long cylindrical body. We introduce the notion of a capacitary potential and its asymptotic approximation in the elongated domain and construct an asymptotic approximation of Green's function in the long rod in Sects. 5.3.2 and 5.3.3.

5.1 Perturbation of a Smooth Exterior Boundary

Consider an example of a bounded domain Ω_ε^- in $\mathbb{R}^3$, as shown in Fig. 5.1. Let γ_ε^- denote the perturbed small part of the boundary, and l be a flat part of the boundary surrounding γ_ε^-, while Γ^- is the remaining unperturbed part of the exterior surface.

Green's function for the Dirichlet–Neumann boundary value problem in Ω_ε^- is introduced as a solution of the following boundary value problem

$$\Delta_x G_\varepsilon(\mathbf{x},\mathbf{y}) + \delta(\mathbf{x}-\mathbf{y}) = 0, \quad \mathbf{x},\mathbf{y} \in \Omega_\varepsilon^-,$$
$$G_\varepsilon(\mathbf{x},\mathbf{y}) = 0, \ \mathbf{x} \in \gamma_\varepsilon^- \cup \Gamma^-, \mathbf{y} \in \Omega_\varepsilon^-,$$
$$\frac{\partial G_\varepsilon}{\partial n_x}(\mathbf{x},\mathbf{y}) = 0, \quad \mathbf{x} \in l, \ \mathbf{y} \in \Omega_\varepsilon^-.$$

To construct an asymptotic approximation of G_ε one also needs model limit domains shown in Fig. 5.1: the unperturbed limit domain Ω^- and the unbounded domain D^- corresponding to boundary layers near the perturbed boundary. Let G_{Ω^-} and g_{D^-} be Green's functions of the corresponding mixed boundary value problems in Ω^- and D^-. By H_{Ω^-} we denote the regular part of G_{Ω^-}. The capacitary potential is introduced as a function P_{γ^-}, harmonic in D^-, which

V. Maz'ya et al., *Green's Kernels and Meso-Scale Approximations in Perforated Domains*, Lecture Notes in Mathematics 2077, DOI 10.1007/978-3-319-00357-3_5,

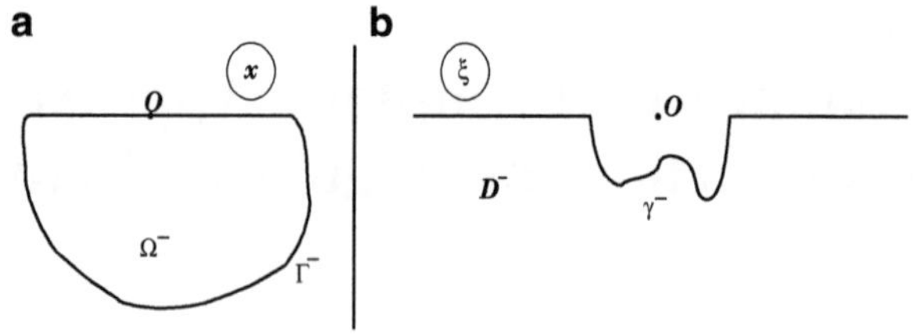

Fig. 5.1 (**a**) Unperturbed domain Ω^-; (**b**) Unbounded model domain D^-

satisfies the homogeneous Neumann condition on $(\partial D^-) \setminus \gamma^-$, equals to 1 on γ^-, and decays at infinity. Then the asymptotic approximation for G_ε takes the form

$$\begin{aligned} G_\varepsilon(\mathbf{x},\mathbf{y}) &= G_{\Omega^-}(\mathbf{x},\mathbf{y}) + \varepsilon^{-1} g_{D^-}(\varepsilon^{-1}\mathbf{x}, \varepsilon^{-1}\mathbf{y}) - (4\pi|\mathbf{x}-\mathbf{y}|)^{-1} \\ &+ H_{\Omega^-}(0,\mathbf{y}) P_{\gamma^-}(\varepsilon^{-1}\mathbf{x}) + H_{\Omega^-}(\mathbf{x},0) P_{\gamma^-}(\varepsilon^{-1}\mathbf{y}) \\ &- H_{\Omega^-}(0,0) P_{\gamma^-}(\varepsilon^{-1}\mathbf{x}) P_{\gamma^-}(\varepsilon^{-1}\mathbf{y}) + O(\varepsilon). \end{aligned} \tag{5.1}$$

For the particular example of the domain in Fig. 5.1, one can make a mirror reflection across the flat part l of the boundary, so that the extended set represents a domain with a small hole. Then the method of images enables one to employ the formula (1.17) and to deduce the asymptotic approximation (5.1). Indeed, other shapes of the perturbed exterior boundaries can be considered: in particular, this may include the case of a domain with a perturbed conical surface outlined below.

5.2 Green's Function for the Dirichlet–Neumann Problem in a Truncated Cone

Consider an example involving a three-dimensional domain shown in Fig. 5.2a. Let K be an infinite cone $\{\mathbf{x} : |\mathbf{x}| > 0, |\mathbf{x}|^{-1}\mathbf{x} \in \Xi\}$, where Ξ is a subdomain of the unit sphere S_1 such that $S_1 \setminus \Xi$ has a positive two-dimensional harmonic capacity. The notations ω and Ω are used for subdomains of K separated from the vertex of K and from infinity by surfaces γ and Γ, respectively, (see Figs. 5.3 and 5.2b). By Ω_ε we denote a domain involving a "small truncation" of the conical part of the boundary, i.e. $\Omega_\varepsilon = \{\mathbf{x} \in \Omega : \varepsilon^{-1}\mathbf{x} \in \omega\}$, where ε stands for a small positive parameter. The conical surface is denoted by l, whereas $\gamma_\varepsilon = \{\mathbf{x} : \varepsilon^{-1}\mathbf{x} \in \gamma\}$ stands for the part of surface near the vertex of the truncated cone, as shown in Fig. 5.2a.

Let G_ε and G_{cone} be Green's functions for the Dirichlet–Neumann problem for $-\Delta$ in Ω_ε and the Neumann problem in K, respectively:

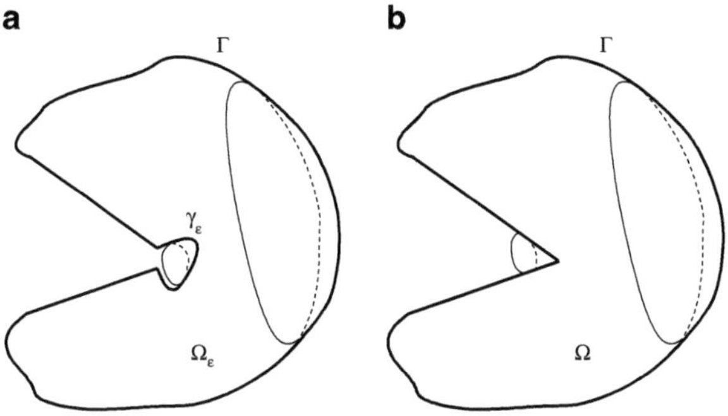

Fig. 5.2 (**a**) A domain with a singularly perturbed conical boundary. (**b**) A limit unperturbed domain

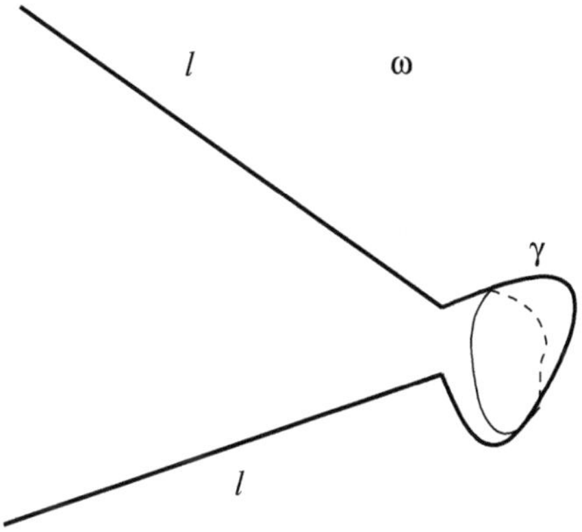

Fig. 5.3 Scaled region in the vicinity of the perturbed boundary

$$\Delta_x G_\varepsilon(\mathbf{x},\mathbf{y}) + \delta(\mathbf{x}-\mathbf{y}) = 0, \;\; \mathbf{x},\mathbf{y} \in \Omega_\varepsilon, \tag{5.2}$$

$$\frac{\partial G_\varepsilon}{\partial n_x}(\mathbf{x},\mathbf{y}) = 0, \;\; \mathbf{x} \in l, \; \mathbf{y} \in \Omega_\varepsilon, \tag{5.3}$$

$$G_\varepsilon(\mathbf{x},\mathbf{y}) = 0, \;\; \mathbf{x} \in \gamma_\varepsilon \cup \Gamma, \; \mathbf{y} \in \Omega_\varepsilon. \tag{5.4}$$

and

$$\begin{aligned} \Delta_x G_{cone}(\mathbf{x},\mathbf{y}) + \delta(\mathbf{x}-\mathbf{y}) &= 0, \;\; \mathbf{x},\mathbf{y} \in K, \\ \frac{\partial G_{cone}}{\partial n_x}(\mathbf{x},\mathbf{y}) &= 0, \;\; \mathbf{x} \in l, \; \mathbf{y} \in K, \\ G_{cone}(\mathbf{x},\mathbf{y}) &\to 0, \;\; |\mathbf{x}| \to \infty, \; \mathbf{y} \in K. \end{aligned}$$

Also the notation G is used for Green's function of the Dirichlet–Neumann problems for $-\Delta$ in Ω, that is $G(\mathbf{x},\mathbf{y}) = G_{cone}(\mathbf{x},\mathbf{y}) - \mathcal{K}(\mathbf{x},\mathbf{y})$, where the harmonic function $\mathcal{K}(\mathbf{x},\mathbf{y})$ is a solution of the boundary value problem

$$\Delta_x \mathcal{K}(\mathbf{x}, \mathbf{y}) = 0, \quad \mathbf{x}, \mathbf{y} \in \Omega,$$
$$\frac{\partial \mathcal{K}}{\partial n_x}(\mathbf{x}, \mathbf{y}) = 0, \quad \mathbf{x} \in l, \ \mathbf{y} \in \Omega,$$
$$\mathcal{K}(\mathbf{x}, \mathbf{y}) = G_{cone}(\mathbf{x}, \mathbf{y}), \quad \mathbf{x} \in \Gamma, \mathbf{y} \in \Omega.$$

We note that

$$\mathcal{K}(0, \mathbf{y}) = (s|\mathbf{y}|)^{-1} - G(0, \mathbf{y}), \text{ and } \mathcal{K}(\mathbf{x}, 0) = (s|\mathbf{x}|)^{-1} - G(\mathbf{x}, 0),$$

where s is the area of $C \bigcap S_1$.

To describe the model fields in the unbounded domain ω, we use the scaled coordinates $\boldsymbol{\xi} = \varepsilon^{-1}\mathbf{x}$, $\boldsymbol{\eta} = \varepsilon^{-1}y$. Let $P(\boldsymbol{\xi})$ be a relative capacitary potential of γ, which solves the boundary value problem

$$\Delta P(\boldsymbol{\xi}) = 0, \quad \boldsymbol{\xi} \in \omega,$$
$$P(\boldsymbol{\xi}) = 1, \quad \boldsymbol{\xi} \in \gamma,$$
$$\frac{\partial P}{\partial n}(\boldsymbol{\xi}) = 0, \quad \boldsymbol{\xi} \in l, \ P(\boldsymbol{\xi}) \to 0 \ \text{ as } |\boldsymbol{\xi}| \to \infty.$$

Green's function $g(\boldsymbol{\xi}, \boldsymbol{\eta})$ for the unbounded domain ω is represented as $g(\boldsymbol{\xi}, \boldsymbol{\eta}) = G_{cone}(\boldsymbol{\xi}, \boldsymbol{\eta}) - \kappa(\boldsymbol{\xi}, \boldsymbol{\eta})$, where $\kappa(\boldsymbol{\xi}, \boldsymbol{\eta})$ is a solution of the model problem

$$\Delta_\xi \kappa(\boldsymbol{\xi}, \boldsymbol{\eta}) = 0, \quad \boldsymbol{\xi}, \boldsymbol{\eta} \in \omega,$$
$$\kappa(\boldsymbol{\xi}, \boldsymbol{\eta}) = G_{cone}(\boldsymbol{\xi}, \boldsymbol{\eta}), \quad \boldsymbol{\xi} \in \gamma, \ \boldsymbol{\eta} \in \omega,$$
$$\frac{\partial \kappa}{\partial n_\xi}(\boldsymbol{\xi}, \boldsymbol{\eta}) = 0, \quad \boldsymbol{\xi} \in l, \ \boldsymbol{\eta} \in \omega,$$
$$\kappa(\boldsymbol{\xi}, \boldsymbol{\eta}) \to 0 \ \text{ as } |\boldsymbol{\xi}| \to \infty, \boldsymbol{\eta} \in \omega.$$

Then the required Green's function $G_\varepsilon(\mathbf{x}, \mathbf{y})$, solving the problem (5.2)–(5.4), is approximated by the uniform asymptotic formula

$$G_\varepsilon(\mathbf{x}, \mathbf{y}) = G(\mathbf{x}, \mathbf{y}) + \varepsilon^{-1} g(\varepsilon^{-1}\mathbf{x}, \varepsilon^{-1}\mathbf{y}) - G_{cone}(\mathbf{x}, \mathbf{y})$$
$$+\mathcal{K}(0, \mathbf{y}) P(\varepsilon^{-1}\mathbf{x}) + \mathcal{K}(\mathbf{x}, 0) P(\varepsilon^{-1}\mathbf{y})$$
$$-\mathcal{K}(0, 0) P(\varepsilon^{-1}\mathbf{y}) P(\varepsilon^{-1}\mathbf{x}) + O(\varepsilon^{\lambda}),$$

where λ is a positive exponent depending on the cone opening.

In the following section we present a result including uniform asymptotic approximations of Green's functions for a mixed boundary value problem for the Laplacian in an elongated domain. Dirichlet boundary conditions are set at the end regions of this domain, whereas the Neumann boundary condition are prescribed on the lateral surface.

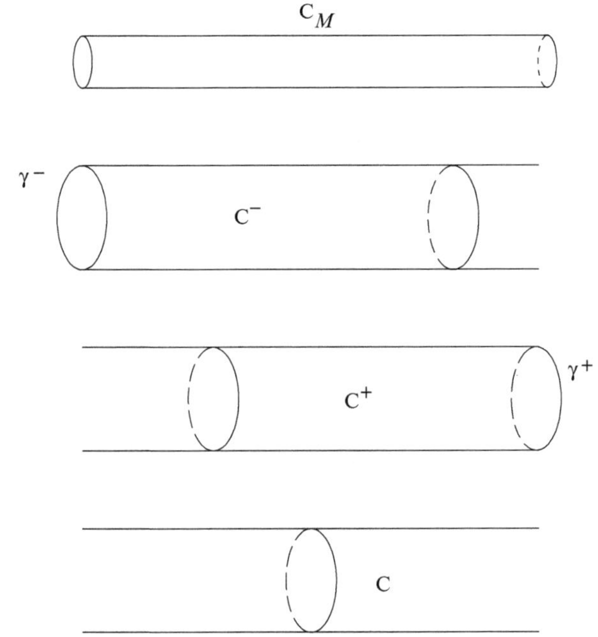

Fig. 5.4 A long rod C_M and associated unbounded model domains

5.3 The Dirichlet–Neumann Problem in a Long Rod

Let C be the infinite cylinder $\{(\mathbf{x}', x_n) : \mathbf{x}' \in \omega, x_n \in \mathbb{R}\}$, where ω is a bounded domain in $\mathbb{R}^{n-1}$ with smooth boundary; here $n \geq 2$. Also let $C^{\pm}$ denote Lipschitz subdomains of C separated from $\pm\infty$ by surfaces $\gamma^{\pm}$, respectively.

Let us introduce a positive number M and the vector $\mathbf{M} = (\mathbf{O}', M)$, where $\mathbf{O}'$ is the origin of $\mathbb{R}^{n-1}$. It is assumed that the ratio (diam $\omega)/M$ is small.

A long rod C_M is defined as follows

$$C_M = \{\mathbf{x} : (\mathbf{x} - \mathbf{M}) \in C^+, \ (\mathbf{x} + \mathbf{M}) \in C^-\},$$

the lateral surface of the rod is denoted by Γ, as shown in Fig. 5.4.

Let $G_M(\mathbf{x}, \mathbf{y})$ denote the fundamental solution for $-\Delta$ in the domain C_M subject to zero Neumann condition on the lateral surface Γ and zero Dirichlet conditions on the end parts $\gamma^{\pm}$ of the boundary of the long rod:

$$\Delta_x G_M(\mathbf{x}, \mathbf{y}) + \delta(\mathbf{x} - \mathbf{y}) = 0, \ \ \mathbf{x}, \mathbf{y} \in C_M,$$

$$\frac{\partial G_M}{\partial n_x}(\mathbf{x}, \mathbf{y}) = 0, \ \ \mathbf{x} \in \Gamma, \ \mathbf{y} \in C_M,$$

$$G_M(\mathbf{x}, \mathbf{y}) = 0, \ \ \mathbf{x} \in \gamma^{\pm}, \ \mathbf{y} \in C_M.$$

In order to obtain an approximation of G_M we also introduce several model problems independent of the cylinder length M.

By $G_\infty(\mathbf{x}, \mathbf{y})$ we denote Green's function of the Neumann problem in C

$$\Delta_x G_\infty(\mathbf{x}, \mathbf{y}) + \delta(\mathbf{x} - \mathbf{y}) = 0, \quad \mathbf{x}, \mathbf{y} \in C,$$

$$\frac{\partial G_\infty}{\partial n_x}(\mathbf{x}, \mathbf{y}) = 0, \quad \mathbf{x} \in \Gamma, \mathbf{y} \in C,$$

$$G_\infty(\mathbf{x}, \mathbf{y}) = -(2|\omega|)^{-1}|x_n - y_n| + O(\exp(-\alpha|x_n - y_n|)) \text{ as } |x_n| \to \infty,$$

where α is a positive constant, and $|\omega|$ is the $(n-1)$-dimensional measure of ω.

Similarly, G^+ and G^- stand for the fundamental solutions for $-\Delta$ in the domains $C^\pm$, with the homogeneous boundary conditions defined as follows

$$\Delta_x G^\pm(\mathbf{x}^\pm, \mathbf{y}^\pm) + \delta(\mathbf{x}^\pm - \mathbf{y}^\pm) = 0, \quad \mathbf{x}^\pm, \mathbf{y}^\pm \in C^\pm,$$

$$G^\pm(\mathbf{x}^\pm, \mathbf{y}^\pm) = 0, \quad \mathbf{x}^\pm \in \gamma^\pm, \mathbf{y}^\pm \in C^\pm,$$

$$\frac{\partial G^\pm}{\partial n_x}(\mathbf{x}^\pm, \mathbf{y}^\pm) = 0, \quad \mathbf{x}^\pm \in \Gamma, \mathbf{y}^\pm \in C^\pm,$$

and it is also assumed that $G^\pm(\mathbf{x}^\pm, \mathbf{y}^\pm)$ are bounded as $x_n^\pm \to \mp\infty$.

5.3.1 Capacitary Potential

The capacitary potential P_M is defined as a solution of the Dirichlet–Neumann boundary value problem in C_M:

$$\Delta P_M(\mathbf{x}) = 0, \quad \mathbf{x} \in C_M, \tag{5.5}$$

$$\frac{\partial P_M}{\partial n}(\mathbf{x}) = 0, \quad \mathbf{x} \in \Gamma, \tag{5.6}$$

$$P_M(\mathbf{x}) = 1, \quad \mathbf{x} \in \gamma^- \text{ and } P_M(\mathbf{x}) = 0, \quad \mathbf{x} \in \gamma^+. \tag{5.7}$$

We shall also use the solutions $\zeta^\pm$ of the homogeneous Dirichlet–Neumann problems in semi-infinite domains $C^\pm$, as follows:

$$\Delta \zeta^\pm(\mathbf{x}^\pm) = 0, \quad \mathbf{x} \in C^\pm, \tag{5.8}$$

$$\frac{\partial \zeta^\pm}{\partial n}(\mathbf{x}^\pm) = 0, \quad \mathbf{x}^\pm \in \Gamma, \tag{5.9}$$

$$\zeta^\pm(\mathbf{x}^\pm) = 0, \quad \mathbf{x} \in \gamma^\pm, \tag{5.10}$$

and

$$\zeta^{\pm}(\mathbf{x}^{\pm}) = \mp x_n^{\pm} + \zeta_{\infty}^{\pm} + O(\exp(-\alpha|x_n^{\pm}|)) \text{ as } |x_n^{\pm}| \to \infty, \tag{5.11}$$

where α is a positive constant, $\mathbf{x}^{\pm} = (\mathbf{x}', x_n \mp M)$ are local coordinates at the ends of the long rod C_M, and $\zeta_{\infty}^{\pm}$ are constant terms that depend on the geometry of the cross-section ω and the end parts $\gamma^{\pm}$ of the boundary of the long rod.

Theorem 5.1. *The following asymptotic formula, uniform with respect to $\mathbf{x} \in C_M$, for the capacitary potential $P_M(\mathbf{x})$ holds:*

$$P_M(\mathbf{x}) = \frac{M + x_n + \zeta_{\infty}^{-} - \zeta^{-}(\mathbf{x}^{-}) + \zeta^{+}(\mathbf{x}^{+})}{2M + \zeta_{\infty}^{+} + \zeta_{\infty}^{-}} + O(\exp(-\alpha M)). \tag{5.12}$$

Here, the functions $\zeta^{\pm}$, variables $\mathbf{x}^{\pm}$ and the constants $\zeta_{\infty}^{\pm}$ are the same as in (5.8)–(5.11), α is a positive constant.

To prove this statement we use the direct substitution of (5.12) into (5.5)–(5.7), which shows that the remainder term is a harmonic function satisfying homogeneous Neumann boundary conditions on the lateral surface of the rod and is exponentially small at the end parts $\gamma^{\pm}$ of the boundary. Then it remains to apply the estimate similar to Lemma 1.3 of Sect. 1.5 in Kozlov, Maz'ya and Movchan [19].

5.3.2 Asymptotic Approximation of Green's Function

Let $H^{\pm}(\mathbf{x}^{\pm}, \mathbf{y}^{\pm})$ be functions defined in semi-infinite domains $C^{\pm}$, and assume that they also satisfy the Dirichlet–Neumann boundary value problems

$$\Delta_x H^{\pm}(\mathbf{x}^{\pm}, \mathbf{y}^{\pm}) = 0, \ \mathbf{x}^{\pm}, \mathbf{y}^{\pm} \in C^{\pm}, \tag{5.13}$$

$$\frac{\partial H^{\pm}}{\partial n_x}(\mathbf{x}^{\pm}, \mathbf{y}^{\pm}) = 0, \ \mathbf{x}^{\pm} \in \Gamma, \ \mathbf{y}^{\pm} \in C^{\pm}, \tag{5.14}$$

$$H^{+}(\mathbf{x}^{+}, \mathbf{y}^{+}) - G_{\infty}(\mathbf{x}, \mathbf{y}) + (2|\omega|)^{-1}\zeta^{\pm}(\mathbf{y}^{\pm}), \ \mathbf{x} \subset \gamma^{\pm}, \ \mathbf{y}^{\pm} \in C^{\pm}, \tag{5.15}$$

and

$$H^{\pm}(\mathbf{x}^{\pm}, \mathbf{y}^{\pm}) \to 0 \text{ as } x_n^{\pm} \to \mp\infty. \tag{5.16}$$

The asymptotic approximation is given by the following statement.

Theorem 5.2. *Green's function $G_M(\mathbf{x}, \mathbf{y})$ is approximated by the asymptotic formula, uniform with respect to $\mathbf{x}, \mathbf{y} \in C_M$*

$$G_M(\mathbf{x},\mathbf{y}) = G_\infty(\mathbf{x},\mathbf{y}) - H^+(\mathbf{x}^+,\mathbf{y}^+) - H^-(\mathbf{x}^-,\mathbf{y}^-)$$

$$-\frac{\mathfrak{A}_M}{|\omega|}\Big(\frac{1}{2} - P_M(\mathbf{x})\Big)\Big(\frac{1}{2} - P_M(\mathbf{y})\Big) + \frac{\mathfrak{A}_M}{4|\omega|} + O(\exp(-\alpha M)), \tag{5.17}$$

where $\mathfrak{A}_M = 2M + \zeta_\infty^+ + \zeta_\infty^-$, *and* α *is a positive constant.*

In the text below we present a formal argument that leads to the asymptotic formula (5.2).

Let

$$G_M(\mathbf{x},\mathbf{y}) = G_\infty(\mathbf{x},\mathbf{y}) - H_M^+(\mathbf{x},\mathbf{y}) - H_M^-(\mathbf{x},\mathbf{y}), \tag{5.18}$$

where the functions $H_M^\pm$ are defined as solutions of the boundary value problems

$$\Delta_x H_M^\pm(\mathbf{x},\mathbf{y}) = 0, \quad \mathbf{x},\mathbf{y} \in C_M,$$

$$\frac{\partial H_M^\pm}{\partial n_x}(\mathbf{x},\mathbf{y}) = 0, \quad \mathbf{x} \in \Gamma, \mathbf{y} \in C_M,$$

$$H_M^\pm(\mathbf{x},\mathbf{y}) = G_\infty(\mathbf{x},\mathbf{y}), \quad \mathbf{x} \in \gamma^\pm, \mathbf{y} \in C_M,$$

$$H_M^\pm(\mathbf{x},\mathbf{y}) = 0, \quad \mathbf{x} \in \gamma^\mp, \ \mathbf{y} \in C_M.$$

We note that the sum $\sum_\pm H_M^\pm$ is symmetric, i.e.

$$H_M^+(\mathbf{x},\mathbf{y}) + H_M^-(\mathbf{x},\mathbf{y}) = H_M^+(\mathbf{y},\mathbf{x}) + H_M^-(\mathbf{y},\mathbf{x}).$$

The functions $H_M^\pm$ can be approximated by the formulae

$$H_M^+(\mathbf{x},\mathbf{y}) = H^+(\mathbf{x}^+,\mathbf{y}^+) - \frac{1}{2|\omega|}\zeta^+(\mathbf{y}^+)$$

$$-P_M(\mathbf{x})\Big(H^+(\mathbf{x}^{+\prime}, -\infty, \mathbf{y}^+) - \frac{1}{2|\omega|}\zeta^+(\mathbf{y}^+)\Big) + h_M^+,$$

and

$$H_M^-(\mathbf{x},\mathbf{y}) = H^-(\mathbf{x}^-,\mathbf{y}^-) - \frac{1}{2|\omega|}\zeta^-(\mathbf{y}^-)$$

$$-P_M(\mathbf{x})\Big(H^-(\mathbf{x}^{-\prime}, +\infty, \mathbf{y}^-) - \frac{1}{2|\omega|}\zeta^-(\mathbf{y}^-)\Big) + h_M^-,$$

with exponentially small remainder terms $h_M^\pm$. Applying Green's formula to the functions $H^\pm$ and $\zeta^\pm$ in the domains $C^\pm$, respectively, we deduce that

$$H^-(\mathbf{x}^{-\prime}, +\infty, \mathbf{y}^-) = -\frac{1}{2|\omega|}\{\zeta^-(\mathbf{y}^-) - (M + y_n + \zeta_-^\infty)\},$$

and

$$H^+(\mathbf{x}^{+'}, -\infty, \mathbf{y}^+) = -\frac{1}{2|\omega|}\{\zeta^+(\mathbf{y}^+) - (M - \mathbf{y}_n + \zeta_+^{\infty})\}.$$

The condition (5.11) yields

$$\lim_{\mathbf{y}_n^- \to +\infty} H^-(\mathbf{y}^{-'}, +\infty, \mathbf{y}^-) = 0,$$

and

$$\lim_{\mathbf{y}_n^+ \to -\infty} H^+(\mathbf{y}^{+'}, -\infty, \mathbf{y}^+) = 0.$$

If $\mathfrak{A} = 2M + \zeta_\infty^+ + \zeta_\infty^-$, then the following identity holds

$$H_M^+(\mathbf{x}, \mathbf{y}) + H_M^-(\mathbf{x}, \mathbf{y}) = H^+(\mathbf{x}^+, \mathbf{y}^+) + H^-(\mathbf{x}^-, \mathbf{y}^-)$$
$$+ \frac{\mathfrak{A}}{|\omega|}\Big(\frac{1}{2} - P_M(\mathbf{x})\Big)\Big(\frac{1}{2} - P_M(\mathbf{y})\Big) - \frac{\mathfrak{A}_M}{4|\omega|}. \tag{5.19}$$

Combining the formulae (5.18) and (5.19) we deduce (5.17).

The direct substitution of (5.17) into (5.14), (5.15) shows that the remainder term is a harmonic function satisfying homogeneous Neumann boundary conditions on the lateral surface of the rod, and it is exponentially small at the end parts $\gamma^\pm$ of the boundary. Applying the estimate similar to Lemma 1.3 of Sect. 1.5 in Kozlov, Maz'ya and Movchan [19] we complete the proof.

Example of Green's functions in model domains. In some cases, Green's functions for model problems required for the above asymptotic approximation can be constructed in a simple form. As an illustration, we suggest an example involving a long rectangular strip. In this case, the function $G_\infty(\mathbf{x}, \mathbf{y})$ is the Neumann function for the Laplacian in the infinite strip $\Pi = \{(x_1, x_2) : -\infty < x_1 < \infty, |x_2| < 1/2\}$, given in the form

$$G_\infty(\mathbf{x}, \mathbf{y}) = \frac{1}{2\pi}\int_{-\infty}^{\infty} \tilde{G}(k, x_2, y_2)\exp(\ ik(x_1 - y_1))dk,$$

where

$$\tilde{G}(k, x_2, y_2) = \frac{\cosh(k(x_2 + y_2)) + \cosh(k)\cosh(k(x_2 - y_2))}{2k\sinh(k)}$$
$$- \begin{cases} (2k)^{-1}\sinh(k(x_2 - y_2)), & x_2 > y_2 \\ -(2k)^{-1}\sinh(k(x_2 - y_2)), & x_2 < y_2. \end{cases}$$

Assuming that the end regions of the rectangular domain are "flat", i.e. they are located on the vertical straight lines $x_1 = \pm M$, we can construct Green's functions $G_\pm$ for semi-infinite strips as follows:

$$G_\pm(\mathbf{x}^\pm, \mathbf{y}^\pm) = G_\infty(\mathbf{x}^\pm, y_1^\pm, y_2^\pm) - G_\infty(\mathbf{x}^\pm, -y_1^\pm, y_2^\pm).$$

These model fields are readily applicable in the asymptotic formula of Theorem 5.2.

5.3.3 *Green's Function G_M Versus Green's Functions for Unbounded Domains*

The result of Sect. 5.3.2 together with definitions of functions G_∞ and $G^\pm$ lead to the following

Theorem 5.3. *The Green's function $G_M(\mathbf{x}, \mathbf{y})$ and the functions $G^\pm$, G_∞ are related by the asymptotic formula*

$$G_M(\mathbf{x}, \mathbf{y}) = \sum_\pm G^\pm(\mathbf{x}^\pm, \mathbf{y}^\pm) - G_\infty(\mathbf{x}, \mathbf{y}) - \frac{1}{2|\omega|} \sum_\pm \Big(\zeta^\pm(\mathbf{x}^\pm) + \zeta^\pm(\mathbf{y}^\pm)\Big)$$

$$- \frac{\mathfrak{A}_M}{|\omega|}\Big(\frac{1}{2} - P_M(\mathbf{x})\Big)\Big(\frac{1}{2} - P_M(\mathbf{y})\Big) + \frac{\mathfrak{A}_M}{4|\omega|} + O(\exp(-\alpha M)) \qquad (5.20)$$

where α is a positive constant independent of M.

Corollary 5.1. *The formula* (5.20) *allows for an equivalent representation involving the model fields $\zeta^\pm$ defined as solutions of the boundary value problems* (5.8)–(5.11)*:*

$$G_M(\mathbf{x}, \mathbf{y}) = \sum_\pm G^\pm(\mathbf{x}^\pm, \mathbf{y}^\pm) - G_\infty(\mathbf{x}, \mathbf{y}) + \frac{1}{4|\omega|}\Big\{\mathfrak{A}_M - 2\sum_\pm \Big(\zeta^\pm(\mathbf{x}^\pm) + \zeta^\pm(\mathbf{y}^\pm)\Big)\Big\}$$

$$- \Big(|\omega|\mathfrak{A}_M\Big)^{-1}\Big(\mathbf{x}_n - \tfrac{1}{2}(\zeta_\infty^+ - \zeta_\infty^-) + \zeta^+(\mathbf{x}^+) - \zeta^-(\mathbf{x}^-)\Big) \qquad (5.21)$$

$$\times\Big(\mathbf{y}_n - \tfrac{1}{2}(\zeta_\infty^+ - \zeta_\infty^-) + \zeta^+(\mathbf{y}^+) - \zeta^-(\mathbf{y}^-)\Big) + O(\exp(-\alpha M)),$$

where α is a positive constant independent of M.

The above formulae can be simplified if we introduce additional constraints on the positions of the points $\mathbf{x}$ and $\mathbf{y}$ within C_M.

When the points $\mathbf{x}$ and $\mathbf{y}$ are "far away" from the ends $\gamma^\pm$ of the long rod the quantities $H^\pm$ become exponentially small, and hence we arrive to the following

Corollary 5.2. *When* $\min\{(\mathbf{x}\pm M)/M, (\mathbf{x}\pm M)/M\} \geq \text{Const}$, *the Green's function* G_M *is approximated by the formula*

$$G_M(\mathbf{x},\mathbf{y}) \sim G_\infty(\mathbf{x},\mathbf{y}) - (|\omega|\mathfrak{A}_M)^{-1}\Big(x_n - \tfrac{1}{2}(\zeta_\infty^+ - \zeta_\infty^-)\Big)\Big(y_n - \tfrac{1}{2}(\zeta_\infty^+ - \zeta_\infty^-)\Big) + \frac{\mathfrak{A}_M}{4|\omega|}, \tag{5.22}$$

as $M \to \infty$.

Another simplified formula for the Green's function can be written for the case when the points $\mathbf{x}$ and $\mathbf{y}$ are sufficiently close to one of the ends of the rod.

Corollary 5.3. *Assume that the points* $\mathbf{x}$ *and* $\mathbf{y}$ *are close to the left end* γ^- *of the long rod* C_M, *i.e.* $\max\{\mathbf{x}+M, \mathbf{y}+M\} \leq \text{Const}$. *Then the function* G_M *is approximated by the formula*

$$G_M(\mathbf{x},\mathbf{y}) \sim G^-(\mathbf{x}^-,\mathbf{y}^-) - |\omega|\frac{G^-(\mathbf{x}^{-\prime}, +\infty, \mathbf{y}^-)G^-(\mathbf{x}^-, \mathbf{y}^{-\prime}, +\infty)}{\mathfrak{A}_M}, \tag{5.23}$$

as $M \to \infty$.

Similar approximation is valid near the other end γ^+ of the long rod.

5.3.4 The Dirichlet–Neumann Problem in a Thin Rod

By rescaling, the above results can be used to find an asymptotic approximation for Green's function $G^{(\varepsilon)}$ in a thin rod rather than the long rod. Let a thin domain be defined by

$$C_\varepsilon = \{\mathbf{x} : \varepsilon^{-1}(\mathbf{x}-\mathbf{a}) \in C^+,\ \varepsilon^{-1}(\mathbf{x}+\mathbf{a}) \in C^-\},$$

where the notations $C^\pm$ are the same as in the beginning of Sect. 5.3 (see Fig. 5.4), $2a$ is the length of the rod, and now ε is a positive small parameter. As above, it is assumed that Green's function is subject to zero Neumann condition on the cylindrical part of C_ε and zero Dirichlet condition on the remaining part of ∂C_ε.

Theorem 5.4. *The following asymptotic formula for* $G^{(\varepsilon)}(\mathbf{x},\mathbf{y})$, *uniform with respect to* $\mathbf{x},\mathbf{y} \in \Omega_\varepsilon$, *holds*

$$G^{(\varepsilon)}(\mathbf{x},\mathbf{y}) = \varepsilon^{2-n}\Big\{G^+(\varepsilon^{-1}(\mathbf{x}-\mathbf{a}), \varepsilon^{-1}(\mathbf{y}-\mathbf{a})) + G^-(\varepsilon^{-1}(\mathbf{x}+\mathbf{a}), \varepsilon^{-1}(\mathbf{y}+\mathbf{a})) - G_\infty(\varepsilon^{-1}\mathbf{x}, \varepsilon^{-1}\mathbf{y})$$

$$-\varepsilon\{2|\omega|^{-1}a+\varepsilon(\zeta_\infty^{+}+\zeta_\infty^{-})\}^{-1}\Big(\frac{x_n}{\varepsilon|\omega|}-\tfrac{1}{2}(\zeta_\infty^{-}-\zeta_\infty^{+})+\zeta^{+}\Big(\frac{\mathbf{x}-\mathbf{a}}{\varepsilon}\Big)-\zeta^{-}\Big(\frac{\mathbf{x}+\mathbf{a}}{\varepsilon}\Big)\Big)$$

$$\times\Big(\frac{y_n}{\varepsilon|\omega|}-\tfrac{1}{2}(\zeta_\infty^{-}-\zeta_\infty^{+})+\zeta^{+}\Big(\frac{\mathbf{y}-\mathbf{a}}{\varepsilon}\Big)-\zeta^{-}\Big(\frac{\mathbf{y}+\mathbf{a}}{\varepsilon}\Big)\Big)$$

$$+\frac{1}{4}\Big((\varepsilon|\omega|)^{-1}2a+\zeta_\infty^{-}+\zeta_\infty^{+}-2\sum_{\pm}\big(\zeta^{\pm}(\varepsilon^{-1}(\mathbf{x}\mp\mathbf{a}))+\zeta^{\pm}(\varepsilon^{-1}(\mathbf{y}\mp\mathbf{a}))\big)\Big)$$

$$+\,O(\exp(-\beta/\varepsilon))\Big\}, \tag{5.24}$$

where β is a positive constant independent of ε.

Part II
Green's Tensors for Vector Elasticity in Bodies with Small Defects

Chapter 6
Green's Tensor for the Dirichlet Boundary Value Problem in a Domain with a Single Inclusion

We consider an elastic domain containing a single small inclusion. The columns of Green's tensor correspond to the displacement vectors produced by unit point forces aligned with the coordinate axes. Governing equations and main definitions are given in Sect. 6.1. Here, we also discuss an application of this tensor, concerning Green's representation for a particular class of problems in elasticity for a domain with a small inclusion. Section 6.2, includes the result on the estimates for the maximum modulus of solutions to the homogeneous Lamé system in a domain containing a small inclusion. In Sect. 6.3, for such a domain, we derive the uniform approximation of Green's tensor in a three-dimensional domain. For the case of a planar singularly perturbed domain we construct the corresponding Green's tensor for the Lamé operator, in Sect. 6.4. Section 6.5 contains corollaries, which show that under certain constraints on the independent variables, the asymptotic formulae for Green's matrices can be simplified.

6.1 Green's Representation for Vector Elasticity

Let $\Omega_\varepsilon \subset \mathbb{R}^n$, $n = 2, 3$ be a domain containing a small inclusion or void dependent upon a small parameter ε. As a simple example, consider the following problem posed in Ω_ε

$$\mu \Delta_{\mathbf{x}} \mathbf{u}(\mathbf{x}) + (\lambda + \mu)\nabla_{\mathbf{x}}(\nabla_{\mathbf{x}} \cdot \mathbf{u}(\mathbf{x})) + \mathbf{f}(\mathbf{x}) = \mathbf{O}\ , \mathbf{x} \in \Omega_\varepsilon, \tag{6.1}$$

$$\mathbf{u}(\mathbf{x}) = \mathbf{O}\ , \quad \mathbf{x} \in \partial\Omega_\varepsilon\ , \tag{6.2}$$

where λ and μ are the Lamé elastic moduli, $\mathbf{O}$ is the zero vector in $\mathbb{R}^n$, $\mathbf{u}(\mathbf{x}) = (u_1(\mathbf{x}), \dots, u_n(\mathbf{x}))^T$, $\mathbf{f} = (f_1(\mathbf{x}), \dots, f_n(\mathbf{x}))^T$ and the components of $\mathbf{f}$ are assumed to be smooth. Then suppose G_ε is the Green's tensor of the Lamé operator, which solves

$$\mu \Delta_{\mathbf{x}} G_\varepsilon(\mathbf{x}, \mathbf{y}) + (\lambda + \mu)\nabla_{\mathbf{x}}(\nabla_{\mathbf{x}} \cdot G_\varepsilon(\mathbf{x}, \mathbf{y})) + \delta(\mathbf{x} - \mathbf{y})I_n = 0I_n, \mathbf{x}, \mathbf{y} \in \Omega_\varepsilon, \tag{6.3}$$

V. Maz'ya et al., *Green's Kernels and Meso-Scale Approximations in Perforated Domains*, Lecture Notes in Mathematics 2077, DOI 10.1007/978-3-319-00357-3_6,

$$G_\varepsilon(\mathbf{x},\mathbf{y}) = 0 I_n\ , \quad \mathbf{x} \in \partial\Omega_\varepsilon, \mathbf{y} \in \Omega_\varepsilon\ . \tag{6.4}$$

where I_n is the $n \times n$ identity matrix. Then the solution of (6.1) and (6.2), can be computed using this Green's tensor in the following way. By applying Betti's formula to the tensor G_ε and vector function $\mathbf{u}$, we immediately obtain

$$\mathbf{u}(\mathbf{x}) = \int_{\Omega_\varepsilon} G_\varepsilon(\mathbf{x},\mathbf{y})\mathbf{f}(\mathbf{y})\, d\mathbf{y}\ . \tag{6.5}$$

We note that a similar formula can be obtained for the case of the mixed problem considered in Chap. 8.

6.1.1 Geometry and Matrix Differential Operators

We now give several notations adopted in the following text. Let Ω be a bounded domain in $\mathbb{R}^n$, $n = 2, 3$, with compact closure $\bar{\Omega}$ and smooth boundary $\partial\Omega$. By ω we denote a contractible domain in $\mathbb{R}^n$ with smooth boundary $\partial\omega$ and compact closure $\bar{\omega}$; its complement being $C\bar{\omega} = \mathbb{R}^n \backslash \bar{\omega}$. We shall assume that both Ω and ω contain the origin $\mathbf{O}$ as an interior point. It is also assumed that the minimum distance between $\mathbf{O}$ and the points of $\partial\Omega$ is equal to 1. In addition the maximum distance between $\mathbf{O}$ and the points of $\partial C\bar{\omega}$ will be taken as 1. We introduce the set $\omega_\varepsilon = \{\mathbf{x} : \varepsilon^{-1}\mathbf{x} \in \omega\}$, where ε is a small positive parameter, and the open set $\Omega_\varepsilon = \Omega\backslash\bar{\omega}_\varepsilon$. The notation B_ϱ stands for the open ball centered at $\mathbf{O}$ with radius ϱ.

In the sequel, along with $\mathbf{x}$ and $\mathbf{y}$, we shall use scaled variables $\boldsymbol{\xi} = \varepsilon^{-1}\mathbf{x}$ and $\boldsymbol{\eta} = \varepsilon^{-1}\mathbf{y}$.

Let $\boldsymbol{\sigma}(u) = [\sigma_{ij}(u)]_{i,j=1}^n$, $(n = 2, 3)$, represent the Cauchy stress tensor, which for an isotropic solid with displacements $u = \{u_k\}_{k=1}^n$ has entries of the form

$$\sigma_{ij}(u) = \lambda\delta_{ij}u_{p,p} + \mu(u_{i,j} + u_{j,i})\ , \tag{6.6}$$

where λ and μ are Lamé elastic moduli.

Here and elsewhere in the text, the repeated indices are regarded as the indices of summation, and $t_i^n(u) = \sigma_{ij}(u)n_j$ are the tractions computed for displacements u, where n_j is the jth component of the unit outward normal.

Also $\boldsymbol{e}(u) = [e_{ij}(u)]_{i,j=1}^n$ denotes the strain tensor, whose entries are given by

$$e_{ij}(u) = 2^{-1}(u_{i,j} + u_{j,i})\ , \tag{6.7}$$

for $n = 2, 3$.

Let $T_n(\partial_{\mathbf{x}})$ be the differential operator of tractions and $\mathbf{u}(\mathbf{x})$ a vector function with k-components. The tractions of this vector function on the boundary are defined by

$$T_n(\partial_{\mathbf{x}})\mathbf{u}(\mathbf{x}) = n_1 T^{(1)}(\partial_{\mathbf{x}})\mathbf{u}(\mathbf{x}) + \cdots + n_k T^{(k)}(\partial_{\mathbf{x}})\mathbf{u}(\mathbf{x})\ , \tag{6.8}$$

where $\boldsymbol{n} = (n_1, \ldots, n_k)^T$ is the unit outward normal. In the two-dimensional case

$$T^{(1)}(\partial_{\mathbf{x}}) = \begin{pmatrix} (\lambda+2\mu)\partial_1 & \lambda\partial_2 \\ \mu\partial_2 & \mu\partial_1 \end{pmatrix}, T^{(2)}(\partial_{\mathbf{x}}) = \begin{pmatrix} \mu\partial_2 & \mu\partial_1 \\ \lambda\partial_1 & (\lambda+2\mu)\partial_2 \end{pmatrix}, \tag{6.9}$$

and in three dimensions we have

$$T^{(1)}(\partial_{\mathbf{x}}) = \begin{pmatrix} (\lambda+2\mu)\partial_1 & \lambda\partial_2 & \lambda\partial_3 \\ \mu\partial_2 & \mu\partial_1 & 0 \\ \mu\partial_3 & 0 & \mu\partial_1 \end{pmatrix}, T^{(2)}(\partial_{\mathbf{x}}) = \begin{pmatrix} \mu\partial_2 & \mu\partial_1 & 0 \\ \lambda\partial_1 & (\lambda+2\mu)\partial_2 & \lambda\partial_3 \\ 0 & \mu\partial_3 & \mu\partial_2 \end{pmatrix},$$

$$T^{(3)}(\partial_{\mathbf{x}}) = \begin{pmatrix} \mu\partial_3 & 0 & \mu\partial_1 \\ 0 & \mu\partial_3 & \mu\partial_2 \\ \lambda\partial_1 & \lambda\partial_2 & (\lambda+2\mu)\partial_3 \end{pmatrix}, \tag{6.10}$$

with $\partial_{\mathbf{x}} = \partial/\partial\mathbf{x}$, $\partial_i = \partial/\partial x_i$.

We shall study the Green's tensor G_ε for the Lamé operator, which is denoted by $L(\partial_{\mathbf{x}}) = [L_{ij}(\partial_{\mathbf{x}})]_{i,j=1}^n$, $n = 2, 3$, whose entries are given by

$$L_{ij}(\partial_{\mathbf{x}}) = \begin{cases} (\lambda+2\mu)\partial_i^2 + \mu\sum_{m=1}^n (1-\delta_{im})\partial_m^2 & \text{for } i = j \\ (\lambda+\mu)\partial_{ij}^2 & \text{for } i \neq j\,, \end{cases} \tag{6.11}$$

and here δ_{im} is the Kronecker delta.

The tensor G_ε is a solution of the following problem in $\Omega_\varepsilon \subset \mathbb{R}^n, n = 2, 3$,

$$L(\partial_{\mathbf{x}})G_\varepsilon(\mathbf{x}, \mathbf{y}) + \delta(\mathbf{x}-\mathbf{y})I_n = 0I_n, \quad \mathbf{x}, \mathbf{y} \in \Omega_\varepsilon\,, \tag{6.12}$$

$$G_\varepsilon(\mathbf{x}, \mathbf{y}) = 0I_n, \quad \mathbf{x} \in \partial\Omega_\varepsilon, \mathbf{y} \in \Omega_\varepsilon\,, \tag{6.13}$$

where I_n, is the $n \times n$ identity matrix.

An important property of Green's tensor is the following symmetry relation

$$G_\varepsilon(\mathbf{x}, \mathbf{y}) = (G_\varepsilon(\mathbf{y}, \mathbf{x}))^T, \quad \text{for } \mathbf{x}, \mathbf{y} \in \Omega_\varepsilon, \mathbf{x} \neq \mathbf{y}\,. \tag{6.14}$$

Betti's Identities. Let $\mathbf{u}(\mathbf{x}) = \{\mathbf{u}_i(\mathbf{x})\}_{i=1}^n$ and $\mathbf{v}(\mathbf{x}) = \{\mathbf{v}_i(\mathbf{x})\}_{i=1}^n$ be real vector functions on a domain $\Omega \subset \mathbb{R}^n, n = 2, 3$. Then *Betti's first identity* can be written as

$$\int_\Omega \mathbf{u}(\mathbf{x}) \cdot L(\partial_{\mathbf{x}})\mathbf{v}(\mathbf{x})\,d\mathbf{x} = -\int_\Omega \text{Trace}(\boldsymbol{\sigma}(\mathbf{u})\boldsymbol{e}(\mathbf{v}))\,d\mathbf{x} + \int_{\partial\Omega} \mathbf{u}(\mathbf{x}) \cdot T_n(\partial_{\mathbf{x}})\mathbf{v}(\mathbf{x})\,dS_{\mathbf{x}}\,,$$

which is a direct consequence of integration by parts. By interchanging $\mathbf{u}$ and $\mathbf{v}$ in the above relation, and subtracting we can obtain *Betti's second identity*

$$\int_{\Omega} \{\mathbf{u}(\mathbf{x}) \cdot L(\partial_{\mathbf{x}})\mathbf{v}(\mathbf{x}) - \mathbf{v}(\mathbf{x}) \cdot L(\partial_{\mathbf{x}})\mathbf{u}(\mathbf{x})\}\, d\mathbf{x}$$
$$= \int_{\partial\Omega} \{\mathbf{u}(\mathbf{x}) \cdot T_n(\partial_{\mathbf{x}})\mathbf{v}(\mathbf{x}) - \mathbf{v}(\mathbf{x}) \cdot T_n(\partial_{\mathbf{x}})\mathbf{u}(\mathbf{x})\}\, dS_{\mathbf{x}} \,.$$

The differential operators and Betti's identities introduced above can also be given an equivalent representation as we will now see.

We now consider the case of two dimensions. Let $\mathbf{D}(\boldsymbol{\xi})$ be the matrix function

$$\mathbf{D}(\boldsymbol{\xi}) = \begin{pmatrix} \xi_1 & 0 & 2^{-1/2}\xi_2 \\ 0 & \xi_2 & 2^{-1/2}\xi_1 \end{pmatrix} \tag{6.15}$$

and $\mathcal{C}$ be the 3×3 symmetric constant matrix:

$$\mathcal{C} = \begin{pmatrix} \lambda + 2\mu & \lambda & 0 \\ \lambda & \lambda + 2\mu & 0 \\ 0 & 0 & 2\mu \end{pmatrix} . \tag{6.16}$$

Then the operator $L(\partial_{\mathbf{x}})$, using the above matrix differential operator (6.15) and matrix of elastic constants (6.16), can be written as

$$L(\partial_{\mathbf{x}}) = \mathbf{D}(\partial_{\mathbf{x}})\mathcal{C}\mathbf{D}(\partial_{\mathbf{x}})^T \,.$$

We also write the differential operator of tractions $T_n(\partial_{\mathbf{x}})$ in this way

$$T_n(\partial_{\mathbf{x}}) = \mathbf{D}(\mathbf{n})\mathcal{C}\mathbf{D}(\partial_{\mathbf{x}})^T \,,$$

where $\mathbf{n}$ is the unit outward normal to the boundary at which this operator is considered.

We set

$$\mathbf{S}(\mathbf{u}) = (\sigma_{11}(\mathbf{u}), \sigma_{22}(\mathbf{u}), 2^{1/2}\sigma_{12}(\mathbf{u}))^T \,,$$

which is known as the *vector of stress*, and by *vector of strain* we mean

$$\mathbf{E}(\mathbf{u}) = (e_{11}(\mathbf{u}), e_{22}(\mathbf{u}), 2^{1/2}e_{12}(\mathbf{u}))^T \,.$$

Under these notations, we have the relations

$$\mathbf{S}(\mathbf{u}) = \mathcal{C}\mathbf{D}(\partial_{\mathbf{x}})^T\mathbf{u}(\mathbf{x}) \quad \text{and } \mathbf{E}(\mathbf{u}) = \mathbf{D}(\partial_{\mathbf{x}})^T\mathbf{u}(\mathbf{x}) \,.$$

Then the second Betti identity takes the form

$$\int_{\Omega} \{\mathbf{u}(\mathbf{x})^T \mathbf{D}(\partial_{\mathbf{x}}) \mathcal{C} \mathbf{D}(\partial_{\mathbf{x}})^T \mathbf{v}(\mathbf{x}) - \mathbf{v}(\mathbf{x})^T \mathbf{D}(\partial_{\mathbf{x}}) \mathcal{C} \mathbf{D}(\partial_{\mathbf{x}})^T \mathbf{u}(\mathbf{x})\} d\mathbf{x}$$
$$= \int_{\partial\Omega} \{\mathbf{u}(\mathbf{x})^T \mathbf{D}(\mathbf{n}) \mathcal{C} \mathbf{D}(\partial_{\mathbf{x}})^T \mathbf{v}(\mathbf{x}) - \mathbf{v}(\mathbf{x})^T \mathbf{D}(\mathbf{n}) \mathcal{C} \mathbf{D}(\partial_{\mathbf{x}})^T \mathbf{u}(\mathbf{x})\} dS_{\mathbf{x}} .$$

Another consequence of the matrix function (6.15) is it allows for the linear Taylor approximation of a vector function $\mathbf{u}$ about a point, say $\mathbf{x} = \mathbf{O}$, to take the form

$$\mathbf{u}(\mathbf{x}) = \mathbf{u}(\mathbf{O}) + \mathbf{r}(\mathbf{x}) \mathbf{r}(\partial_{\mathbf{x}})^T \mathbf{u}(\mathbf{O}) + \mathbf{D}(\mathbf{x}) \mathbf{D}(\partial_{\mathbf{x}})^T \mathbf{u}(\mathbf{O}) + O(|\mathbf{x}|^2) , \qquad (6.17)$$

where $\mathbf{r}(\mathbf{x}) = [x_2, -x_1]^T$ and the first two terms in (6.17) correspond to a rigid body displacement and rotation, for which the strain vector E is the null vector.

6.2 Estimates for the Maximum Modulus of Solutions of Elasticity Problems in Domains with Small Inclusions

In order to obtain the estimates for the remainders in the representations for G_ε in Sects. 6.3 and 6.4, we need an auxiliary result concerning an estimate for the maximum modulus of solutions for Lamé equation in domains with small inclusions. In what follows we shall formulate and prove such a result.

Let $\mathbf{u}$ be the displacement vector which satisfies the Dirichlet boundary value problem in the domain $\Omega_\varepsilon \subset \mathbb{R}^n$,

$$L(\partial_{\mathbf{x}}) \mathbf{u}(\mathbf{x}) := \mu \Delta \mathbf{u}(\mathbf{x}) + (\lambda + \mu) \nabla (\nabla \cdot \mathbf{u}(\mathbf{x})) = \mathbf{O} , \quad \mathbf{x} \in \Omega_\varepsilon , \qquad (6.18)$$

$$\mathbf{u}(\mathbf{x}) = \boldsymbol{\varphi}_\varepsilon(\mathbf{x}) , \quad \mathbf{x} \in \partial\omega_\varepsilon , \qquad (6.19)$$

$$\mathbf{u}(\mathbf{x}) = \boldsymbol{\psi}(\mathbf{x}) , \quad \mathbf{x} \in \partial\Omega , \qquad (6.20)$$

where $\partial_{\mathbf{x}} = \partial/\partial\mathbf{x}$, $\mathbf{O}$ is the zero vector, $\boldsymbol{\varphi}_\varepsilon(\mathbf{x}) = \boldsymbol{\varphi}(\varepsilon^{-1}\mathbf{x})$, and we assume that $\boldsymbol{\varphi}_\varepsilon$ and $\boldsymbol{\psi}$ are continuous vector functions.

In this section, we prove the following.

Lemma 6.1. *There exists a unique solution* $\mathbf{u} \in C(\bar{\Omega}_\varepsilon)$ *of problem (6.18)–(6.20) which satisfies the estimate*

$$\max_{\bar{\Omega}_\varepsilon} |\mathbf{u}(\mathbf{x})| \leq \text{Const} \max\{\|\boldsymbol{\varphi}_\varepsilon\|_{C(\partial\omega_\varepsilon)} , \|\boldsymbol{\psi}\|_{C(\partial\Omega)}\} . \qquad (6.21)$$

We consider the cases when the dimension n is equal to 3 or 2.

The proof of the theorem involves auxiliary statements related to model domains Ω and $C\bar{\omega} = \mathbb{R}^n \backslash \bar{\omega}$.

6.2.1 The Maximum Principle in Ω

Let $\mathbf{u}$ solve the Dirichlet boundary value problem in Ω

$$L(\partial_{\mathbf{x}})\,\mathbf{u}(\mathbf{x}) = \mathbf{O}\,, \quad \mathbf{x} \in \Omega\,, \tag{6.22}$$

$$\mathbf{u}(\mathbf{x}) = \boldsymbol{\psi}(\mathbf{x})\,, \quad \mathbf{x} \in \partial\Omega\,, \tag{6.23}$$

where $\boldsymbol{\psi}$ is continuous on $\partial\Omega$.

The following assertion is essentially due to Fichera [8], who proved its analogue for the three-dimensional case. The same argument works for the case of a planar domain and is even simpler.

Lemma 6.2. (Fichera's maximum principle, see [8]) *There exists a unique solution* $\mathbf{u} \in C(\bar{\Omega})$ *of problem (6.22), (6.23). This solution satisfies the estimate*

$$\|\mathbf{u}\|_{C(\bar{\Omega})} \le A_\Omega \|\boldsymbol{\psi}\|_{C(\partial\Omega)}\,, \tag{6.24}$$

where A_Ω is a constant coefficient.

6.2.2 The Maximum Principle in $C\bar{\omega}$

Let ω be a domain containing the origin with compact closure and smooth boundary $\partial\omega$. Without loss of generality we assume that $\operatorname{diam}\omega = 1$. Let $\mathbf{v}(\boldsymbol{\xi})$ be a solution of the Dirichlet boundary value problem in the unbounded domain $C\bar{\omega}$:

$$L(\partial_{\boldsymbol{\xi}})\mathbf{v}(\boldsymbol{\xi}) = \mathbf{O}\,, \quad \boldsymbol{\xi} \in C\bar{\omega}\,, \tag{6.25}$$

$$\mathbf{v}(\boldsymbol{\xi}) = \boldsymbol{\varphi}(\boldsymbol{\xi})\,, \quad \boldsymbol{\xi} \in \partial\omega\,, \tag{6.26}$$

$$|\mathbf{v}| \to 0 \text{ as } |\boldsymbol{\xi}| \to \infty\,, \tag{6.27}$$

when $n = 3$.

For the two-dimensional case ($n = 2$), the formulation (6.25)–(6.27) has to be supplied with the orthogonality conditions for the right-hand side $\boldsymbol{\varphi}$:

$$\int_{\partial\omega} \boldsymbol{\varphi}(\boldsymbol{\xi}) \cdot T_n(\partial_{\boldsymbol{\xi}})\zeta^{(j)}(\boldsymbol{\xi})\,ds = 0\,, \quad j = 1, 2\,, \tag{6.28}$$

which guarantees the decay of the solution $\mathbf{v}$ at infinity. The vector functions $\zeta^{(j)}$ are solutions of the model problem

$$L(\partial_{\boldsymbol{\xi}})\zeta^{(j)}(\boldsymbol{\xi}) = \mathbf{O}\,, \quad \boldsymbol{\xi} \in C\bar{\omega}\,, \tag{6.29}$$

$$\zeta^{(j)}(\boldsymbol{\xi}) = \mathbf{O}\,, \quad \boldsymbol{\xi} \in \partial C\bar{\omega}\,, \tag{6.30}$$

$$\zeta^{(j)}(\boldsymbol{\xi}) \sim -\gamma^{(j)}(\boldsymbol{\xi}, \mathbf{O}) \quad \text{as} \quad |\boldsymbol{\xi}| \to \infty\,, \tag{6.31}$$

where $\gamma^{(j)}$ are the columns of the fundamental solution γ for the Lamé operator in an infinite plane and T_n denotes the matrix differential operator of tractions

$$T_n(\partial_\xi)\zeta^{(j)}(\xi) = \begin{pmatrix} \sigma_{11}(\zeta^{(j)})n_1 + \sigma_{12}(\zeta^{(j)})n_2 \\ \sigma_{12}(\zeta^{(j)})n_1 + \sigma_{22}(\zeta^{(j)})n_2 \end{pmatrix}$$

where $\mathbf{n} = (n_1, n_2)^T$ is the unit outward normal on $\partial C\bar{\omega}$. We shall also use the notation $\mathfrak{N}$ for the 2×2 matrix function:

$$\mathfrak{N}(\xi) = \{T_n(\partial_\xi)\zeta^{(1)}(\xi), T_n(\partial_\xi)\zeta^{(2)}(\xi)\} . \tag{6.32}$$

Lemma 6.3. *There exists a unique solution in $C(\mathbb{R}^n \backslash \bar{\omega})$ of the problem (6.25)–(6.27) ((6.25)–(6.28) for $n = 2$). This solution satisfies the estimate*

$$\sup_{\xi \in C\bar{\omega}} \{|\xi||u(\xi)|\} \le A_{C\bar{\omega}} \|\varphi\|_{C(\partial\omega)} . \tag{6.33}$$

Proof. By Lemma 6.2 there exists a unique solution $\mathbf{U} \in C(\bar{B}_3 \backslash \omega)$ of the Dirichlet problem

$$L(\partial_\xi)\mathbf{U}(\xi) = \mathbf{O} \quad \text{in } B_3 \backslash \bar{\omega} , \tag{6.34}$$

$$\mathbf{U}(\xi) = \mathbf{O} \quad \text{on } \partial B_3 , \tag{6.35}$$

$$\mathbf{U}(\xi) = \varphi(\xi) \quad \text{on } \partial\omega , \tag{6.36}$$

where B_3 is the ball of radius 3 centered at the origin.

This solution satisfies the estimate

$$\|\mathbf{U}\|_{C(\bar{B}_3 \backslash \omega)} \le A \, \|\varphi\|_{C(\partial\omega)} . \tag{6.37}$$

It suffices to prove the lemma assuming that φ is smooth, with the general case being settled by approximation. Owing to the classical elliptic theory and smoothness of both $\partial\omega$ and φ, there exists a unique variational solution $\mathbf{v} \in C(\mathbb{R}^n \backslash \bar{\omega})$.

Let

$$\mathbf{w} = \mathbf{v} - \eta\mathbf{U} , \tag{6.38}$$

where $\eta \in C_0^\infty(B_3)$ and $\eta = 1$ on B_2. The vector function $\eta\mathbf{U}$ is extended by zero outside B_3. Obviously,

$$\mathrm{Tr}_{\partial\omega}\mathbf{w} = \mathbf{O} , \tag{6.39}$$

and

$$\mathbf{w} = O(|\xi|^{-1}) \quad \text{as } |\xi| \to \infty . \tag{6.40}$$

Furthermore,

$$L(\partial_\xi)\mathbf{w} = -[L(\partial_\xi), \eta] \, \mathbf{U} , \tag{6.41}$$

so that $L(\partial_{\xi})\mathbf{w} \in C_0^\infty(\mathbb{R}^n \backslash \bar{\omega})$ and $\text{supp}\, L(\partial_{\xi})\mathbf{w} \subset \bar{B}_3 \backslash B_2$. By Betti's formula and Korn's inequality we obtain

$$\|\mathbf{w}\|_{W_2^1(B_3\backslash\bar{\omega})} \leq \text{Const} \left(\int_{B_3\backslash\bar{B}_2} |\mathbf{U}|^2 \, d\mathbf{x} \right)^{1/2} . \tag{6.42}$$

This along with (6.37) gives

$$\|\mathbf{w}\|_{L_2(B_2\backslash\bar{B}_{3/2})} \leq \text{Const}\, \|\boldsymbol{\varphi}\|_{C(\partial\omega)} . \tag{6.43}$$

By the local regularity estimate for solutions of $L(\partial_{\xi})\mathbf{w} = \mathbf{O}$ we have

$$\|\mathbf{w}\|_{C(\partial B_{7/4})} \leq \text{Const}\, \|\boldsymbol{\varphi}\|_{C(\partial\omega)} . \tag{6.44}$$

This and (6.37), (6.38) imply

$$\|\mathbf{v}\|_{C(\partial B_{7/4})} \leq \text{Const}\, \|\boldsymbol{\varphi}\|_{C(\partial\omega)} . \tag{6.45}$$

Applying Fichera's maximum principle (see Lemma 6.2) for the domain $B_{7/4} \backslash \bar{\omega}$ we find

$$\|\mathbf{v}\|_{C(\bar{B}_{7/4}\backslash\omega)} \leq \text{Const}\, \|\boldsymbol{\varphi}\|_{C(\partial\omega)} . \tag{6.46}$$

Let $\tau \in C_0^\infty(B_{7/4})$, and $\tau = 1$ on $B_{5/4}$. Then

$$- L(\partial_{\xi})((1 - \tau)\mathbf{v}) := \mathbf{f}\, , \tag{6.47}$$

where

$$\mathbf{f} \in C_0^\infty(\mathbb{R}^n) \quad \text{and} \quad \text{supp}\, \mathbf{f} \subset B_{7/4} \backslash \bar{B}_{5/4} . \tag{6.48}$$

We have

$$(1 - \tau)\mathbf{v} = \mathfrak{G} * \mathbf{f}\, , \tag{6.49}$$

where $\mathfrak{G}$ is the fundamental solution of the Lamé operator.

Now, (6.49) implies directly that

$$|\boldsymbol{\xi}||1 - \tau(\boldsymbol{\xi})||\mathbf{v}(\boldsymbol{\xi})| \leq \text{Const}\, \|\mathbf{v}\|_{L_2(B_{7/4}\backslash\bar{B}_{5/4})} , \tag{6.50}$$

in the three-dimensional case. For $n = 2$, we notice that the condition that $(1 - \tau)\mathbf{v}$ vanishes at infinity results in the self-balanced condition for $\mathbf{f}$. Therefore, the logarithmic and homogeneous of order zero terms in the asymptotics of $\mathfrak{G}$ disappear. Referring to (6.46) we obtain for $\boldsymbol{\xi} \in \mathbb{R}^n \backslash \bar{B}_{7/4}$

$$|\boldsymbol{\xi}||\mathbf{v}(\boldsymbol{\xi})| \leq \text{Const}\, \|\boldsymbol{\varphi}\|_{C(\partial\omega)}\,, \tag{6.51}$$

and using (6.46) once more we complete the proof of (6.33). □

6.2.3 The Operator Notations

We introduce the operators P_Ω and $P_{C\bar{\omega}}$ in such a way that the solutions $\mathbf{u}$, $\mathbf{v}$ of problems (6.22), (6.23) and (6.25)–(6.27) are represented in the form

$$\mathbf{u} = P_\Omega(\boldsymbol{\psi})\,, \quad \mathbf{v} = P_{C\bar{\omega}}(\boldsymbol{\varphi})\,. \tag{6.52}$$

The notation $P_{C\bar{\omega}}(\boldsymbol{\varphi})(\boldsymbol{\xi}) = P_{C\bar{\omega}_\varepsilon}(\boldsymbol{\varphi})(\mathbf{x})$ will also be used.

In the case of $n = 2$, we will also need the approximation $\mathcal{P}_\varepsilon$ of the capacitary potential:

$$\begin{aligned}\mathcal{P}_\varepsilon = {} & G(\mathbf{x},\mathbf{O})D(\log\varepsilon)\\ & + P_{C\bar{\omega}_\varepsilon}(I_2 - \text{Tr}_{\partial\omega_\varepsilon} G(\mathbf{x},\mathbf{O})D(\log\varepsilon))\\ & - P_\Omega(\text{Tr}_{\partial\Omega} P_{C\bar{\omega}_\varepsilon}(I_2 - \text{Tr}_{\partial\omega_\varepsilon} G(\mathbf{x},\mathbf{O})D(\log\varepsilon)))\,,\end{aligned}$$

where G is the Green's tensor in Ω, $D(\log\varepsilon)$ is the 2×2 matrix defined by

$$D = -\frac{1}{K_1}\begin{pmatrix} K_2\log\varepsilon - \zeta_{22}^\infty + H_{22}(\mathbf{O},\mathbf{O}) & \zeta_{12}^\infty - H_{12}(\mathbf{O},\mathbf{O}) \\ \zeta_{21}^\infty - H_{21}(\mathbf{O},\mathbf{O}) & K_2\log\varepsilon - \zeta_{11}^\infty + H_{11}(\mathbf{O},\mathbf{O}) \end{pmatrix}, \tag{6.53}$$

with

$$\begin{aligned}K_1 = {} & \left(K_2\log\varepsilon - \zeta_{11}^\infty + H_{11}(\mathbf{O},\mathbf{O})\right)\left(K_2\log\varepsilon - \zeta_{22}^\infty + H_{22}(\mathbf{O},\mathbf{O})\right)\\ & -(H_{12}(\mathbf{O},\mathbf{O}) - \zeta_{12}^\infty)(H_{21}(\mathbf{O},\mathbf{O}) - \zeta_{21}^\infty)\,,\end{aligned} \tag{6.54}$$

$$K_2 = \frac{\lambda + 3\mu}{4\pi\mu(\lambda+2\mu)}\,, \tag{6.55}$$

and $H = [H_{ij}]_{i,j=1}^2$ is the regular part of Green's tensor for the domain Ω,

$$\zeta^\infty = [\zeta_{ij}^\infty]_{i,j=1}^2 = \lim_{|\boldsymbol{\eta}|\to\infty}\lim_{|\boldsymbol{\xi}|\to\infty}\{\gamma(\boldsymbol{\eta},\mathbf{O}) + g(\boldsymbol{\eta},\boldsymbol{\xi})\}\,, \tag{6.56}$$

where g is Green's tensor for the unbounded domain $C\bar{\omega}$.

By direct substitution, we can verify that

$$L(\partial_{\mathbf{x}})\,\mathcal{P}_\varepsilon(\mathbf{x}) = 0I_2\,, \quad \mathbf{x} \in \Omega_\varepsilon\,, \tag{6.57}$$

$$\mathcal{P}_\varepsilon(\mathbf{x}) = 0I_2\,, \quad \mathbf{x} \in \partial\Omega\,, \tag{6.58}$$

$$\mathcal{P}_\varepsilon(\mathbf{x}) = I_2 + O(\varepsilon)\,, \quad \mathbf{x} \in \partial\omega_\varepsilon\,. \tag{6.59}$$

The Proof of Lemma 6.1 for n = 2

First, consider the case when the *homogeneous boundary condition is set on* $\partial\Omega$, so that

$$L(\partial_{\mathbf{x}})\,\mathbf{u}(\mathbf{x}) = \mathbf{O}\,, \quad \mathbf{x} \in \Omega_\varepsilon\,, \tag{6.60}$$

$$\mathbf{u}(\mathbf{x}) = \boldsymbol{\varphi}_\varepsilon(\mathbf{x})\,, \quad \mathbf{x} \in \partial\omega_\varepsilon\,, \tag{6.61}$$

$$\mathbf{u}(\mathbf{x}) = \mathbf{O}\,, \quad \mathbf{x} \in \partial\Omega\,. \tag{6.62}$$

We are looking for the solution in the form

$$\begin{aligned}\mathbf{u} = {} & P_{C\bar{\omega}_\varepsilon}(\mathbf{g}_\varepsilon - A_{\mathbf{g}}) + \mathcal{P}_\varepsilon A_{\mathbf{g}} \\ & - P_\Omega(\mathrm{Tr}_{\partial\Omega} P_{C\bar{\omega}_\varepsilon}(\mathbf{g}_\varepsilon - A_{\mathbf{g}}))\,,\end{aligned} \tag{6.63}$$

where $\mathbf{g}_\varepsilon(\mathbf{x}) = \mathbf{g}(\varepsilon^{-1}\mathbf{x})$ is an unknown vector function and the constant vector $A_{\mathbf{g}}$ is determined by

$$A_{\mathbf{g}} = -\int_{\partial\omega} \mathfrak{N}^T(\boldsymbol{\xi})\mathbf{g}(\boldsymbol{\xi})\,dS_{\boldsymbol{\xi}}\,, \tag{6.64}$$

here the matrix $\mathfrak{N}$ is the same as in (6.32). We note that

$$\int_{\partial\omega_\varepsilon} \|\mathfrak{N}\|\,dS_{\mathbf{x}} < C\,, \tag{6.65}$$

where C is independent of ε and $\|\mathfrak{N}\|$ is the norm of the matrix $\mathfrak{N}$.

Evaluating the trace of (6.63) on $\partial\omega_\varepsilon$ we obtain

$$\boldsymbol{\varphi}_\varepsilon = \mathbf{g}_\varepsilon + S_\varepsilon\mathbf{g}_\varepsilon\,, \tag{6.66}$$

where the operator S_ε is defined by

$$\begin{aligned}S_\varepsilon\mathbf{g}_\varepsilon = {} & \mathrm{Tr}_{\partial\omega_\varepsilon}(\mathcal{P}_\varepsilon - I_2)A_{\mathbf{g}} \\ & - \mathrm{Tr}_{\partial\omega_\varepsilon} P_\Omega(\mathrm{Tr}_{\partial\Omega} P_{C\bar{\omega}_\varepsilon}(\mathbf{g}_\varepsilon - A_{\mathbf{g}}))\,.\end{aligned}$$

By (6.64), (6.65) and (6.59)

$$\|\mathrm{Tr}_{\partial\omega_\varepsilon}(\mathcal{P}_\varepsilon - I_2)A_{\mathbf{g}}\|_{C(\partial\omega_\varepsilon)} \le \mathrm{Const}\,\varepsilon\|\mathbf{g}_\varepsilon\|_{C(\partial\omega_\varepsilon)} . \tag{6.67}$$

Lemma 6.3 implies

$$|\mathbf{x}|\,|P_{C\bar{\omega}_\varepsilon}(\mathbf{g}_\varepsilon - A_{\mathbf{g}})(\mathbf{x})| \le \mathrm{Const}\,\varepsilon\,\|\mathbf{g}_\varepsilon\|_{C(\partial\omega_\varepsilon)} , \tag{6.68}$$

for all $\mathbf{x} \in \Omega_\varepsilon$.

Combining (6.67) and (6.68) we conclude

$$\|S_\varepsilon\|_{C(\partial\omega_\varepsilon)\to C(\partial\omega_\varepsilon)} \le \mathrm{Const}\,\varepsilon . \tag{6.69}$$

It follows from (6.66) that $\mathbf{g}_\varepsilon = (I + S_\varepsilon)^{-1}\boldsymbol{\varphi}_\varepsilon$, and then we deduce

$$\|\mathbf{g}_\varepsilon\|_{C(\partial\omega_\varepsilon)} \le \mathrm{Const}\,\|\boldsymbol{\varphi}_\varepsilon\|_{C(\partial\omega_\varepsilon)} .$$

Due to (6.68) and Lemma 6.2 we obtain

$$\max_{\bar{\Omega}_\varepsilon}|\mathbf{u}| \le \mathrm{Const}\,\|\mathbf{g}_\varepsilon\|_{C(\partial\omega_\varepsilon)} \le \mathrm{Const}\,\|\boldsymbol{\varphi}_\varepsilon\|_{C(\partial\omega_\varepsilon)} . \tag{6.70}$$

Second, we consider *the case of the inhomogeneous boundary condition on* $\partial\Omega$

$$L(\partial_{\mathbf{x}})\,\mathbf{u}(\mathbf{x}) = \mathbf{O} , \quad \mathbf{x} \in \Omega_\varepsilon , \tag{6.71}$$

$$\mathbf{u}(\mathbf{x}) = \boldsymbol{\psi}(\mathbf{x}) , \quad \mathbf{x} \in \partial\Omega , \tag{6.72}$$

$$\mathbf{u}(\mathbf{x}) = \mathbf{O} , \quad \mathbf{x} \in \partial\omega_\varepsilon . \tag{6.73}$$

The solution is sought in the form

$$\mathbf{u} = P_\Omega\boldsymbol{\psi} + \mathbf{v} , \tag{6.74}$$

where the second term $\mathbf{v}$ is defined as a solution of the problem, which is similar to (6.60)–(6.62), with the boundary condition on $\partial\omega_\varepsilon$ being replaced by

$$\mathbf{v}(\mathbf{x}) = -(\mathrm{Tr}_{\partial\omega_\varepsilon}P_\Omega\boldsymbol{\psi})(\mathbf{x}) , \quad \mathbf{x} \in \partial\omega_\varepsilon .$$

According to the result of first part of the proof (6.70), we have

$$\max_{\bar{\Omega}_\varepsilon}|\mathbf{v}| \le \mathrm{Const}\,\max_{\partial\omega_\varepsilon}|\mathrm{Tr}_{\partial\omega_\varepsilon}P_\Omega\boldsymbol{\psi}|$$
$$\le \mathrm{Const}\,\|\boldsymbol{\psi}\|_{C(\partial\Omega)} . \tag{6.75}$$

It follows from Lemma 6.2 that

$$\max_{\bar{\Omega}_\varepsilon} |P_\Omega \boldsymbol{\psi}| \leq \text{Const} \, \|\boldsymbol{\psi}\|_{C(\partial\Omega)} \, . \tag{6.76}$$

Combining (6.74), (6.75) and (6.76) we deduce

$$\max_{\bar{\Omega}_\varepsilon} |\mathbf{u}| \leq \text{Const} \, \|\boldsymbol{\psi}\|_{C(\partial\Omega)} \, .$$

This completes the proof for the case $n = 2$.

The Proof of Lemma 6.1 for n = 3

First, we address the formulation (6.60)–(6.62), where Ω_ε is a domain in $\mathbb{R}^3$, and the inhomogeneous boundary condition is specified on $\partial\omega_\varepsilon$.

The solution is sought in the form

$$\mathbf{u} = P_{C\bar{\omega}_\varepsilon} \mathbf{g}_\varepsilon - P_\Omega (\text{Tr}_{\partial\Omega} P_{C\bar{\omega}_\varepsilon} \mathbf{g}_\varepsilon) \, , \tag{6.77}$$

with $\mathbf{g}_\varepsilon = \mathbf{g}(\varepsilon^{-1}\mathbf{x})$ being an unknown function. Evaluating the trace of (6.77) on $\partial\omega_\varepsilon$ we obtain

$$\boldsymbol{\varphi}_\varepsilon = \mathbf{g}_\varepsilon + S_\varepsilon \mathbf{g}_\varepsilon \, ,$$

where $S_\varepsilon \mathbf{g}_\varepsilon = -\text{Tr}_{\partial\omega_\varepsilon} P_\Omega (\text{Tr}_{\partial\Omega} P_{C\bar{\omega}_\varepsilon} \mathbf{g}_\varepsilon)$.

Since $\|\text{Tr}_{\partial\Omega} P_{C\bar{\omega}_\varepsilon} \mathbf{g}_\varepsilon\|_{C(\partial\Omega)} \leq \text{Const} \, \varepsilon \|\mathbf{g}_\varepsilon\|_{C(\partial\omega_\varepsilon)}$ it follows from Lemma 6.2 that

$$\|S_\varepsilon\|_{C(\partial\omega_\varepsilon) \to C(\partial\omega_\varepsilon)} \leq \text{Const} \, \varepsilon \, .$$

Hence

$$\mathbf{g}_\varepsilon = (I + S_\varepsilon)^{-1} \boldsymbol{\varphi}_\varepsilon \, ,$$

and the following estimate holds

$$\|\mathbf{g}_\varepsilon\|_{C(\partial\omega_\varepsilon)} \leq \text{Const} \, \|\boldsymbol{\varphi}_\varepsilon\|_{C(\partial\omega_\varepsilon)} \, .$$

Applying Lemmas 6.2 and 6.3 we conclude

$$\max_{\bar{\Omega}_\varepsilon} |\mathbf{u}| \leq \text{Const} \, \|\mathbf{g}_\varepsilon\|_{C(\partial\omega_\varepsilon)} \leq \text{Const} \, \|\boldsymbol{\varphi}_\varepsilon\|_{C(\partial\omega_\varepsilon)} \, .$$

The case when an inhomogeneous boundary condition is set on $\partial\Omega$ is treated similarly to the proof of Sect. 6.2.3.

The proof of the theorem is complete. □

6.3 Green's Tensor for a Three-Dimensional Domain with a Small Inclusion

In this section, we derive a uniform asymptotic approximation of the Green's tensor $G_\varepsilon(\mathbf{x},\mathbf{y})$ in a three-dimensional domain with a small inclusion, as described in Sect. 6.1.1 (see (6.12) and (6.13)). Before formulating the asymptotic representation, we list model domains and associated model problems required for the asymptotic algorithm.

6.3.1 Green's Matrices for Model Domains in Three Dimensions

Let $G(\mathbf{x},\mathbf{y}) = [G^{(1)}(\mathbf{x},\mathbf{y}), G^{(2)}(\mathbf{x},\mathbf{y}), G^{(3)}(\mathbf{x},\mathbf{y})]$ and $g(\boldsymbol{\xi},\boldsymbol{\eta}) = [g^{(1)}(\boldsymbol{\xi},\boldsymbol{\eta}),$ $g^{(2)}(\boldsymbol{\xi},\boldsymbol{\eta}),$ $g^{(3)}(\boldsymbol{\xi},\boldsymbol{\eta})]$ denote Green's tensors in the sets Ω and $C\bar{\omega} = \mathbb{R}^3\backslash\bar{\omega}$, respectively, for the Lamé operator given by (6.11) for the case of three dimensions. The tensor G solves the following problem

$$L(\partial_{\mathbf{x}})G(\mathbf{x},\mathbf{y}) + \delta(\mathbf{x}-\mathbf{y})I_3 = 0I_3\ , \quad \mathbf{x},\mathbf{y} \in \Omega\ , \tag{6.78}$$

$$G(\mathbf{x},\mathbf{y}) = 0I_3\ , \quad \mathbf{x} \in \partial\Omega, \mathbf{y} \in \Omega\ , \tag{6.79}$$

and the tensor g is solution of

$$L(\partial_{\boldsymbol{\xi}})g(\boldsymbol{\xi},\boldsymbol{\eta}) + \delta(\boldsymbol{\xi}-\boldsymbol{\eta})I_3 = 0I_3\ , \quad \boldsymbol{\xi},\boldsymbol{\eta} \in C\bar{\omega}\ , \tag{6.80}$$

$$g(\boldsymbol{\xi},\boldsymbol{\eta}) = 0I_3\ , \quad \boldsymbol{\xi} \in \partial C\bar{\omega}, \boldsymbol{\eta} \in C\bar{\omega}\ , \tag{6.81}$$

$$g(\boldsymbol{\xi},\boldsymbol{\eta}) \to 0I_3 \quad \text{as} \quad |\boldsymbol{\xi}| \to \infty\ . \tag{6.82}$$

From (6.78), (6.79), we have that G satisfies the symmetry relation

$$G(\mathbf{x},\mathbf{y}) = G^T(\mathbf{y},\mathbf{x}) \quad \mathbf{x},\mathbf{y} \in \Omega, \mathbf{x} \neq \mathbf{y}\ , \tag{6.83}$$

and in the unbounded domain $C\bar{\omega}$ the Green's function g satisfies

$$g(\boldsymbol{\xi},\boldsymbol{\eta}) = g^T(\boldsymbol{\eta},\boldsymbol{\xi}), \quad \boldsymbol{\xi},\boldsymbol{\eta} \in C\bar{\omega}, \boldsymbol{\xi} \neq \boldsymbol{\eta}\ . \tag{6.84}$$

We represent $G(\mathbf{x},\mathbf{y})$ and $g(\boldsymbol{\xi},\boldsymbol{\eta})$ as

$$G(\mathbf{x},\mathbf{y}) = \Gamma(\mathbf{x},\mathbf{y}) - H(\mathbf{x},\mathbf{y})\ , \tag{6.85}$$

and

$$g(\boldsymbol{\xi},\boldsymbol{\eta}) = \Gamma(\boldsymbol{\xi},\boldsymbol{\eta}) - h(\boldsymbol{\xi},\boldsymbol{\eta})\ , \tag{6.86}$$

where $\Gamma(\mathbf{x},\mathbf{y}) = [\Gamma_{ij}(\mathbf{x},\mathbf{y})]$, $i,j = 1,2,3$, is the fundamental solution of the Lamé operator in three dimensions, whose entries are given by

$$\Gamma_{ij}(\mathbf{x},\mathbf{y}) = (8\pi\mu(\lambda+2\mu)|\mathbf{x}-\mathbf{y}|)^{-1}((\lambda+\mu)(x_i-y_i)(x_j-y_j)|\mathbf{x}-\mathbf{y}|^{-2} \quad (6.87)$$

$$+(\lambda+3\mu)\delta_{ij}) ,$$

and H, h are the regular parts of G, g respectively.

6.3.2 The Elastic Capacitary Potential Matrix

By $P(\boldsymbol{\xi}) = [P^{(1)}(\boldsymbol{\xi}), P^{(2)}(\boldsymbol{\xi}), P^{(3)}(\boldsymbol{\xi})]$, we mean the elastic capacitary potential matrix of the set ω, whose columns satisfy

$$L(\partial_{\boldsymbol{\xi}})P^{(j)}(\boldsymbol{\xi}) = \mathbf{O} \quad \text{in } C\bar{\omega} , \quad (6.88)$$

$$P^{(j)}(\boldsymbol{\xi}) = \mathbf{e}^{(j)} \quad \text{on } \partial C\bar{\omega} , \quad (6.89)$$

$$P^{(j)}(\boldsymbol{\xi}) \to \mathbf{O} \quad \text{as} \quad |\boldsymbol{\xi}| \to \infty , \quad (6.90)$$

for $j = 1,2,3$, where $\mathbf{e}^{(j)}$ is a basis vector, whose j^{th} entry is equal to 1, and all other entries are zero.

Lemma 6.4. *The columns $P^{(j)}$, $j = 1,2,3$, of the elastic capacitary potential satisfy the inequality*

$$\sup_{\boldsymbol{\xi}\in C\bar{\omega}} \{|\boldsymbol{\xi}||P^{(j)}(\boldsymbol{\xi})|\} \le \text{Const} . \quad (6.91)$$

Proof. The proof follows directly from the maximum principle for unbounded domains (cf. Lemma 6.3). □

In the sequel, we will need the following lemma, which is a reformulation of that by Kondratiev and Oleinik, in [18] (p. 78).

Lemma 6.5. *Suppose the columns $u^{(j)}(\boldsymbol{\xi})$ of the matrix $u(\boldsymbol{\xi})$ are solutions of*

$$L(\partial_{\boldsymbol{\xi}})u^{(j)}(\boldsymbol{\xi}) = \mathbf{O} , \quad \textit{in } C\bar{\omega} ,$$

and that $|u^{(j)}(\boldsymbol{\xi})| \le \text{Const}\,(1 + |\boldsymbol{\xi}|)^k$, $k \ge 0$, for $j = 1,2,3$.

Then for $|\boldsymbol{\xi}| > 2$

$$u^{(j)}(\boldsymbol{\xi}) = \mathscr{P}_k^{(j)}(\boldsymbol{\xi}) + \Gamma(\boldsymbol{\xi},\mathbf{O})C^{(j)} + O(|\boldsymbol{\xi}|^{-2}) , \quad (6.92)$$

where $\mathscr{P}_k^{(j)}(\boldsymbol{\xi}) = \{\mathscr{P}_i^{(j,k)}(\boldsymbol{\xi})\}_{i=1}^3$, $\mathscr{P}_i^{(j,k)}(\boldsymbol{\xi})$ are polynomials of order not greater than k, $C^{(j)} = \{C_i^{(j)}\}_{i=1}^3$, where $C_i^{(j)}$ are constants.

6.3.2.1 Properties of the Elastic Capacity Matrix

Let $B = [B_{ij}]$, $i, j = 1, 2, 3$, be a constant matrix that we shall call the elastic capacity matrix of the set ω. In the present subsection, we will discuss some properties of the elastic capacity matrix. The aim of this subsection is to show that upper and lower elastic capacity (obtained from the maximum and minimum eigenvalues of B, respectively) are equivalent to electrostatic capacity.

Throughout we will need the following Lemma related to the asymptotic behaviour of P.

Lemma 6.6. *If $|\boldsymbol{\xi}| > 2$, then for $P^{(j)}$ the following estimate holds*

$$|P^{(j)}(\boldsymbol{\xi}) - B_{ij}\Gamma^{(i)}(\boldsymbol{\xi}, \mathbf{O})| \leq \text{Const}\, |\boldsymbol{\xi}|^{-2}\,, \tag{6.93}$$

for $j = 1, 2, 3$, and i being the index of summation, where $\Gamma^{(i)}$ are columns of the fundamental solution for the Lamé operator and B_{ij} are entries of the elastic capacity matrix B of the set ω.

Proof. By Lemma 6.4, it is sufficient to take $P(\boldsymbol{\xi}) = O(|\boldsymbol{\xi}|^{-1})$, then from Lemma 6.5, for $|\boldsymbol{\xi}| > 2$ the columns $P^{(j)}(\boldsymbol{\xi})$ can be written in the following way

$$P^{(j)}(\boldsymbol{\xi}) = \Gamma(\boldsymbol{\xi}, \mathbf{O})C^{(j)} + O(|\boldsymbol{\xi}|^{-2})\,. \tag{6.94}$$

Then taking $C^{(j)} = B^{(j)}$ we obtain (6.93). □

We also use the electrostatic potential $\mathcal{P}$ of the set ω, with electrostatic capacity cap ω, as a solution of the problem

$$\Delta_{\boldsymbol{\xi}}\mathcal{P}(\boldsymbol{\xi}) = 0\,, \quad \boldsymbol{\xi} \in C\bar{\omega}\,, \tag{6.95}$$

$$\mathcal{P}(\boldsymbol{\xi}) = 1\,, \quad \boldsymbol{\xi} \in \partial\omega\,, \tag{6.96}$$

$$\mathcal{P}(\boldsymbol{\xi}) \to 0 \quad \text{as} \quad |\boldsymbol{\xi}| \to \infty\,. \tag{6.97}$$

The electrostatic energy for a scalar function u in a domain $T \subset \mathbb{R}^n$ is defined as

$$\mathcal{E}(u, T) = \int_T |\nabla u|^2\, d\mathbf{x}\,. \tag{6.98}$$

It is well known that for the function $\mathcal{P}$, we have for the energy functional $\mathcal{E}$ in $C\bar{\omega}$

$$\mathcal{E}(\mathcal{P}, C\bar{\omega}) = \int_{C\bar{\omega}} |\nabla\mathcal{P}|^2\, d\boldsymbol{\xi} = \text{cap}\,\omega\,. \tag{6.99}$$

In contrast, the elastic energy functional for a vector $\mathbf{u}$ in the domain T is given by

$$\mathscr{E}(\mathbf{u}, T) = 2^{-1}\int_T e_{ij}(\mathbf{u})\sigma_{ij}(\mathbf{u})\, d\mathbf{x}\,, \tag{6.100}$$

also we define the elastic energy matrix $E = [E_{ij}]_{i,j=1}^3$ for a matrix $\mathcal{A}$ in the domain T with entries

$$E_{ij}(\mathcal{A}, T) = 2^{-1} \int_T e_{st}(\mathcal{A}^{(i)}) \sigma_{st}(\mathcal{A}^{(j)})\, d\mathbf{x} , \tag{6.101}$$

where $\mathcal{A}^{(i)}, i = 1, 2, 3$ are the columns of the matrix $\mathcal{A}$. Clearly, the diagonal entries E_{11}, E_{22} and E_{33} give the elastic energy for the vectors $\mathcal{A}^{(i)}, i = 1, 2, 3$ respectively.

We shall show that the elastic energy matrix can be represented in terms of the elastic capacity matrix B of the set ω, by considering the entries of elastic energy matrix for the matrix function P, defined as a solution of (6.88)–(6.90).

Lemma 6.7. *(i) For the elastic capacitary potential P, we have*

$$E(P(\boldsymbol{\xi}), C\bar{\omega}) = 2^{-1} B , \tag{6.102}$$

where B is the elastic capacity matrix of the set ω and
(ii) the matrix B is symmetric.

Proof. (*i*) We take a ball $B_R = \{\boldsymbol{\xi} : |\boldsymbol{\xi}| < R\}$ with sufficiently large radius R. We consider the component E_{jk} of the elastic energy matrix in the domain $B_R \backslash \bar{\omega}$ as follows

$$\begin{aligned} E_{jk}(P(\boldsymbol{\xi}), B_R \backslash \bar{\omega}) &= 2^{-1} \int_{B_R \backslash \bar{\omega}} e_{st}(P^{(j)}(\boldsymbol{\xi})) \sigma_{st}(P^{(k)}(\boldsymbol{\xi}))\, d\boldsymbol{\xi} \\ &= 2^{-1} \int_{\partial(B_R \backslash \bar{\omega})} P^{(j)}(\boldsymbol{\xi}) \cdot T_n(\partial_{\boldsymbol{\xi}}) P^{(k)}(\boldsymbol{\xi})\, dS_{\boldsymbol{\xi}} , \end{aligned} \tag{6.103}$$

where we have used Betti's formula and the fact that the columns of P satisfy the homogeneous Lamé equation. Noting the boundary condition (6.89), the preceding equation may be written as

$$\begin{aligned} E_{jk}(P(\boldsymbol{\xi}), B_R \backslash \bar{\omega}) = 2^{-1} \Big\{ &\int_{\partial B_R} P^{(j)}(\boldsymbol{\xi}) \cdot T_n(\partial_{\boldsymbol{\xi}}) P^{(k)}(\boldsymbol{\xi})\, dS_{\boldsymbol{\xi}} \\ &+ \int_{\partial \omega} \mathbf{e}^{(j)} \cdot T_n(\partial_{\boldsymbol{\xi}}) P^{(k)}(\boldsymbol{\xi})\, dS_{\boldsymbol{\xi}} \Big\} . \end{aligned} \tag{6.104}$$

Applying Betti's formula once more to the vectors $\mathbf{e}^{(j)}$ and $P^{(k)}(\boldsymbol{\xi})$ in the domain $B_R \backslash \bar{\omega}$, we have

$$\begin{aligned} E_{jk}(P(\boldsymbol{\xi}), B_R \backslash \bar{\omega}) = 2^{-1} \Big\{ &\int_{\partial B_R} P^{(j)}(\boldsymbol{\xi}) \cdot T_n(\partial_{\boldsymbol{\xi}}) P^{(k)}(\boldsymbol{\xi})\, dS_{\boldsymbol{\xi}} \\ &- \int_{\partial B_R} \mathbf{e}^{(j)} \cdot T_n(\partial_{\boldsymbol{\xi}}) P^{(k)}(\boldsymbol{\xi})\, dS_{\boldsymbol{\xi}} \Big\} , \end{aligned} \tag{6.105}$$

which holds for all R. Using the asymptotic representation for P given in Lemma 6.6, we pass to the limit as $R \to \infty$, yielding

$$E_{jk}(P(\boldsymbol{\xi}), C\bar{\omega}) = -2^{-1} \lim_{R\to\infty} \int_{\partial B_R} B_{rk}\sigma_{jp}(\Gamma^{(r)}(\boldsymbol{\xi}, \mathbf{O}))n_p \, dS_{\boldsymbol{\xi}}$$

$$= 2^{-1} B_{jk} , \tag{6.106}$$

where (6.106) has been obtained via Betti's formula applied to the vectors $\mathbf{e}^{(j)}$ and $\Gamma^{(r)}(\boldsymbol{\xi}, \mathbf{O})$ in B_R. Thus we have proved relation (6.102).

(*ii*) Now we prove the symmetry of the matrix B. Again using Lemma 6.6, we take the limit in (6.104) as $R \to \infty$, then comparing to (6.106), we have

$$\int_{\partial\omega} \mathbf{e}^{(j)} \cdot T_n(\partial_{\boldsymbol{\xi}}) P^{(k)}(\boldsymbol{\xi}) \, dS_{\boldsymbol{\xi}} = B_{jk} . \tag{6.107}$$

Then, interchanging the indices k and j, and subtracting the result from (6.107) gives

$$B_{jk} - B_{kj} = \int_{\partial\omega} \{\mathbf{e}^{(j)} \cdot T_n(\partial_{\boldsymbol{\xi}}) P^{(k)}(\boldsymbol{\xi}) - \mathbf{e}^{(k)} \cdot T_n(\partial_{\boldsymbol{\xi}}) P^{(j)}(\boldsymbol{\xi})\} \, dS_{\boldsymbol{\xi}} . \tag{6.108}$$

Recalling that on $\partial\omega$ we have $P^{(j)}(\boldsymbol{\xi}) = \mathbf{e}^{(j)}$, for $j = 1, 2, 3$, we see that the right-hand side is the result of application of the Betti formula to vectors $P^{(j)}(\boldsymbol{\xi})$ and $P^{(k)}(\boldsymbol{\xi})$ in $C\bar{\omega}$. Namely in (6.108) we have

$$B_{jk} - B_{kj} = \int_{C\bar{\omega}} \{P^{(j)}(\boldsymbol{\xi}) \cdot L(\partial_{\boldsymbol{\xi}}) P^{(k)}(\boldsymbol{\xi}) - P^{(k)}(\boldsymbol{\xi}) \cdot L(\partial_{\boldsymbol{\xi}}) P^{(j)}(\boldsymbol{\xi})\} \, d\boldsymbol{\xi} . \tag{6.109}$$

Since the columns of P are solutions to the homogeneous Lamé equation the right-hand side in (6.109) is zero and

$$B_{jk} = B_{kj} ,$$

i.e. the elastic capacity matrix B is symmetric. □

Next we prove that the elastic capacity matrix B represents a tensor.

Lemma 6.8. *The elastic capacity matrix is a Cartesian tensor of rank* 2.

Proof. Consider the boundary value problem in Ω_ε:

$$L(\partial_{\mathbf{x}})\mathbf{u}_\varepsilon(\mathbf{x}) = \mathbf{O} , \quad \mathbf{x} \in \Omega_\varepsilon ,$$

$$\mathbf{u}_\varepsilon(\mathbf{x}) = \boldsymbol{\psi}(\mathbf{x}) , \quad \mathbf{x} \in \partial\Omega , \tag{6.110}$$

$$\mathbf{u}_\varepsilon(\mathbf{x}) = \mathbf{O} , \quad \mathbf{x} \in \partial\omega_\varepsilon . \tag{6.111}$$

The solution $\mathbf{u}_\varepsilon$ can be approximated by

$$\mathbf{u}_\varepsilon(\mathbf{x}) = \mathbf{u}(\mathbf{x}) - (P(\boldsymbol{\xi}) - \varepsilon H(\mathbf{x}, \mathbf{O})B)\mathbf{u}(\mathbf{O}) + O(\varepsilon^2|\mathbf{x}|^{-2}) , \tag{6.112}$$

where $\mathbf{u}$ is the solution of the unperturbed problem

$$L(\partial_{\mathbf{x}})\mathbf{u}(\mathbf{x}) = \mathbf{O}\,, \quad \mathbf{x} \in \Omega\,,$$
$$\mathbf{u}(\mathbf{x}) = \boldsymbol{\psi}(\mathbf{x})\,, \quad \mathbf{x} \in \partial\Omega\,. \tag{6.113}$$

Next we show how the elastic capacity matrix can be used to determine the increment in elastic energy when a small rigid inclusion is introduced in to the body Ω. This energy increment is defined by

$$\delta\mathscr{E} = \mathscr{E}(\mathbf{u}_\varepsilon, \Omega_\varepsilon) - \mathscr{E}(\mathbf{u}, \Omega)\,.$$

Using this definition of $\partial\mathscr{E}$ along with (6.100), we can apply integration by parts and derive

$$\delta\mathscr{E} = \frac{1}{2}\int_{\partial\Omega_\varepsilon} \mathbf{u}_\varepsilon(\mathbf{x}) \cdot T_n(\partial_{\mathbf{x}})\mathbf{u}_\varepsilon(\mathbf{x})dS_{\mathbf{x}} - \frac{1}{2}\int_{\partial\Omega} \mathbf{u}(\mathbf{x}) \cdot T_n(\partial_{\mathbf{x}})\mathbf{u}(\mathbf{x})dS_{\mathbf{x}}\,, \tag{6.114}$$

where in obtaining the above expression we have used the fact that $\mathbf{u}_\varepsilon$ and $\mathbf{u}$ satisfy the homogenous Lamé equation in Ω_ε and Ω, respectively. The integral over $\partial\omega_\varepsilon$ in (6.114) is zero due to (6.111). Then from (6.110) and (6.113) we have

$$\delta\mathscr{E} = \frac{1}{2}\int_{\partial\Omega} \boldsymbol{\psi}(\mathbf{x}) \cdot T_n(\partial_{\mathbf{x}})(\mathbf{u}_\varepsilon(\mathbf{x}) - \mathbf{u}(\mathbf{x}))dS_{\mathbf{x}}\,.$$

The approximation (6.112), Lemma 6.6 and (6.113) then allow this increment to be approximated by

$$\delta\mathscr{E} = -\frac{1}{2}\varepsilon\int_{\partial\Omega} \mathbf{u}(\mathbf{x}) \cdot T_n(\partial_{\mathbf{x}})(G(\mathbf{x}, \mathbf{O})B\mathbf{u}(\mathbf{O}))dS_{\mathbf{x}} + O(\varepsilon^2)\,, \tag{6.115}$$

where (6.85) has also been used.

One more application of the Betti formula in Ω to the vectors $\mathbf{u}$ and $G(\mathbf{x}, \mathbf{O})B\mathbf{u}(\mathbf{O})$, which satisfies the homogeneous displacement condition on $\partial\Omega$, results in the asymptotic approximation for $\delta\mathscr{E}$

$$\delta\mathscr{E} = -\frac{1}{2}\varepsilon\int_{\Omega} \mathbf{u}(\mathbf{x}) \cdot L(\partial_{\mathbf{x}})(G(\mathbf{x}, \mathbf{O})B\mathbf{u}(\mathbf{O}))d\mathbf{x} + O(\varepsilon^2)\,,$$

or by (6.78) we can assert

$$\delta\mathscr{E} = \frac{1}{2}\varepsilon\mathbf{u}(\mathbf{O}) \cdot B\mathbf{u}(\mathbf{O}) + O(\varepsilon^2)\,.$$

Now since the energy increment is a scalar quantity, it is also a Cartesian tensor of rank 0. The displacement vector $\mathbf{u}$ is a Cartesian tensor of rank 1. Therefore, after applying a rotation of coordinates, the above formula implies that the elastic capacity matrix is a Cartesian tensor of rank 2. The proof is complete. □

6.3.2.2 Upper and Lower Elastic Capacity Versus Electrostatic Capacity

Let S denote set of vector functions $\mathbf{u}$, such that

$$L(\partial_\xi)\mathbf{u}(\xi) = \mathbf{O} \quad \text{in } C\bar{\omega} \,, \tag{6.116}$$

$$\mathbf{u}(\xi) = \mathbf{c} \quad \text{on } \partial C\bar{\omega} \,, \tag{6.117}$$

$$\mathbf{u}(\xi) \to \mathbf{O} \quad \text{as} \quad |\xi| \to \infty \,, \tag{6.118}$$

and for $|\xi| > 2$ has the asymptotic representation

$$\mathbf{u}(\xi) = \Gamma(\xi, \mathbf{O})B\mathbf{c} + O(|\xi|^{-2}) \,, \tag{6.119}$$

where $\mathbf{c} = \{c_j\}_{j=1}^3$ is a constant vector with $|\mathbf{c}| = 1$.

We define the lower elastic capacity, of the set $C\bar{\omega}$, to be

$$\underline{\text{cap}}_{elast}\,\omega = \inf_{\substack{\mathbf{u}\in S\\ \mathbf{c},|\mathbf{c}|=1}} \mathscr{E}(\mathbf{u}, C\bar{\omega}) \,, \tag{6.120}$$

and upper elastic capacity as

$$\overline{\text{cap}}_{elast}\,\omega = \sup_{\substack{\mathbf{u}\in S\\ \mathbf{c},|\mathbf{c}|=1}} \mathscr{E}(\mathbf{u}, C\bar{\omega}) \,. \tag{6.121}$$

The following Lemma shows that upper and lower elastic capacity are equivalent to electrostatic capacity.

Lemma 6.9. *For the upper and lower capacities the following inequalities hold*

$$\overline{\text{cap}}_{elast}\,\omega \le k_2 \,\text{cap}\; \omega \,, \tag{6.122}$$

$$k_1 \,\text{cap}\; \omega \le \underline{\text{cap}}_{elast}\,\omega \,, \tag{6.123}$$

where $k_1 = 2^{-1}\mu$ *and* $k_2 = 2^{-1}(\lambda + 2\mu)$. *(From which it follows*

$$\overline{\text{cap}}_{elast}\,\omega \le k_3 \,\underline{\text{cap}}_{elast}\,\omega \,, \tag{6.124}$$

where $k_3 = k_2/k_1$.*)*

In order that we prove the preceding Lemma, we shall need the following auxiliary inequality

Lemma 6.10. *For any vector function* $\mathbf{v}$ *in* $C\bar{\omega}$, *constant on* $\partial\omega$, *the elastic energy functional* $\mathscr{E}$ *satisfies the inequality*

$$k_1 \int_{C\bar{\omega}} \|\nabla \mathbf{v}\|^2 \, d\xi \le \mathscr{E}(\mathbf{v}, C\bar{\omega}) \le k_2 \int_{C\bar{\omega}} \|\nabla \mathbf{v}\|^2 \, d\xi \,. \tag{6.125}$$

Proof. We take an arbitrary vector function $\mathbf{v}$ such that $\mathbf{v}|_{\partial\omega} = \mathbf{b}$, where $\mathbf{b}$ is a constant vector, and consider the elastic energy for this in the domain $C\bar{\omega}$

$$\mathscr{E}(\mathbf{v}, C\bar{\omega}) = 2^{-1} \int_{C\bar{\omega}} e_{ij}(\mathbf{v})\sigma_{ij}(\mathbf{v})\, d\boldsymbol{\xi} \ . \tag{6.126}$$

We may rewrite this in the following way

$$2\mathscr{E}(\mathbf{v}, C\bar{\omega}) = \mu \int_{C\bar{\omega}} \|\nabla \mathbf{v}\|^2\, d\boldsymbol{\xi} + (\lambda + \mu) \int_{C\bar{\omega}} (\nabla \cdot \mathbf{v})^2\, d\boldsymbol{\xi} \ . \tag{6.127}$$

Extending $\mathbf{v}$ by $\mathbf{b}$ over the domain ω, we have using Parseval's identity and the Schwarz inequality,

$$\int_{C\bar{\omega}} (\nabla \cdot \mathbf{v})^2\, d\boldsymbol{\xi} = \int_{\mathbb{R}^3} |\mathcal{F}(\nabla \cdot \mathbf{v})|^2\, d\boldsymbol{\nu} \le \int_{\mathbb{R}^3} |\boldsymbol{\nu}|^2 |\mathcal{F}(\mathbf{v})|^2\, d\boldsymbol{\nu} = \int_{C\bar{\omega}} \|\nabla \mathbf{v}\|^2\, d\boldsymbol{\xi} \ , \tag{6.128}$$

where $\mathcal{F}$ is the Fourier transform and $\boldsymbol{\nu} = (\nu_1, \nu_2, \nu_3)$ is the Fourier transform variable.

Thus using (6.128) in (6.127) we deduce that

$$\mathscr{E}(\mathbf{v}, C\bar{\omega}) \le 2^{-1}(\lambda + 2\mu) \int_{C\bar{\omega}} \|\nabla \mathbf{v}\|^2\, d\boldsymbol{\xi} \ . \tag{6.129}$$

It is clear from (6.127) that

$$\mathscr{E}(\mathbf{v}, C\bar{\omega}) \ge 2^{-1}\mu \int_{C\bar{\omega}} \|\nabla \mathbf{v}\|^2\, d\boldsymbol{\xi} \ . \tag{6.130}$$

Hence from (6.129) and (6.130) we have

$$2^{-1}\mu \int_{C\bar{\omega}} \|\nabla \mathbf{v}\|^2\, d\boldsymbol{\xi} \le \mathscr{E}(\mathbf{v}, C\bar{\omega}) \le 2^{-1}(\lambda + 2\mu) \int_{C\bar{\omega}} \|\nabla \mathbf{v}\|^2\, d\boldsymbol{\xi} \ . \tag{6.131}$$

□

Now we are in a position to prove Lemma 6.9.

Proof of Lemma 6.9. We first take $\mathbf{u} \in S$, and consider the elastic energy for this vector function in the domain $B_R \backslash \bar{\omega}$. Repeating the same procedure as in the proof (6.102) we obtain for the vector $\mathbf{u}$, that

$$\mathscr{E}(\mathbf{u}, C\bar{\omega}) = 2^{-1}(\mathbf{c}, B\mathbf{c}) \ . \tag{6.132}$$

Let α be an eigenvalue of the matrix B and $\mathbf{c}$ the corresponding eigenvector, i.e.

$$B\mathbf{c} = \alpha\mathbf{c} \ , \quad \text{where } |\mathbf{c}| = 1 \ . \tag{6.133}$$

From (6.133), we obtain that $\alpha = (\mathbf{c}, B\mathbf{c})$ and this means that for (6.132), we have

$$\mathscr{E}(\mathbf{u}, C\bar{\omega}) = 2^{-1}\alpha\ . \tag{6.134}$$

Moreover, by the definition of upper and lower elastic capacity (6.121), (6.120) we have that upper and lower elastic capacity are the maximum, minimum eigenvalues, respectively, of the elastic capacity matrix $2^{-1}B$.

We shall obtain the inequality (6.122) first. Let the vector $\mathbf{u}^{(1)}$ be sought in the form $\mathbf{u}^{(1)} = \mathcal{P}(\boldsymbol{\xi})\mathbf{c}$ where $\mathcal{P}$ is the electrostatic potential. Considering the right-hand side of (6.125) for $\mathbf{u}^{(1)}$ in $C\bar{\omega}$, we obtain

$$\int_{C\bar{\omega}} \|\nabla \mathbf{u}^{(1)}\|^2\, d\boldsymbol{\xi} = \sum_{j=1}^{3} \int_{C\bar{\omega}} c_j^2 |\nabla \mathcal{P}|^2\, d\boldsymbol{\xi} = \text{cap}\ \omega\ , \tag{6.135}$$

since the function $\mathcal{P}$ minimises the electrostatic energy functional and $|\mathbf{c}| = 1$. Applying now the upper inequality of (6.125) of Lemma 6.10 to the vector function $\mathbf{u}^{(1)}$ we have

$$\mathscr{E}(\mathbf{u}, C\bar{\omega}) \le \mathscr{E}(\mathbf{u}^{(1)}, C\bar{\omega}) \le k_2\, \text{cap}\ \omega\ , \quad \mathbf{u} \in S\ . \tag{6.136}$$

Then taking the supremum on the left-hand side with respect to $\mathbf{c}$, with $|\mathbf{c}| = 1$, we arrive at

$$\overline{\text{cap}}_{elast}\omega \le k_2\, \text{cap}\ \omega\ , \tag{6.137}$$

which proves (6.122).

Next, we take a vector function $\mathbf{u}^{(2)} \in S$, with boundary condition $\mathbf{u}^{(2)} = \mathbf{c}^{(2)}$ on $C\bar{\omega}$ that minimises the elastic energy in $\mathbf{u}$ and $\mathbf{c}$. Applying the lower inequality of (6.125) to $\mathbf{u}^{(2)}$, we have

$$k_1 \int_{C\bar{\omega}} \|\nabla \mathbf{u}^{(2)}\|^2\, d\boldsymbol{\xi} \le \underline{\text{cap}}_{elast}\omega\ . \tag{6.138}$$

However the vector $\mathbf{u}^{(2)}$ is not a minimizer of the Dirichlet integral (we have seen that $\mathbf{u}^{(1)}$ is such a vector). Thus

$$k_1\, \text{cap}\ \omega = k_1 \int_{C\bar{\omega}} \|\nabla \mathbf{u}^{(1)}\|^2\, d\boldsymbol{\xi} \le k_1 \int_{C\bar{\omega}} \|\nabla \mathbf{u}^{(2)}\|^2\, d\boldsymbol{\xi} \le \underline{\text{cap}}_{elast}\omega\ , \tag{6.139}$$

completing the proof of (6.123).

Combining inequalities (6.122) and (6.123), we arrive at the proof of (6.124). □

Hence from Lemma 6.9 we have the elastic capacity and the electrostatic capacity are equivalent.

6.3.3 Asymptotic Estimates for the Regular Part h of Green's Tensor in an Unbounded Domain

We now give an auxiliary result concerning an asymptotic estimate for the tensor h, which we shall make use of in the algorithm for the case of three-dimensional elasticity.

Lemma 6.11. *For all $\boldsymbol{\eta} \in C\bar{\omega}$ and $\boldsymbol{\xi}$ with $|\boldsymbol{\xi}| > 2$ the estimate holds*

$$|h^{(j)}(\boldsymbol{\xi},\boldsymbol{\eta}) - \Gamma(\boldsymbol{\xi},\mathbf{O})P^{T(j)}(\boldsymbol{\eta})| \le \text{Const}\,|\boldsymbol{\xi}|^{-2}|\boldsymbol{\eta}|^{-1}\,, \tag{6.140}$$

where $j = 1,2,3$.

Proof. From the definition of $h(\boldsymbol{\xi},\boldsymbol{\eta})$ in (6.86), the columns of $h(\boldsymbol{\xi},\boldsymbol{\eta})$ satisfy

$$L(\partial_{\boldsymbol{\xi}})h^{(j)}(\boldsymbol{\xi},\boldsymbol{\eta}) = \mathbf{O} \quad \boldsymbol{\xi},\boldsymbol{\eta} \in C\bar{\omega}\,, \tag{6.141}$$

$$h^{(j)}(\boldsymbol{\xi},\boldsymbol{\eta}) = \Gamma^{(j)}(\boldsymbol{\xi},\boldsymbol{\eta})\,, \quad \boldsymbol{\xi} \in \partial C\bar{\omega} \text{ and } \boldsymbol{\eta} \in C\bar{\omega}\,, \tag{6.142}$$

$$h^{(j)}(\boldsymbol{\xi},\boldsymbol{\eta}) \to \mathbf{O} \quad \text{as } |\boldsymbol{\xi}| \to \infty \text{ and } \boldsymbol{\eta} \in C\bar{\omega}\,, \tag{6.143}$$

for $j = 1,2,3$.

From Lemma 6.5, we see that $g^{(i)}(\boldsymbol{\xi},\boldsymbol{\eta})$, $i = 1,2,3$ for $\boldsymbol{\xi}$ with sufficiently large modulus, can be approximated by a linear combination of columns of the fundamental solution

$$|\boldsymbol{\xi}|(g^{(i)}(\boldsymbol{\xi},\boldsymbol{\eta}) - C_{ji}(\boldsymbol{\eta})\Gamma^{(j)}(\boldsymbol{\xi},\mathbf{O})) \xrightarrow{|\boldsymbol{\xi}|\to\infty} \mathbf{O}\,. \tag{6.144}$$

We now apply Betti's formula to the vectors $g^{(k)}(\boldsymbol{\xi},\boldsymbol{\eta})$ and $\mathbf{e}^{(l)} - P^{(l)}(\boldsymbol{\xi})$, $k,l = 1,2,3$, in the domain $B_R\backslash\bar{\omega}$ where $B_R = \{\boldsymbol{\xi} : |\boldsymbol{\xi}| < R\}$ is a ball with sufficiently large radius R. Recalling $P^{(j)}(\boldsymbol{\xi}) = \mathbf{e}^{(j)}$ and $g^{(k)}(\boldsymbol{\xi},\boldsymbol{\eta}) = \mathbf{O}$ when $\boldsymbol{\xi} \in \partial C\bar{\omega}$, we have

$$\begin{aligned}&\int_{B_R\backslash\bar{\omega}} e_{ij}(g^{(k)}(\boldsymbol{\xi},\boldsymbol{\eta}))\sigma_{ij}(P^{(l)}(\boldsymbol{\xi}))\,d\boldsymbol{\xi}\\ &= P_{kl}(\boldsymbol{\eta}) - \delta_{kl} - \int_{\partial B_R}(\delta_{il} - P_{il}(\boldsymbol{\xi}))\sigma_{ij}(g^{(k)}(\boldsymbol{\xi},\boldsymbol{\eta}))n_j\,dS_{\boldsymbol{\xi}}\,,\end{aligned} \tag{6.145}$$

and

$$\int_{B_R\backslash\bar{\omega}} e_{ij}(g^{(k)}(\boldsymbol{\xi},\boldsymbol{\eta}))\sigma_{ij}(P^{(l)}(\boldsymbol{\xi}))\,d\boldsymbol{\xi} = \int_{\partial B_R} g_{ik}(\boldsymbol{\xi},\boldsymbol{\eta})\sigma_{ij}(P^{(l)}(\boldsymbol{\xi}))n_j\,dS_{\boldsymbol{\xi}}\,, \tag{6.146}$$

for $k,l = 1,2,3$.

Then from (6.145), (6.146) we have

$$\delta_{kl} - P_{kl}(\boldsymbol{\eta}) = -\int_{\partial B_R} \{(\delta_{il} - P_{il}(\boldsymbol{\xi}))\sigma_{ij}(g^{(k)}(\boldsymbol{\xi},\boldsymbol{\eta}))n_j + g_{ik}(\boldsymbol{\xi},\boldsymbol{\eta})\sigma_{ij}(P^{(l)}(\boldsymbol{\xi}))n_j\}\, dS_{\boldsymbol{\xi}} . \tag{6.147}$$

Using the asymptotic representation for g given in (6.144) and that for P given in Lemma 6.6, we take the limit in (6.147) as $R \to \infty$ and obtain

$$\delta_{kl} - P_{kl}(\boldsymbol{\eta}) = -\lim_{R\to\infty}\int_{\partial B_R} C_{rk}(\boldsymbol{\eta})\sigma_{lj}(\Gamma^{(r)}(\boldsymbol{\xi},\mathbf{O}))n_j\, dS_{\boldsymbol{\xi}} . \tag{6.148}$$

Computing the above integral, by applying integration by parts to $\mathbf{e}^{(l)}$ and $\Gamma^{(r)}(\boldsymbol{\xi},\mathbf{O})$ in B_R, yields

$$\delta_{kl} - P_{kl}(\boldsymbol{\eta}) = C_{lk}(\boldsymbol{\eta}) , \tag{6.149}$$

or equivalently in the form of matrices

$$I_3 - P^T(\boldsymbol{\eta}) = C(\boldsymbol{\eta}) . \tag{6.150}$$

Let $|\boldsymbol{\xi}| > 2$. Then for $\boldsymbol{\eta} \in \partial C\bar{\omega}$

$$|h^{(j)}(\boldsymbol{\xi},\boldsymbol{\eta}) - \Gamma(\boldsymbol{\xi},\mathbf{O})P^{T(j)}(\boldsymbol{\eta})| = |h^{(j)}(\boldsymbol{\xi},\boldsymbol{\eta}) - \Gamma^{(j)}(\boldsymbol{\xi},\mathbf{O})|$$

$$= |\Gamma^{(j)}(\boldsymbol{\xi},\boldsymbol{\eta}) - \Gamma^{(j)}(\boldsymbol{\xi},\mathbf{O})| \le \text{Const}\, |\boldsymbol{\eta}||\boldsymbol{\xi}|^{-2} \le \text{Const}\, |\boldsymbol{\xi}|^{-2} , \tag{6.151}$$

here we have used that for $\boldsymbol{\eta} \in \partial C\bar{\omega}$, $|\boldsymbol{\eta}| \le 1$. By Lemma 6.3 for functions satisfying the Lamé equation in $\boldsymbol{\eta}$, we have from (6.151) that

$$|h^{(j)}(\boldsymbol{\xi},\boldsymbol{\eta}) - \Gamma(\boldsymbol{\xi},\mathbf{O})P^{T(j)}(\boldsymbol{\eta})| \le \text{Const}\, |\boldsymbol{\xi}|^{-2}|\boldsymbol{\eta}|^{-1} , \tag{6.152}$$

for $\boldsymbol{\eta} \in C\bar{\omega}$ and $|\boldsymbol{\xi}| > 2$. □

6.3.4 *A Uniform Asymptotic Formula for Green's Function G_ε in Three Dimensions*

Now we present the main result concerning the uniform approximation of Green's tensor G_ε in the case of three-dimensions.

Theorem 6.1. *Green's tensor $G_\varepsilon(\mathbf{x},\mathbf{y})$ for the Lamé operator in $\Omega_\varepsilon \subset \mathbb{R}^3$ admits the representation*

$$
\begin{aligned}
G_\varepsilon(\mathbf{x},\mathbf{y}) = G(\mathbf{x},\mathbf{y}) + \varepsilon^{-1}g(\varepsilon^{-1}\mathbf{x},\varepsilon^{-1}\mathbf{y}) - \Gamma(\mathbf{x},\mathbf{y}) + P(\varepsilon^{-1}\mathbf{x})H(\mathbf{O},\mathbf{y})\\
+H(\mathbf{x},\mathbf{O})P^T(\varepsilon^{-1}\mathbf{y}) - P(\varepsilon^{-1}\mathbf{x})H(\mathbf{O},\mathbf{O})P^T(\varepsilon^{-1}\mathbf{y})\\
-\varepsilon H(\mathbf{x},\mathbf{O})BH(\mathbf{O},\mathbf{y}) + O(\varepsilon^2(\min\{|\mathbf{x}|,|\mathbf{y}|\})^{-1})\ ,
\end{aligned}
\tag{6.153}
$$

uniformly with respect to $\mathbf{x}$, $\mathbf{y} \in \Omega_\varepsilon$.

Proof. As in Maz'ya and Movchan [24], we first present a formal argument concerning the structure of $G_\varepsilon(\mathbf{x},\mathbf{y})$, then give a rigorous proof of the remainder estimate.

Formal Argument

Let G_ε be represented in the form

$$
G_\varepsilon(\mathbf{x},\mathbf{y}) = \Gamma(\mathbf{x},\mathbf{y}) - H_\varepsilon(\mathbf{x},\mathbf{y}) - h_\varepsilon(\mathbf{x},\mathbf{y})\ , \tag{6.154}
$$

where the columns of $H_\varepsilon(\mathbf{x},\mathbf{y}) = [H_\varepsilon^{(j)}(\mathbf{x},\mathbf{y})]$, $h_\varepsilon(\mathbf{x},\mathbf{y}) = [h_\varepsilon^{(j)}(\mathbf{x},\mathbf{y})]$, $j = 1,2,3$, satisfy the Dirichlet problems

$$
\begin{aligned}
L(\partial_{\mathbf{x}})H_\varepsilon^{(j)}(\mathbf{x},\mathbf{y}) = \mathbf{O}\ , \quad \mathbf{x},\mathbf{y} \in \Omega_\varepsilon\ ,\\
H_\varepsilon^{(j)}(\mathbf{x},\mathbf{y}) = \Gamma^{(j)}(\mathbf{x},\mathbf{y})\ , \quad \mathbf{x} \in \partial\Omega, \mathbf{y} \in \Omega_\varepsilon\ ,\\
H_\varepsilon^{(j)}(\mathbf{x},\mathbf{y}) = \mathbf{O}\ , \quad \mathbf{x} \in \partial\omega_\varepsilon, \mathbf{y} \in \Omega_\varepsilon\ ,
\end{aligned}
$$

and

$$
\begin{aligned}
L(\partial_{\mathbf{x}})h_\varepsilon^{(j)}(\mathbf{x},\mathbf{y}) = \mathbf{O}\ , \quad \mathbf{x},\mathbf{y} \in \Omega_\varepsilon\ ,\\
h_\varepsilon^{(j)}(\mathbf{x},\mathbf{y}) = \Gamma^{(j)}(\mathbf{x},\mathbf{y})\ , \quad \mathbf{x} \in \partial\omega_\varepsilon, \mathbf{y} \in \Omega_\varepsilon\ ,\\
h_\varepsilon^{(j)}(\mathbf{x},\mathbf{y}) = \mathbf{O}\ , \quad \mathbf{x} \in \partial\Omega, \mathbf{y} \in \Omega_\varepsilon\ .
\end{aligned}
\tag{6.155}
$$

From (6.154), it is enough to approximate the columns of H_ε and h_ε, to obtain the asymptotic formula for G_ε.

Approximation of $H_\varepsilon(\mathbf{x},\mathbf{y})$

Consider $H_\varepsilon(\mathbf{x},\mathbf{y}) - H(\mathbf{x},\mathbf{y})$, which satisfies the homogeneous Lamé equation and has zero boundary value when $\mathbf{x} \in \partial\Omega, \mathbf{y} \in \Omega_\varepsilon$. When $\mathbf{x} \in \partial\omega_\varepsilon$, the leading part of $H_\varepsilon(\mathbf{x},\mathbf{y}) - H(\mathbf{x},\mathbf{y})$ is given by $-H(\mathbf{O},\mathbf{y})$. We extend $-H(\mathbf{O},\mathbf{y})$ onto $C\bar{\omega}_\varepsilon$ as a tensor that satisfies the homogeneous Lamé equation in variable $\mathbf{x}$, in the form $-P(\varepsilon^{-1}\mathbf{x})H(\mathbf{O},\mathbf{y})$, whose leading order part is $-\varepsilon\Gamma(\mathbf{x},\mathbf{O})BH(\mathbf{O},\mathbf{y})$ for $\mathbf{x} \in \partial\Omega, \mathbf{y} \in \Omega_\varepsilon$. Thus

$$H_\varepsilon(\mathbf{x},\mathbf{y}) - H(\mathbf{x},\mathbf{y}) = -P(\varepsilon^{-1}\mathbf{x})H(\mathbf{O},\mathbf{y}) + \varepsilon H(\mathbf{x},\mathbf{O})BH(\mathbf{O},\mathbf{y})$$
$$+\mathfrak{H}_\varepsilon(\mathbf{x},\mathbf{y}) , \quad \mathbf{x},\mathbf{y} \in \Omega_\varepsilon , \tag{6.156}$$

where $\mathfrak{H}_\varepsilon(\mathbf{x},\mathbf{y})$ is the remainder term produced by this approximation.

Approximation of $h_\varepsilon(\mathbf{x},\mathbf{y})$

Using the definition of h and (6.155) of h_ε, we have

$$h_\varepsilon(\mathbf{x},\mathbf{y}) - \varepsilon^{-1}h(\varepsilon^{-1}\mathbf{x},\varepsilon^{-1}\mathbf{y}) = \mathbf{O} \quad \text{for } \mathbf{x} \in \partial\omega_\varepsilon . \tag{6.157}$$

Then from Lemma 6.11, we have

$$h_\varepsilon(\mathbf{x},\mathbf{y}) - \varepsilon^{-1}h(\varepsilon^{-1}\mathbf{x},\varepsilon^{-1}\mathbf{y}) = -\Gamma(\mathbf{x},\mathbf{O})P^T(\eta) + O(\varepsilon^2(|\mathbf{x}|^2|\mathbf{y}|)^{-1}) ,$$

for $\mathbf{x} \in \partial\Omega, \mathbf{y} \in \Omega_\varepsilon$. The tensor that satisfies the homogeneous Lamé equation in $\mathbf{x}$ and has boundary data $\Gamma(\mathbf{x},\mathbf{O})P^T(\eta)$ when $\mathbf{x} \in \partial\Omega$ is

$$H(\mathbf{x},\mathbf{O})P^T(\eta) .$$

Thus, we have

$$h_\varepsilon(\mathbf{x},\mathbf{y}) - \varepsilon^{-1}h(\varepsilon^{-1}\mathbf{x},\varepsilon^{-1}\mathbf{y}) = -H(\mathbf{x},\mathbf{O})P^T(\eta) + \chi_\varepsilon(\mathbf{x},\mathbf{y}) ,$$

where $\chi_\varepsilon(\mathbf{x},\mathbf{y})$ is the remainder. For $\mathbf{x} \in \partial\omega_\varepsilon$, $\chi_\varepsilon(\mathbf{x},\mathbf{y}) = H(\mathbf{x},\mathbf{O})P^T(\eta)$. Since the components of $H(\mathbf{x},\mathbf{O})$ are smooth for $\mathbf{x},\mathbf{y} \in \Omega$, we may approximate the latter by $H(\mathbf{O},\mathbf{O})P^T(\eta)$. However this matrix is not necessarily small. Making an extension of $H(\mathbf{O},\mathbf{O})P^T(\eta)$ to a matrix which satisfies the homogeneous Lamé equation for $\mathbf{x} \in C\bar{\omega}_\varepsilon$, and is small for $\mathbf{x} \in \partial\Omega, \mathbf{y} \in \Omega_\varepsilon$, we have

$$\chi_\varepsilon(\mathbf{x},\mathbf{y}) = P(\varepsilon^{-1}\mathbf{x})H(\mathbf{O},\mathbf{O})P^T(\varepsilon^{-1}\mathbf{y}) + \mathfrak{h}_\varepsilon(\mathbf{x},\mathbf{y}) ,$$

where $\mathfrak{h}_\varepsilon(\mathbf{x},\mathbf{y})$ is the new remainder. Hence we may now assume the asymptotic representation

$$h_\varepsilon(\mathbf{x},\mathbf{y}) - \varepsilon^{-1}h(\varepsilon^{-1}\mathbf{x},\varepsilon^{-1}\mathbf{y}) = -H(\mathbf{x},\mathbf{O})P^T(\varepsilon^{-1}\mathbf{y})$$
$$+P(\varepsilon^{-1}\mathbf{x})H(\mathbf{O},\mathbf{O})P^T(\varepsilon^{-1}\mathbf{y})$$
$$+\mathfrak{h}_\varepsilon(\mathbf{x},\mathbf{y}) , \tag{6.158}$$

for $\mathbf{x},\mathbf{y} \in \Omega_\varepsilon$.

Combined Formula

Combining (6.156) and (6.158) in (6.154), yields

$$\begin{aligned} G_\varepsilon(\mathbf{x},\mathbf{y}) = {} & \Gamma(\mathbf{x},\mathbf{y}) - H(\mathbf{x},\mathbf{y}) + P(\varepsilon^{-1}\mathbf{x})H(\mathbf{O},\mathbf{y}) \\ & -\varepsilon H(\mathbf{x},\mathbf{O})BH(\mathbf{O},\mathbf{y}) - \varepsilon^{-1}h(\varepsilon^{-1}\mathbf{x},\varepsilon^{-1}\mathbf{y}) \\ & +H(\mathbf{x},\mathbf{O})P^T(\varepsilon^{-1}\mathbf{y}) - P(\varepsilon^{-1}\mathbf{x})H(\mathbf{O},\mathbf{O})P^T(\varepsilon^{-1}\mathbf{y}) \\ & +R_\varepsilon(\mathbf{x},\mathbf{y})\,, \end{aligned} \tag{6.159}$$

where $R_\varepsilon(\mathbf{x},\mathbf{y})$ is the sum of the remainders $\mathfrak{H}_\varepsilon(\mathbf{x},\mathbf{y})$ and $\mathfrak{h}_\varepsilon(\mathbf{x},\mathbf{y})$, which we shall estimate. Recalling the definition of G and g from (6.85) and (6.86), the preceding expression is equivalent to

$$\begin{aligned} G_\varepsilon(\mathbf{x},\mathbf{y}) = {} & G(\mathbf{x},\mathbf{y}) + \varepsilon^{-1}g(\varepsilon^{-1}\mathbf{x},\varepsilon^{-1}\mathbf{y}) - \Gamma(\mathbf{x},\mathbf{y}) \\ & +P(\varepsilon^{-1}\mathbf{x})H(\mathbf{O},\mathbf{y}) + H(\mathbf{x},\mathbf{O})P^T(\varepsilon^{-1}\mathbf{y}) \\ & -P(\varepsilon^{-1}\mathbf{x})H(\mathbf{O},\mathbf{O})P^T(\varepsilon^{-1}\mathbf{y}) - \varepsilon H(\mathbf{x},\mathbf{O})BH(\mathbf{O},\mathbf{y}) \\ & +R_\varepsilon(\mathbf{x},\mathbf{y})\,. \end{aligned} \tag{6.160}$$

Next we give a rigorous proof of (6.153).

Proof of Theorem 6.1

The columns of $R_\varepsilon(\mathbf{x},\mathbf{y})$ solve the problem

$$L(\partial_\mathbf{x})R_\varepsilon^{(j)}(\mathbf{x},\mathbf{y}) = \mathbf{O} \quad \mathbf{x},\mathbf{y} \in \Omega_\varepsilon\,, \tag{6.161}$$

$$\begin{aligned} R_\varepsilon^{(j)}(\mathbf{x},\mathbf{y}) = {} & \varepsilon^{-1}h^{(j)}(\varepsilon^{-1}\mathbf{x},\varepsilon^{-1}\mathbf{y}) - H(\mathbf{x},\mathbf{O})P^{T(j)}(\varepsilon^{-1}\mathbf{y}) \\ & -P(\varepsilon^{-1}\mathbf{x})H^{(j)}(\mathbf{O},\mathbf{y}) + P(\varepsilon^{-1}\mathbf{x})H(\mathbf{O},\mathbf{O})P^{T(j)}(\varepsilon^{-1}\mathbf{y}) \\ & +\varepsilon H(\mathbf{x},\mathbf{O})BH^{(j)}(\mathbf{O},\mathbf{y})\,, \quad \mathbf{x} \in \partial\Omega, \mathbf{y} \in \Omega_\varepsilon\,, \end{aligned} \tag{6.162}$$

$$\begin{aligned} R_\varepsilon^{(j)}(\mathbf{x},\mathbf{y}) = {} & H^{(j)}(\mathbf{x},\mathbf{y}) - H^{(j)}(\mathbf{O},\mathbf{y}) - H(\mathbf{x},\mathbf{O})P^{T(j)}(\varepsilon^{-1}\mathbf{y}) \\ & +H(\mathbf{O},\mathbf{O})P^{T(j)}(\varepsilon^{-1}\mathbf{y}) + \varepsilon H(\mathbf{x},\mathbf{O})BH^{(j)}(\mathbf{O},\mathbf{y})\,, \\ & \mathbf{x} \in \partial\omega_\varepsilon, \mathbf{y} \in \Omega_\varepsilon\,. \end{aligned} \tag{6.163}$$

Both $H^{(j)}(\mathbf{x},\mathbf{O})$ and $H^{(j)}(\mathbf{O},\mathbf{y})$ are columns of H (see (6.85)), and $H^{(j)}(\mathbf{x},\mathbf{O})$ has bounded components on $\partial\Omega$. They are also bounded for $\mathbf{x} \in \partial\omega_\varepsilon$, $\mathbf{y} \in \Omega_\varepsilon$. The components of the term $\varepsilon H(\mathbf{x},\mathbf{O})BH^{(j)}(\mathbf{O},\mathbf{y})$ are bounded by Const ε in (6.162) and (6.163). Since the components of $H(\mathbf{x},\mathbf{y})$ are smooth for $\mathbf{x},\mathbf{y} \in \Omega$ and by Lemma 6.4 the entries of the tensor $P(\boldsymbol{\xi})$ are bounded, from (6.163) we have

$$|H^{(j)}(\mathbf{x},\mathbf{y}) - H^{(j)}(\mathbf{O},\mathbf{y}) - (H(\mathbf{x},\mathbf{O}) - H(\mathbf{O},\mathbf{O}))P^{T(j)}(\boldsymbol{\eta})| \leq \text{Const}\,\varepsilon\,, \quad (6.164)$$

for $\mathbf{x} \in \partial\omega_\varepsilon, \mathbf{y} \in \Omega_\varepsilon$. Thus when $\mathbf{x} \in \partial\omega_\varepsilon$ and $\mathbf{y} \in \Omega_\varepsilon$

$$|R_\varepsilon^{(j)}(\mathbf{x},\mathbf{y})| \leq \text{Const}\,\varepsilon\,,$$

for $j = 1,2,3$.

Next we estimate $|R_\varepsilon^{(j)}(\mathbf{x},\mathbf{y})|$ when $\mathbf{x} \in \partial\Omega, \mathbf{y} \in \Omega_\varepsilon$. By Lemma 6.4, the columns of capacitary potential satisfy the following inequality

$$|P^{(j)}(\varepsilon^{-1}\mathbf{x})| \leq \text{Const}\,\varepsilon|\mathbf{x}|^{-1}\,, \quad j = 1,2,3, \quad \text{for } \mathbf{x} \in \Omega_\varepsilon\,. \quad (6.165)$$

Now, (6.93) of Lemma 6.6 and the definition of $H(\mathbf{x},\mathbf{y})$ imply

$$\begin{aligned} &|\varepsilon H(\mathbf{x},\mathbf{O})BH^{(j)}(\mathbf{O},\mathbf{y}) - P(\varepsilon^{-1}\mathbf{x})H^{(j)}(\mathbf{O},\mathbf{y})| \\ &= |(\Gamma(\varepsilon^{-1}\mathbf{x},\mathbf{O})B - P(\varepsilon^{-1}\mathbf{x}))H^{(j)}(\mathbf{O},\mathbf{y})| \leq \text{Const}\,\varepsilon^2\,, \end{aligned} \quad (6.166)$$

for $\mathbf{x} \in \partial\Omega, \mathbf{y} \in \Omega_\varepsilon$. We also have, using Lemma 6.11, the following estimate

$$\begin{aligned} &|\varepsilon^{-1}h^{(j)}(\varepsilon^{-1}\mathbf{x},\varepsilon^{-1}\mathbf{y}) - H(\mathbf{x},\mathbf{O})P^{T(j)}(\varepsilon^{-1}\mathbf{y})| \\ &= \varepsilon^{-1}|h^{(j)}(\varepsilon^{-1}\mathbf{x},\varepsilon^{-1}\mathbf{y}) - \Gamma(\boldsymbol{\xi},\mathbf{O})P^{T(j)}(\varepsilon^{-1}\mathbf{y})| \\ &\leq \text{Const}\,\varepsilon^2|\mathbf{x}|^{-2}|\mathbf{y}|^{-1} \leq \text{Const}\,\varepsilon^2|\mathbf{y}|^{-1}\,, \quad \mathbf{x} \in \partial\Omega, \mathbf{y} \in \Omega_\varepsilon\,, \end{aligned} \quad (6.167)$$

where we have used the estimate (6.140) and for $\mathbf{x} \in \partial\Omega$, $|\mathbf{x}| \geq 1$. Combining (6.165), (6.166) and (6.167) in (6.162) we obtain

$$|R_\varepsilon^{(j)}(\mathbf{x},\mathbf{y})| \leq \text{Const}\,\varepsilon^2|\mathbf{y}|^{-1} \quad \text{for } \mathbf{x} \in \partial\Omega, \mathbf{y} \in \Omega_\varepsilon\,, \quad (6.168)$$

for $j = 1,2,3$.

Therefore, by Lemma 6.1, we have

$$|R_\varepsilon^{(j)}(\mathbf{x},\mathbf{y})| \leq \text{Const}\,\max\left\{\varepsilon^2|\mathbf{x}|^{-1}, \varepsilon^2|\mathbf{y}|^{-1}\right\}\,, \quad (6.169)$$

for $j = 1,2,3$, and $\mathbf{x},\mathbf{y} \in \Omega_\varepsilon$. Thus,

$$|R_\varepsilon^{(j)}(\mathbf{x},\mathbf{y})| \leq \text{Const}\,\varepsilon^2(\min\{|\mathbf{x}|,|\mathbf{y}|\})^{-1}\,. \quad (6.170)$$

The proof is complete. □

6.4 Green's Tensor for a Planar Domain with a Small Inclusion

Now we present the uniform approximation of the tensor $G_\varepsilon(\mathbf{x},\mathbf{y})$ for the case of a planar domain with a small inclusion, defined in Sect. 6.1.1. We once again introduce model domains and governing equations needed for the study related to this case.

6.4.1 *Green's Kernels for Model Domains in Two Dimensions*

Let $G(\mathbf{x},\mathbf{y}) = [G^{(1)}(\mathbf{x},\mathbf{y}), G^{(2)}(\mathbf{x},\mathbf{y})]$ and $g(\boldsymbol{\xi},\boldsymbol{\eta}) = [g^{(1)}(\boldsymbol{\xi},\boldsymbol{\eta}), g^{(2)}(\boldsymbol{\xi},\boldsymbol{\eta})]$ denote Green's tensor in the bounded domain Ω and $C\bar{\omega} = \mathbb{R}^2\backslash\bar{\omega}$, respectively, for the Lamé operator given by (6.11) in two dimensions. The tensor G is a solution the following problem

$$L(\partial_{\mathbf{x}})G(\mathbf{x},\mathbf{y}) + \delta(\mathbf{x}-\mathbf{y})I_2 = 0I_2\,, \quad \mathbf{x},\mathbf{y} \in \Omega\,, \tag{6.171}$$

$$G(\mathbf{x},\mathbf{y}) = 0I_2\,, \quad \mathbf{x} \in \partial\Omega, \mathbf{y} \in \Omega\,, \tag{6.172}$$

and the tensor g solves

$$L(\partial_{\boldsymbol{\xi}})g(\boldsymbol{\xi},\boldsymbol{\eta}) + \delta(\boldsymbol{\xi}-\boldsymbol{\eta})I_2 = 0I_2\,, \quad \boldsymbol{\xi},\boldsymbol{\eta} \in C\bar{\omega}\,, \tag{6.173}$$

$$g(\boldsymbol{\xi},\boldsymbol{\eta}) = 0I_2\,, \quad \boldsymbol{\xi} \in \partial C\bar{\omega}, \boldsymbol{\eta} \in C\bar{\omega}\,, \tag{6.174}$$

$$|g^{(j)}(\boldsymbol{\xi},\boldsymbol{\eta})| \text{ is bounded as } |\boldsymbol{\xi}| \to \infty\,, \boldsymbol{\eta} \in C\bar{\omega} \text{ for } j = 1,2\,. \tag{6.175}$$

We have from (6.171), (6.172), that G has the following symmetry property

$$G(\mathbf{x},\mathbf{y}) = (G(\mathbf{y},\mathbf{x}))^T \quad \mathbf{x},\mathbf{y} \in \Omega, \mathbf{x} \neq \mathbf{y}\,, \tag{6.176}$$

and from (6.173)–(6.175) the Green's function g satisfies

$$g(\boldsymbol{\xi},\boldsymbol{\eta}) = (g(\boldsymbol{\eta},\boldsymbol{\xi}))^T, \quad \boldsymbol{\xi},\boldsymbol{\eta} \in C\bar{\omega}, \boldsymbol{\xi} \neq \boldsymbol{\eta}\,. \tag{6.177}$$

We represent $G(\mathbf{x},\mathbf{y})$ as

$$G(\mathbf{x},\mathbf{y}) = \gamma(\mathbf{x},\mathbf{y}) - H(\mathbf{x},\mathbf{y})\,, \tag{6.178}$$

and $g(\boldsymbol{\xi},\boldsymbol{\eta})$ as

$$g(\boldsymbol{\xi},\boldsymbol{\eta}) = \gamma(\boldsymbol{\xi},\boldsymbol{\eta}) - h(\boldsymbol{\xi},\boldsymbol{\eta})\,, \tag{6.179}$$

where H and h are the regular parts of G and g respectively, and $\gamma(\mathbf{x},\mathbf{y}) = [\gamma_{ij}(\mathbf{x},\mathbf{y})]_{i,j=1}^2$, is the fundamental solution of the Lamé operator in two dimensions, with components

$$\gamma_{ij}(\mathbf{x},\mathbf{y}) = K_2(-\log|\mathbf{x}-\mathbf{y}|\delta_{ij} \\ +(\lambda+\mu)(\lambda+3\mu)^{-1}(x_i-y_i)(x_j-y_j)|\mathbf{x}-\mathbf{y}|^{-2}) , \quad (6.180)$$

for $i, j = 1, 2$, where

$$K_2 = \frac{\lambda+3\mu}{4\pi\mu(\lambda+2\mu)} . \quad (6.181)$$

We introduce the tensor ζ as

$$\zeta(\boldsymbol{\eta}) = \lim_{|\boldsymbol{\xi}|\to\infty} g(\boldsymbol{\eta},\boldsymbol{\xi}) , \quad (6.182)$$

and the constant matrix

$$\zeta^{\infty} = \lim_{|\boldsymbol{\eta}|\to\infty} \{\zeta(\boldsymbol{\eta}) + \gamma(\boldsymbol{\eta},\mathbf{O})\} , \quad (6.183)$$

where it will be shown that ζ^{∞} is a symmetric matrix.

6.4.2 *Auxiliary Properties of the Regular Part h of Green's Tensor for an Unbounded Planar Domain and the Tensor ζ*

In the present subsection, we shall formulate and prove an asymptotic representation for the regular part h of Green's tensor g, in the unbounded domain. For this we shall need the following Lemma which is the two-dimensional analogue of Lemma 6.5, and is a reformulation of that by Kondratiev and Oleinik [18] (p. 78).

Lemma 6.12. *Suppose the columns $u^{(j)}(\boldsymbol{\xi})$ of the matrix $u(\boldsymbol{\xi})$ are solutions of*

$$L(\partial_{\boldsymbol{\xi}})u^{(j)}(\boldsymbol{\xi}) = \mathbf{O} , \quad in\ C\bar{\omega} ,$$

and that $|u^{(j)}(\boldsymbol{\xi})| \le \text{Const}\,(1+|\boldsymbol{\xi}|)^k$, $k \ge 0$, for $j = 1, 2$.

Then for $|\boldsymbol{\xi}| > 2$ the representation holds

$$u^{(j)}(\boldsymbol{\xi}) = \mathscr{P}_k^{(j)}(\boldsymbol{\xi}) + \gamma(\boldsymbol{\xi},\mathbf{O})C^{(j)} + O(|\boldsymbol{\xi}|^{-1}) , \quad (6.184)$$

where $\mathscr{P}_k^{(j)}(\boldsymbol{\xi}) = \{\mathscr{P}_i^{(j,k)}(\boldsymbol{\xi})\}_{i=1}^2$, $\mathscr{P}_i^{(j,k)}(\boldsymbol{\xi})$ are polynomials of order not greater than k, $C^{(j)} = \{C_i^{(j)}\}_{i=1}^2$, where $C_i^{(j)}$ are constants.

We now formulate a result related to the approximation of the regular part of Green's tensor g needed for the asymptotic algorithm.

Lemma 6.13. *Let $|\boldsymbol{\xi}| > 2$, $\boldsymbol{\eta} \in C\bar{\omega}$. Then the columns of the regular part $h^{(j)}(\boldsymbol{\xi},\boldsymbol{\eta})$ of Green's tensor in $C\bar{\omega}$ admit the asymptotic representation*

$$h^{(j)}(\boldsymbol{\xi},\boldsymbol{\eta}) = \gamma^{(j)}(\boldsymbol{\xi},\mathbf{O}) - \zeta^{T(j)}(\boldsymbol{\eta}) + O(|\boldsymbol{\xi}|^{-1})\,. \tag{6.185}$$

Proof. By definition of g (cf. (6.173)–(6.175)), the columns $h^{(j)}$ of the regular part satisfy

$$L(\partial_{\boldsymbol{\xi}})h^{(j)}(\boldsymbol{\xi},\boldsymbol{\eta}) = \mathbf{O}\,, \quad \boldsymbol{\xi},\boldsymbol{\eta} \in C\bar{\omega}\,, \tag{6.186}$$

$$h^{(j)}(\boldsymbol{\xi},\boldsymbol{\eta}) = \gamma^{(j)}(\boldsymbol{\xi},\boldsymbol{\eta})\,, \quad \boldsymbol{\xi} \in \partial C\bar{\omega}, \boldsymbol{\eta} \in C\bar{\omega}\,, \tag{6.187}$$

with the following condition at infinity

$$h^{(j)}(\boldsymbol{\xi},\boldsymbol{\eta}) \sim \gamma^{(j)}(\boldsymbol{\xi},\mathbf{O}) - \zeta^{T(j)}(\boldsymbol{\eta})\,, \quad \text{as } |\boldsymbol{\xi}| \to \infty, \boldsymbol{\eta} \in C\bar{\omega}\,, \tag{6.188}$$

for $j = 1, 2$.

Setting $U^{(j)}(\boldsymbol{\xi},\boldsymbol{\eta}) = h^{(j)}(\boldsymbol{\xi},\boldsymbol{\eta}) - \gamma^{(j)}(\boldsymbol{\xi},\mathbf{O})$, we have that $U^{(j)}$ solves

$$L(\partial_{\boldsymbol{\xi}})U^{(j)}(\boldsymbol{\xi},\boldsymbol{\eta}) = \mathbf{O}\,, \quad \boldsymbol{\xi},\boldsymbol{\eta} \in C\bar{\omega}\,, \tag{6.189}$$

$$U^{(j)}(\boldsymbol{\xi},\boldsymbol{\eta}) = \gamma^{(j)}(\boldsymbol{\xi},\boldsymbol{\eta}) - \gamma^{(j)}(\boldsymbol{\xi},\mathbf{O})\,, \quad \boldsymbol{\xi} \in \partial C\bar{\omega}, \boldsymbol{\eta} \in C\bar{\omega}\,, \tag{6.190}$$

and by (6.182)

$$U^{(j)}(\boldsymbol{\xi},\boldsymbol{\eta}) \sim -\zeta^{T(j)}(\boldsymbol{\eta})\,, \quad \text{as } |\boldsymbol{\xi}| \to \infty, \boldsymbol{\eta} \in C\bar{\omega}\,. \tag{6.191}$$

Consulting Lemma 6.12, we see that for $|\boldsymbol{\xi}| > 2$ the following representation for $U^{(j)}$ holds

$$U^{(j)}(\boldsymbol{\xi},\boldsymbol{\eta}) = K^{(j)} + \gamma(\boldsymbol{\xi},\mathbf{O})C^{(j)} + O(|\boldsymbol{\xi}|^{-1})\,. \tag{6.192}$$

where $K^{(j)}$ and $C^{(j)}$ are vector functions of $\boldsymbol{\eta}$ only.

Then, in order that condition (6.191) be satisfied we must take $K^{(j)} = -\zeta^{T(j)}(\boldsymbol{\eta})$ and $C^{(j)} = \mathbf{O}$. Thus, recalling the definition of $U^{(j)}$, we obtain (6.185). □

We also have the following asymptotic representation of the tensor ζ.

Lemma 6.14. *For $|\boldsymbol{\xi}| > 2$, the representation for $\zeta^{(j)}$, $j = 1, 2$, holds*

$$\zeta^{(j)}(\boldsymbol{\xi}) = -\gamma^{(j)}(\boldsymbol{\xi},\mathbf{O}) + \zeta^{(\infty,j)} + O(|\boldsymbol{\xi}|^{-1})\,. \tag{6.193}$$

Proof. By the definition of $\zeta(\boldsymbol{\xi})$, the columns $\zeta^{(j)}(\boldsymbol{\xi})$ are solutions of

$$L(\partial_{\boldsymbol{\xi}})\zeta^{(j)}(\boldsymbol{\xi}) = \mathbf{O}\,, \quad \boldsymbol{\xi} \in C\bar{\omega}\,, \tag{6.194}$$

$$\zeta^{(j)}(\boldsymbol{\xi}) = \mathbf{O}\,, \quad \boldsymbol{\xi} \in \partial C\bar{\omega}\,, \tag{6.195}$$

$$\zeta^{(j)}(\boldsymbol{\xi}) \sim -\gamma^{(j)}(\boldsymbol{\xi},\mathbf{O}) + \zeta^{(\infty,j)} \quad \text{as} \quad |\boldsymbol{\xi}| \to \infty\,, \tag{6.196}$$

for $j = 1, 2$, where $\zeta^{(\infty,j)}$ are the columns of ζ^{∞} and the preceding boundary value problem is consistent with (6.182), (6.183).

Setting $U^{(j)} = \zeta^{(j)}(\boldsymbol{\xi}) + \gamma^{(j)}(\boldsymbol{\xi}, \mathbf{O})$, and in the same way as in the proof of the previous lemma, we deduce (6.193). □

From the previous two lemmas we also prove

Lemma 6.15. *The matrix ζ^{∞} is symmetric.*

Proof. Consider the matrix function

$$\Sigma(\boldsymbol{\xi}.\boldsymbol{\eta}) = \gamma(\boldsymbol{\xi}, \mathbf{O}) + \gamma(\boldsymbol{\eta}, \mathbf{O}) - h(\boldsymbol{\xi}, \boldsymbol{\eta}) . \tag{6.197}$$

Let $|\boldsymbol{\xi}| > 2$ and $|\boldsymbol{\eta}| > 2$. Using Lemmas 6.13 and 6.14

$$\begin{aligned}\Sigma(\boldsymbol{\xi}, \boldsymbol{\eta}) &= \gamma(\boldsymbol{\eta}, \mathbf{O}) + \zeta^T(\boldsymbol{\eta}) + O(|\boldsymbol{\xi}|^{-1}) \\ &= (\zeta^{\infty})^T + O(|\boldsymbol{\xi}|^{-1} + |\boldsymbol{\eta}|^{-1}) .\end{aligned} \tag{6.198}$$

In a similar way, we also can derive

$$\begin{aligned}\Sigma(\boldsymbol{\eta}, \boldsymbol{\xi}) &= \gamma(\boldsymbol{\xi}, \mathbf{O}) + \zeta^T(\boldsymbol{\xi}) + O(|\boldsymbol{\eta}|^{-1}) \\ &= (\zeta^{\infty})^T + O(|\boldsymbol{\xi}|^{-1} + |\boldsymbol{\eta}|^{-1}) .\end{aligned} \tag{6.199}$$

From (6.177) and (6.179), the symmetry condition $h(\boldsymbol{\xi}, \boldsymbol{\eta}) = (h(\boldsymbol{\eta}, \boldsymbol{\xi}))^T$ for $\boldsymbol{\xi}, \boldsymbol{\eta} \in C\bar{\omega}$ also holds. This condition and the definition (6.197) of Σ also leads to the relation

$$\Sigma(\boldsymbol{\xi}, \boldsymbol{\eta}) = (\Sigma(\boldsymbol{\eta}, \boldsymbol{\xi}))^T , \quad \boldsymbol{\xi}, \boldsymbol{\eta} \in C\bar{\omega} . \tag{6.200}$$

Taking the limit as $|\boldsymbol{\xi}|$ and $|\boldsymbol{\eta}|$ both tend to infinity, (6.198)–(6.200) imply that $\zeta^{\infty} = (\zeta^{\infty})^T$. Thus ζ^{∞} is symmetric. □

6.4.3 A Uniform Asymptotic Approximation of an Elastic Capacitary Potential Matrix

Let $P_{\varepsilon}(\mathbf{x}) = [P_{\varepsilon}^{(1)}(\mathbf{x}), P_{\varepsilon}^{(2)}(\mathbf{x})]$ denote the elastic capacitary potential of the set ω_{ε}, whose columns are a solution of the following problem

$$L(\partial_{\mathbf{x}})P_{\varepsilon}^{(j)}(\mathbf{x}) = \mathbf{O}, \quad \mathbf{x} \in \Omega_{\varepsilon} , \tag{6.201}$$

$$P_{\varepsilon}^{(j)}(\mathbf{x}) = \mathbf{O} , \quad \mathbf{x} \in \partial\Omega , \tag{6.202}$$

$$P_{\varepsilon}^{(j)}(\mathbf{x}) = \mathbf{e}^{(j)} , \quad \mathbf{x} \in \partial\omega_{\varepsilon} , \tag{6.203}$$

for $j = 1, 2$.

Lemma 6.16. *The asymptotic approximation of $P_\varepsilon(\mathbf{x})$ is given by the formula*

$$P_\varepsilon(\mathbf{x}) = (G(\mathbf{x},\mathbf{O}) - \zeta(\boldsymbol{\xi}) - \gamma(\boldsymbol{\xi},\mathbf{O}) + \zeta^\infty)D + p(\mathbf{x}) , \tag{6.204}$$

where D is the matrix given by (6.53)-(6.55) and $p(\mathbf{x}) = [p^{(1)}(\mathbf{x}), p^{(2)}(\mathbf{x})]$ is such that

$$|p^{(j)}(\mathbf{x})| \le \text{Const}\,\varepsilon|\log\varepsilon|^{-1}, \quad j = 1,2 , \tag{6.205}$$

uniformly with respect to $\mathbf{x} \in \Omega_\varepsilon$.

Proof. Let $\varepsilon \to 0$, then $\Omega_\varepsilon \to \Omega\backslash\{\mathbf{O}\}$. In this limit domain, it is suitable to approximate the columns $P_\varepsilon^{(j)}(\mathbf{x})$ of the elastic capacitary potential, by $V^{(j)}(\mathbf{x})$, which solves the boundary value problem

$$L(\partial_{\mathbf{x}})V^{(j)}(\mathbf{x}) + D_{ij}\delta(\mathbf{x})\mathbf{e}^{(i)} = \mathbf{O}, \quad \mathbf{x} \in \Omega , \tag{6.206}$$

$$V^{(j)}(\mathbf{x}) = \mathbf{O} , \quad \mathbf{x} \in \partial\Omega , \tag{6.207}$$

for $j = 1,2$. Let $V^{(j)}(\mathbf{x})$ be sought in the form

$$V^{(j)}(\mathbf{x}) = D_{1j}G^{(1)}(\mathbf{x},\mathbf{O}) + D_{2j}G^{(2)}(\mathbf{x},\mathbf{O}) , j = 1,2 . \tag{6.208}$$

The representation of $V^{(j)}(\mathbf{x})$ by (6.208) does not satisfy the boundary conditions on $\partial\omega_\varepsilon$. Therefore, we construct a boundary layer $M^{(j)}(\boldsymbol{\xi})$, which is a solution of

$$L(\partial_{\boldsymbol{\xi}})M^{(j)}(\boldsymbol{\xi}) = \mathbf{O}, \quad \boldsymbol{\xi} \in C\bar{\omega} , \tag{6.209}$$

$$M^{(j)}(\boldsymbol{\xi}) = \mathbf{e}^{(j)} - D_{1j}G^{(1)}(\mathbf{x},\mathbf{O}) - D_{2j}G^{(2)}(\mathbf{x},\mathbf{O}) , \quad \boldsymbol{\xi} \in \partial\omega , \tag{6.210}$$

$$M^{(j)}(\boldsymbol{\xi}) \to \mathbf{O} \text{ as } |\boldsymbol{\xi}| \to \infty , \tag{6.211}$$

for $j = 1,2$.

Since ω_ε is a small inclusion, we may rewrite the boundary condition (6.210) for $M^{(j)}(\boldsymbol{\xi})$ by considering $G^{(j)}(\mathbf{x},\mathbf{O})$, $j = 1,2$ as follows. Using

$$G^{(j)}(\mathbf{x},\mathbf{O}) = \gamma^{(j)}(\mathbf{x},\mathbf{O}) - H^{(j)}(\mathbf{x},\mathbf{O}) , \quad j = 1,2 , \tag{6.212}$$

where $\gamma^{(j)}$ is the jth column of $\gamma = \{\gamma_{ij}\}_{i,j=1}^2$ and the fact the components of $H^{(j)}(\mathbf{x},\mathbf{O})$ are smooth functions for $\mathbf{x}, \mathbf{y} \in \Omega$, on $\partial\omega_\varepsilon$ we may expand these about $\mathbf{O}$, to give

$$\begin{aligned} G^{(j)}(\mathbf{x},\mathbf{O}) = &-K_2\log\varepsilon\,\mathbf{e}^{(j)} + \gamma^{(j)}(\boldsymbol{\xi},\mathbf{O}) - H^{(j)}(\mathbf{O},\mathbf{O}) \\ &+O(\varepsilon) , \quad j = 1,2 , \end{aligned} \tag{6.213}$$

where K_2 is the constant given in (6.181).

Then using (6.213) we have from (6.210)

$$M^{(j)}(\boldsymbol{\xi}) = \mathbf{e}^{(j)} + D_{1j}\left(K_2 \log\varepsilon\,\mathbf{e}^{(1)} - \gamma^{(1)}(\boldsymbol{\xi},\mathbf{O}) + H^{(1)}(\mathbf{O},\mathbf{O})\right)$$
$$+D_{2j}\left(K_2 \log\varepsilon\,\mathbf{e}^{(2)} - \gamma^{(2)}(\boldsymbol{\xi},\mathbf{O}) + H^{(2)}(\mathbf{O},\mathbf{O})\right)$$
$$+O(\varepsilon)\,, \tag{6.214}$$

for $\boldsymbol{\xi} \in \partial\omega$.

The vectors $\zeta^{(j)}(\boldsymbol{\xi})$ satisfy (6.194)–(6.196). Setting

$$\Upsilon^{(j)}(\boldsymbol{\xi}) = \zeta^{(j)}(\boldsymbol{\xi}) + \gamma^{(j)}(\boldsymbol{\xi},\mathbf{O}) - \zeta^{(\infty,j)}\,, j = 1,2\,, \tag{6.215}$$

we have that $\Upsilon^{(j)}(\boldsymbol{\xi})$ satisfies

$$L(\partial_{\boldsymbol{\xi}})\Upsilon^{(j)}(\boldsymbol{\xi}) = \mathbf{O}, \quad \boldsymbol{\xi} \in C\bar{\omega}\,, \tag{6.216}$$

$$\Upsilon^{(j)}(\boldsymbol{\xi}) = \gamma^{(j)}(\boldsymbol{\xi},\mathbf{O}) - \zeta^{(\infty,j)}\,, \quad \boldsymbol{\xi} \in \partial C\bar{\omega}\,, \tag{6.217}$$

$$\Upsilon^{(j)}(\boldsymbol{\xi}) \to \mathbf{O} \quad \text{as} \quad |\boldsymbol{\xi}| \to \infty\,, \tag{6.218}$$

for $j = 1,2$.

Substituting the boundary condition (6.217), for $\Upsilon^{(j)}(\boldsymbol{\xi})$ on $\partial C\bar{\omega}$, into (6.214) we have

$$M^{(j)}(\boldsymbol{\xi}) = \mathbf{e}^{(j)} + D_{1j}\left(K_2 \log\varepsilon\,\mathbf{e}^{(1)} - (\Upsilon^{(1)}(\boldsymbol{\xi}) + \zeta^{(\infty,1)}) + H^{(1)}(\mathbf{O},\mathbf{O})\right)$$
$$+D_{2j}\left(K_2 \log\varepsilon\,\mathbf{e}^{(2)} - (\Upsilon^{(2)}(\boldsymbol{\xi}) + \zeta^{(\infty,2)}) + H^{(2)}(\mathbf{O},\mathbf{O})\right)$$
$$+O(\varepsilon)\,, \tag{6.219}$$

for $\boldsymbol{\xi} \in \partial C\bar{\omega}$. The boundary layer $M^{(j)}(\boldsymbol{\xi})$ is sought in the form

$$M^{(j)}(\boldsymbol{\xi}) = -D_{1j}\Upsilon^{(1)}(\boldsymbol{\xi}) - D_{2j}\Upsilon^{(2)}(\boldsymbol{\xi}) + W^{(j)}(\boldsymbol{\xi})\,, \quad j = 1,2\,, \tag{6.220}$$

where $W^{(j)}(\boldsymbol{\xi})$ is a solution of

$$L(\partial_{\boldsymbol{\xi}})W^{(j)}(\boldsymbol{\xi}) = \mathbf{O}, \quad \boldsymbol{\xi} \in C\bar{\omega}\,, \tag{6.221}$$

$$W^{(j)}(\boldsymbol{\xi}) = \mathbf{e}^{(j)} + D_{1j}\left(K_2 \log\varepsilon\,\mathbf{e}^{(1)} - \zeta^{(\infty,1)} + H^{(1)}(\mathbf{O},\mathbf{O})\right)$$
$$+D_{2j}\left(K_2 \log\varepsilon\,\mathbf{e}^{(2)} - \zeta^{(\infty,2)} + H^{(2)}(\mathbf{O},\mathbf{O})\right)\,, \tag{6.222}$$

for $\boldsymbol{\xi} \in \partial C\bar{\omega}$, and

$$W^{(j)}(\boldsymbol{\xi}) \to \mathbf{O} \quad \text{as} \quad |\boldsymbol{\xi}| \to \infty\,. \tag{6.223}$$

We choose $D = [D_{ij}]$, $i, j = 1, 2$ as follows,

$$D = [D^{(1)}, D^{(2)}] = -A^{-1} , \tag{6.224}$$

where $A = [A_{ij}]_{i,j=1}^2$, whose entries are given by

$$A_{ij} = K_2 \log \varepsilon \, \delta_{ij} - \zeta_{ij}^{\infty} + H_{ij}(\mathbf{O}, \mathbf{O}) , \quad i, j = 1, 2 . \tag{6.225}$$

Choosing D as in (6.224) we have from (6.221)–(6.223), $W^{(j)}(\boldsymbol{\xi}) \equiv \mathbf{O}$, $j = 1, 2$, and the form of the constant matrix D (given by (6.53)–(6.55)) has been proved.

Combining (6.208) and (6.220) in

$$P_\varepsilon^{(j)}(\mathbf{x}) = V^{(j)}(\mathbf{x}) + M^{(j)}(\boldsymbol{\xi}) + p^{(j)}(\mathbf{x}) ,$$

where $p^{(j)}(\mathbf{x})$ is the remainder term, we have (6.204).

Estimating the Remainder Term

The remainder $p(\mathbf{x}) = [p^{(1)}(\mathbf{x}), p^{(2)}(\mathbf{x})]$ satisfies

$$L(\partial_{\mathbf{x}}) p(\mathbf{x}) = 0 I_2 , \quad \mathbf{x} \in \Omega_\varepsilon , \tag{6.226}$$

$$p(\mathbf{x}) = (\zeta(\boldsymbol{\xi}) + \gamma(\boldsymbol{\xi}, \mathbf{O}) - \zeta^{\infty}) D , \quad \mathbf{x} \in \partial\Omega , \tag{6.227}$$

$$p(\mathbf{x}) = I_2 - (-K_2 \log \varepsilon I_2 + \zeta^{\infty} - H(\mathbf{x}, \mathbf{O}))\, D , \quad \mathbf{x} \in \partial\omega_\varepsilon . \tag{6.228}$$

For the boundary condition on $\partial\omega_\varepsilon$, using (6.224) and (6.225)

$$p(\mathbf{x}) = (H(\mathbf{x}, \mathbf{O}) - H(\mathbf{O}, \mathbf{O})) D , \quad \mathbf{x} \in \partial\omega_\varepsilon . \tag{6.229}$$

Since the components of $H(\mathbf{x}, \mathbf{O})$ are smooth for $\mathbf{x}$, $\mathbf{y} \in \Omega$

$$H(\mathbf{x}, \mathbf{O}) - H(\mathbf{O}, \mathbf{O}) = O(\varepsilon) , \text{ as } \mathbf{x} \in \partial\omega_\varepsilon .$$

Next we consider the matrix D. Comparing to (6.54) we have $K_1^{-1} = (\det A)^{-1}$, is of $O((\log \varepsilon)^{-2})$, from which we see $D = O(|\log \varepsilon|^{-1})$. Thus we have the right-hand side of (6.228) is $O(\varepsilon |\log \varepsilon|^{-1})$.

Using Lemma 6.14

$$\zeta(\boldsymbol{\xi}) + \gamma(\boldsymbol{\xi}, \mathbf{O}) - \zeta^{\infty} = O(\varepsilon) , \quad \text{for } \mathbf{x} \in \partial\Omega , \tag{6.230}$$

and therefore again we have the right-hand side of (6.227) is $O(\varepsilon |\log \varepsilon|^{-1})$.

Thus by the Lemma 6.1 we have

$$p(\mathbf{x}) = O(\varepsilon |\log \varepsilon|^{-1}) \quad \text{for} \quad \mathbf{x} \in \Omega_\varepsilon . \qquad \square$$

6.4.4 A Uniform Asymptotic Formula for Green's Tensor G_ε in Two Dimensions

We are now in a position to formulate and prove the result concerning the uniform approximation of the tensor G_ε for the case of two dimensions.

Theorem 6.2. *Green's tensor G_ε for the Lamé operator in $\Omega_\varepsilon \subset \mathbb{R}^2$ admits the representation*

$$G_\varepsilon(\mathbf{x},\mathbf{y}) = G(\mathbf{x},\mathbf{y}) + g(\boldsymbol{\xi},\boldsymbol{\eta}) - \gamma(\boldsymbol{\xi},\boldsymbol{\eta}) + P_\varepsilon(\mathbf{x}) A P_\varepsilon^T(\mathbf{y}) - \zeta^T(\boldsymbol{\eta}) - \zeta(\boldsymbol{\xi}) + \zeta^\infty + O(\varepsilon)\,, \tag{6.231}$$

which is uniform with respect to $(\mathbf{x},\mathbf{y}) \in \Omega_\varepsilon \times \Omega_\varepsilon$.

Proof. Let G_ε be given by

$$G_\varepsilon(\mathbf{x},\mathbf{y}) = \gamma(\mathbf{x},\mathbf{y}) - H_\varepsilon(\mathbf{x},\mathbf{y}) - h_\varepsilon(\mathbf{x},\mathbf{y})\,, \tag{6.232}$$

where the columns of $H_\varepsilon(\mathbf{x},\mathbf{y})$ and $h_\varepsilon(\mathbf{x},\mathbf{y})$ are solutions of the boundary value problems

$$L(\partial_\mathbf{x}) H_\varepsilon^{(j)}(\mathbf{x},\mathbf{y}) = \mathbf{O}\,, \quad \mathbf{x},\mathbf{y} \in \Omega_\varepsilon\,, \tag{6.233}$$

$$H_\varepsilon^{(j)}(\mathbf{x},\mathbf{y}) = \gamma^{(j)}(\mathbf{x},\mathbf{y})\,, \quad \mathbf{x} \in \partial\Omega, \mathbf{y} \in \Omega_\varepsilon\,, \tag{6.234}$$

$$H_\varepsilon^{(j)}(\mathbf{x},\mathbf{y}) = \mathbf{O}\,, \quad \mathbf{x} \in \partial\omega_\varepsilon, \mathbf{y} \in \Omega_\varepsilon\,, \tag{6.235}$$

and

$$L(\partial_\mathbf{x}) h_\varepsilon^{(j)}(\mathbf{x},\mathbf{y}) = \mathbf{O}\,, \quad \mathbf{x},\mathbf{y} \in \Omega_\varepsilon\,, \tag{6.236}$$

$$h_\varepsilon^{(j)}(\mathbf{x},\mathbf{y}) = \mathbf{O}\,, \quad \mathbf{x} \in \partial\Omega, \mathbf{y} \in \Omega_\varepsilon\,, \tag{6.237}$$

$$h_\varepsilon^{(j)}(\mathbf{x},\mathbf{y}) = \gamma^{(j)}(\mathbf{x},\mathbf{y})\,, \quad \mathbf{x} \in \partial\omega_\varepsilon, \mathbf{y} \in \Omega_\varepsilon\,, \tag{6.238}$$

for $j = 1, 2$.

The Approximation of $H_\varepsilon(\mathbf{x},\mathbf{y})$

Let $H_\varepsilon^{(j)}(\mathbf{x},\mathbf{y})$ be represented in the form

$$H_\varepsilon^{(j)}(\mathbf{x},\mathbf{y}) = S_{1j}(\mathbf{y},\log\varepsilon) G^{(1)}(\mathbf{x},\mathbf{O}) + S_{2j}(\mathbf{y},\log\varepsilon) G^{(2)}(\mathbf{x},\mathbf{O}) + H^{(j)}(\mathbf{x},\mathbf{y}) + R_\varepsilon^{(j)}(\mathbf{x},\mathbf{y},\log\varepsilon)\,, \tag{6.239}$$

where $S_{ij}(\mathbf{y},\log\varepsilon)$, $i,j = 1,2$ are to be determined. In (6.239), the term $R_\varepsilon^{(j)}(\mathbf{x},\mathbf{y},\log\varepsilon)$ satisfies the boundary value problem

$$L(\partial_{\mathbf{x}})R_\varepsilon^{(j)}(\mathbf{x},\mathbf{y},\log\varepsilon) = \mathbf{O}\,, \mathbf{x},\mathbf{y} \in \Omega_\varepsilon\,, \tag{6.240}$$

$$R_\varepsilon^{(j)}(\mathbf{x},\mathbf{y},\log\varepsilon) = \mathbf{O}\,, \quad \mathbf{x} \in \partial\Omega, \mathbf{y} \in \Omega_\varepsilon\,, \tag{6.241}$$

$$R_\varepsilon^{(j)}(\mathbf{x},\mathbf{y},\log\varepsilon) = -S_{1j}G^{(1)}(\mathbf{x},\mathbf{O}) - S_{2j}G^{(2)}(\mathbf{x},\mathbf{O}) - H^{(j)}(\mathbf{x},\mathbf{y})\,,$$
$$\mathbf{x} \in \partial\omega_\varepsilon, \mathbf{y} \in \Omega_\varepsilon\,, \tag{6.242}$$

and is approximated by $R^{(j)}(\boldsymbol{\xi},\mathbf{y},\log\varepsilon)$, which is a solution of

$$L(\partial_{\boldsymbol{\xi}})R^{(j)}(\boldsymbol{\xi},\mathbf{y},\log\varepsilon) = \mathbf{O}\,, \boldsymbol{\xi} \in C\bar{\omega}\,, \tag{6.243}$$

$$\begin{aligned} R^{(j)}(\boldsymbol{\xi},\mathbf{y},\log\varepsilon) = {} & S_{1j}\left(K_2\log\varepsilon\,\mathbf{e}^{(1)} - \gamma^{(1)}(\boldsymbol{\xi},\mathbf{O}) + H^{(1)}(\mathbf{O},\mathbf{O})\right) \\ & + S_{2j}\left(K_2\log\varepsilon\,\mathbf{e}^{(2)} - \gamma^{(2)}(\boldsymbol{\xi},\mathbf{O}) + H^{(2)}(\mathbf{O},\mathbf{O})\right) \\ & - H^{(j)}(\mathbf{O},\mathbf{y})\,, \qquad \boldsymbol{\xi} \in \partial C\bar{\omega}\,, \end{aligned} \tag{6.244}$$

$$R^{(j)}(\boldsymbol{\xi},\mathbf{y},\log\varepsilon) \to \mathbf{O} \text{ as } |\boldsymbol{\xi}| \to \infty\,, \tag{6.245}$$

where $\mathbf{y} \in \Omega_\varepsilon$. We represent the solution of (6.243)–(6.244) as

$$\begin{aligned} R^{(j)}(\boldsymbol{\xi},\mathbf{y},\log\varepsilon) = {} & S_{1j}\left(K_2\log\varepsilon\,\mathbf{e}^{(1)} - \gamma^{(1)}(\boldsymbol{\xi},\mathbf{O}) + H^{(1)}(\mathbf{O},\mathbf{O}) - \zeta^{(1)}(\boldsymbol{\xi})\right) \\ & + S_{2j}\left(K_2\log\varepsilon\,\mathbf{e}^{(2)} - \gamma^{(2)}(\boldsymbol{\xi},\mathbf{O}) + H^{(2)}(\mathbf{O},\mathbf{O}) - \zeta^{(2)}(\boldsymbol{\xi})\right) \\ & - H^{(j)}(\mathbf{O},\mathbf{y})\,. \end{aligned} \tag{6.246}$$

Now, using the condition (6.196) of $\zeta(\boldsymbol{\xi})$ at infinity, in (6.246), we deduce that in order that (6.245) be satisfied we must choose the columns of S as follows

$$S(\mathbf{y},\log\varepsilon) = [S^{(1)}(\mathbf{y},\log\varepsilon), S^{(2)}(\mathbf{y},\log\varepsilon)] = -DH(\mathbf{O},\mathbf{y})\,, \tag{6.247}$$

where the entries of D are given by (6.53)–(6.55).

Combining (6.246), (6.247) in (6.239), we have

$$\begin{aligned} H_\varepsilon^{(j)}(\mathbf{x},\mathbf{y}) \sim {} & S_{1j}G^{(1)}(\mathbf{x},\mathbf{O}) + S_{2j}G^{(2)}(\mathbf{x},\mathbf{O}) \\ & + S_{1j}\left(K_2\log\varepsilon\,\mathbf{e}^{(1)} - \gamma^{(1)}(\boldsymbol{\xi},\mathbf{O}) + H^{(1)}(\mathbf{O},\mathbf{O}) - \zeta^{(1)}(\boldsymbol{\xi})\right) \\ & + S_{2j}\left(K_2\log\varepsilon\,\mathbf{e}^{(2)} - \gamma^{(2)}(\boldsymbol{\xi},\mathbf{O}) + H^{(2)}(\mathbf{O},\mathbf{O}) - \zeta^{(2)}(\boldsymbol{\xi})\right) \\ & - H^{(j)}(\mathbf{O},\mathbf{y}) + H^{(j)}(\mathbf{x},\mathbf{y}) \end{aligned}$$

or using Lemma 6.16

$$H_\varepsilon^{(j)}(\mathbf{x},\mathbf{y}) = -P_\varepsilon(\mathbf{x})H^{(j)}(\mathbf{O},\mathbf{y}) + H^{(j)}(\mathbf{x},\mathbf{y}) + \mathfrak{H}_\varepsilon^{(j)}(\mathbf{x},\mathbf{y})\,. \tag{6.248}$$

Here $\mathfrak{H}_\varepsilon^{(j)}(\mathbf{x},\mathbf{y})$ satisfies

$$L(\partial_{\mathbf{x}})\mathfrak{H}_\varepsilon^{(j)}(\mathbf{x},\mathbf{y}) = \mathbf{O}\,, \quad \mathbf{x},\mathbf{y} \in \Omega_\varepsilon\,, \tag{6.249}$$

$$\mathfrak{H}_\varepsilon^{(j)}(\mathbf{x},\mathbf{y}) = H^{(j)}(\mathbf{O},\mathbf{y}) - H^{(j)}(\mathbf{x},\mathbf{y})\,, \quad \mathbf{x} \in \partial\omega_\varepsilon\,, \mathbf{y} \in \Omega_\varepsilon\,, \tag{6.250}$$

$$\mathfrak{H}_\varepsilon^{(j)}(\mathbf{x},\mathbf{y}) = \mathbf{O}\,, \quad \mathbf{x} \in \partial\Omega\,, \mathbf{y} \in \Omega_\varepsilon\,, \tag{6.251}$$

where the right-hand side of the boundary condition (6.250) is $O(\varepsilon)$, uniformly with respect to $\mathbf{x} \in \partial\omega_\varepsilon$ and $\mathbf{y} \in \Omega_\varepsilon$.

Using Lemma 6.1 we obtain $\mathfrak{H}_\varepsilon(\mathbf{x},\mathbf{y}) = O(\varepsilon)$ for $\mathbf{x},\mathbf{y} \in \Omega_\varepsilon$.

The Approximation of $h_\varepsilon(\mathbf{x},\mathbf{y})$

Now we shall proceed to approximate h_ε. The columns of $h_\varepsilon(\mathbf{x},\mathbf{y})$ satisfy the homogeneous Dirichlet condition on $\partial\Omega$ and for $\mathbf{x} \in \partial\omega_\varepsilon$ we rewrite the boundary condition (6.238) as

$$h_\varepsilon^{(j)}(\mathbf{x},\mathbf{y}) = -K_2 \log\varepsilon\, \mathbf{e}^{(j)} + \gamma^{(j)}(\boldsymbol{\xi},\boldsymbol{\eta})\,, \quad \mathbf{x} \in \partial\omega_\varepsilon\,, \mathbf{y} \in \Omega_\varepsilon\,.$$

Let $h_\varepsilon^{(j)}(\mathbf{x},\mathbf{y})$ be sought in the form

$$h_\varepsilon^{(j)}(\mathbf{x},\mathbf{y}) = -K_2 \log\varepsilon\, \mathbf{e}^{(j)} + h^{(j)}(\boldsymbol{\xi},\boldsymbol{\eta}) + \chi_\varepsilon^{(j)}(\mathbf{x},\mathbf{y})\,, \tag{6.252}$$

where the vector field $\chi_\varepsilon^{(j)}(\mathbf{x},\mathbf{y})$ satisfies

$$L(\partial_{\mathbf{x}})\chi_\varepsilon^{(j)}(\mathbf{x},\mathbf{y}) = \mathbf{O}\,, \quad \mathbf{x},\mathbf{y} \in \Omega_\varepsilon\,, \tag{6.253}$$

$$\chi_\varepsilon^{(j)}(\mathbf{x},\mathbf{y}) = \mathbf{O}\,, \quad \mathbf{x} \in \partial\omega_\varepsilon, \mathbf{y} \in \Omega_\varepsilon\,, \tag{6.254}$$

$$\chi_\varepsilon^{(j)}(\mathbf{x},\mathbf{y}) = K_2 \log\varepsilon\, \mathbf{e}^{(j)} - h^{(j)}(\boldsymbol{\xi},\boldsymbol{\eta})\,, \quad \mathbf{x} \in \partial\Omega, \mathbf{y} \in \Omega_\varepsilon\,. \tag{6.255}$$

Using Lemma 6.13, we rewrite (6.255) as

$$\chi_\varepsilon^{(j)}(\mathbf{x},\mathbf{y}) = -\gamma^{(j)}(\mathbf{x},\mathbf{O}) + \zeta^{T(j)}(\boldsymbol{\eta}) + O(\varepsilon)\,, \quad \mathbf{x} \in \partial\Omega, \mathbf{y} \in \Omega_\varepsilon\,. \tag{6.256}$$

From the definition of $H(\mathbf{x},\mathbf{y})$ and the elastic capacitary potential we write $\chi_\varepsilon^{(j)}(\mathbf{x},\mathbf{y})$ as

$$\chi_\varepsilon^{(j)}(\mathbf{x},\mathbf{y}) = -H^{(j)}(\mathbf{x},\mathbf{O}) + (I_2 - P_\varepsilon(\mathbf{x}))\zeta^{T(j)}(\boldsymbol{\eta}) + \mathfrak{h}_\varepsilon^{(j)}(\mathbf{x},\mathbf{y})\,, \mathbf{x},\mathbf{y} \in \Omega_\varepsilon\,, \tag{6.257}$$

where $\mathfrak{h}_\varepsilon^{(j)}(\mathbf{x},\mathbf{y})$ satisfies the homogeneous Lamé equation; by Lemma 6.13 is $O(\varepsilon)$ for $\mathbf{x} \in \partial\Omega$, $\mathbf{y} \in \Omega_\varepsilon$ and

$$\mathfrak{h}_\varepsilon^{(j)}(\mathbf{x},\mathbf{y}) = H^{(j)}(\mathbf{x},\mathbf{O}) = H^{(j)}(\mathbf{O},\mathbf{O}) + O(\varepsilon)\,, \tag{6.258}$$

for $\mathbf{x} \in \partial\omega_\varepsilon$, $\mathbf{y} \in \Omega_\varepsilon$. Therefore, using the elastic capacitary potential $P_\varepsilon(\mathbf{x})$, we write

$$\mathfrak{h}_\varepsilon^{(j)}(\mathbf{x},\mathbf{y}) = P_\varepsilon(\mathbf{x})H^{(j)}(\mathbf{O},\mathbf{O}) + O(\varepsilon)\,, \tag{6.259}$$

which is uniform with respect to $\mathbf{x}, \mathbf{y} \in \Omega_\varepsilon$, by Lemma 6.1.

Collecting now (6.257), (6.259) in (6.252) we have

$$\begin{aligned} h_\varepsilon^{(j)}(\mathbf{x},\mathbf{y}) = {} & h^{(j)}(\boldsymbol{\xi},\boldsymbol{\eta}) - K_2 \log\varepsilon\, \mathbf{e}^{(j)} \\ & -H^{(j)}(\mathbf{x},\mathbf{O}) + (I_2 - P_\varepsilon(\mathbf{x}))\zeta^{T(j)}(\boldsymbol{\eta}) \\ & +P_\varepsilon(\mathbf{x})H^{(j)}(\mathbf{O},\mathbf{O}) + O(\varepsilon)\,. \end{aligned} \tag{6.260}$$

Combined Formula

Substituting (6.248), (6.260) in (6.232) we have the columns of Green's tensor for the Lamé operator in the domain Ω_ε

$$\begin{aligned} G_\varepsilon^{(j)}(\mathbf{x},\mathbf{y}) = {} & \gamma^{(j)}(\mathbf{x},\mathbf{y}) - H^{(j)}(\mathbf{x},\mathbf{y}) - h^{(j)}(\boldsymbol{\xi},\boldsymbol{\eta}) \\ & +K_2\log\varepsilon\,\mathbf{e}^{(j)} + H^{(j)}(\mathbf{x},\mathbf{O}) - \zeta^{T(j)}(\boldsymbol{\eta}) \\ & -P_\varepsilon(\mathbf{x})(H^{(j)}(\mathbf{O},\mathbf{O}) - \zeta^{T(j)}(\boldsymbol{\eta}) - H^{(j)}(\mathbf{O},\mathbf{y})) + O(\varepsilon) \\ = {} & \gamma^{(j)}(\mathbf{x},\mathbf{y}) - H^{(j)}(\mathbf{x},\mathbf{y}) - h^{(j)}(\boldsymbol{\xi},\boldsymbol{\eta}) + K_2\log\varepsilon\,\mathbf{e}^{(j)} \\ & +(I_2 - P_\varepsilon(\mathbf{x}))(H^{(j)}(\mathbf{O},\mathbf{O}) - \zeta^{T(j)}(\boldsymbol{\eta}) - H^{(j)}(\mathbf{O},\mathbf{y})) \\ & +H^{(j)}(\mathbf{x},\mathbf{O}) + H^{(j)}(\mathbf{O},\mathbf{y}) - H^{(j)}(\mathbf{O},\mathbf{O}) + O(\varepsilon)\,. \end{aligned} \tag{6.261}$$

Using the relation

$$H(\mathbf{O},\mathbf{O}) - \zeta^T(\boldsymbol{\eta}) - H(\mathbf{O},\mathbf{y}) = A(I_2 - P_\varepsilon^T(\mathbf{y})) + O(\varepsilon)\,, \tag{6.262}$$

obtained from the approximation of P_ε in Lemma 6.16, we have

$$\begin{aligned} G_\varepsilon^{(j)}(\mathbf{x},\mathbf{y}) = {} & \gamma^{(j)}(\mathbf{x},\mathbf{y}) - H^{(j)}(\mathbf{x},\mathbf{y}) - h^{(j)}(\boldsymbol{\xi},\boldsymbol{\eta}) \\ & +K_2\log\varepsilon\,\mathbf{e}^{(j)} + (I_2 - P_\varepsilon(\mathbf{x}))A(\mathbf{e}^{(j)} - P_\varepsilon^{T(j)}(\mathbf{y})) \\ & +H^{(j)}(\mathbf{x},\mathbf{O}) + H^{(j)}(\mathbf{O},\mathbf{y}) - H^{(j)}(\mathbf{O},\mathbf{O}) + O(\varepsilon) \\ = {} & \gamma^{(j)}(\mathbf{x},\mathbf{y}) - H^{(j)}(\mathbf{x},\mathbf{y}) - h^{(j)}(\boldsymbol{\xi},\boldsymbol{\eta}) \\ & +P_\varepsilon(\mathbf{x})AP_\varepsilon^{T(j)}(\mathbf{y}) - \zeta^{T(j)}(\boldsymbol{\eta}) - \zeta^{(j)}(\boldsymbol{\xi}) \\ & +\zeta^{(\infty,j)} + O(\varepsilon)\,, \end{aligned} \tag{6.263}$$

which is (6.231). The proof is complete. □

6.5 Simplified Asymptotic Formulae Subject to Constraints on Independent Variables for Green's Tensors in Domains with a Single Inclusion

It is now of interest to see how the asymptotic formulae obtained in Theorems 6.1 and 6.2, simplify under constraints on the points $\mathbf{x}$, $\mathbf{y} \in \Omega_\varepsilon$, where $\Omega_\varepsilon \subset \mathbb{R}^n$, $n = 2, 3$. We consider two situations, the first is when these points are located outside a small neighborhood of the inclusion, the second is when the points are in the vicinity of the inclusion.

We now turn to the case of three dimensions.

Corollary 6.1. (a) *Let* $\mathbf{x}$ *and* $\mathbf{y}$ *be points of* $\Omega_\varepsilon \subset \mathbb{R}^3$, *such that*

$$\min\{|\mathbf{x}|, |\mathbf{y}|\} > 2\varepsilon \,. \tag{6.264}$$

Then $G_\varepsilon(\mathbf{x}, \mathbf{y})$ *admits the representation*

$$G_\varepsilon(\mathbf{x}, \mathbf{y}) = G(\mathbf{x}, \mathbf{y}) - \varepsilon G(\mathbf{x}, \mathbf{O}) B G(\mathbf{O}, \mathbf{y}) + O(\varepsilon^2 (|\mathbf{x}||\mathbf{y}| \min\{|\mathbf{x}|, |\mathbf{y}|\})^{-1}) \,. \tag{6.265}$$

(b) *If* $\max\{|\mathbf{x}|, |\mathbf{y}|\} < 1/2$, *then*

$$\begin{aligned} G_\varepsilon(\mathbf{x}, \mathbf{y}) &= \varepsilon^{-1} g(\varepsilon^{-1}\mathbf{x}, \varepsilon^{-1}\mathbf{y}) - (I_3 - P(\varepsilon^{-1}\mathbf{x})) H(\mathbf{O}, \mathbf{O}) (I_3 - P^T(\varepsilon^{-1}\mathbf{y})) \\ &\quad + O(\max\{|\mathbf{x}|, |\mathbf{y}|\}) \,. \end{aligned} \tag{6.266}$$

Both (6.265) and (6.266) are uniform with respect to $\mathbf{x}, \mathbf{y} \in \Omega_\varepsilon$.

Proof. a) We may rewrite (6.153) as follows

$$\begin{aligned} G_\varepsilon(\mathbf{x}, \mathbf{y}) = {} & G(\mathbf{x}, \mathbf{y}) - \varepsilon^{-1} h(\varepsilon^{-1}\mathbf{x}, \varepsilon^{-1}\mathbf{y}) \\ & + P(\varepsilon^{-1}\mathbf{x}) H(\mathbf{O}, \mathbf{y}) + H(\mathbf{x}, \mathbf{O}) P^T(\varepsilon^{-1}\mathbf{y}) \\ & - P(\varepsilon^{-1}\mathbf{x}) H(\mathbf{O}, \mathbf{O}) P^T(\varepsilon^{-1}\mathbf{y}) - \varepsilon H(\mathbf{x}, \mathbf{O}) B H(\mathbf{O}, \mathbf{y}) \\ & + O\left(\varepsilon^2 (\min\{|\mathbf{x}|, |\mathbf{y}|\})^{-1}\right) \,. \end{aligned} \tag{6.267}$$

From Lemma 6.6, we have for $|\mathbf{x}| > 2\,\varepsilon$

$$P(\varepsilon^{-1}\mathbf{x}) = \varepsilon \Gamma(\mathbf{x}, \mathbf{O}) B + O\left(\varepsilon^2 |\mathbf{x}|^{-2}\right) \,. \tag{6.268}$$

Also, by Lemma 6.11 we have

$$\begin{aligned} \varepsilon^{-1} h(\varepsilon^{-1}\mathbf{x}, \varepsilon^{-1}\mathbf{y}) &= \varepsilon^{-1} \Gamma(\varepsilon^{-1}\mathbf{x}, \mathbf{O}) P^T(\varepsilon^{-1}\mathbf{y}) + O\left(\varepsilon^2 (|\mathbf{x}|^2 |\mathbf{y}|)^{-1}\right) \\ &= \varepsilon^{-1} \Gamma(\varepsilon^{-1}\mathbf{x}, \mathbf{O}) B \Gamma(\varepsilon^{-1}\mathbf{y}, \mathbf{O}) \\ &\quad + O\left(\varepsilon^2 (|\mathbf{x}||\mathbf{y}| \min\{|\mathbf{x}|, |\mathbf{y}|\})^{-1}\right) \,. \end{aligned} \tag{6.269}$$

By substitution of (6.268) and (6.269) into (6.267) we have

$$\begin{aligned} G_\varepsilon(\mathbf{x},\mathbf{y}) = {} & G(\mathbf{x},\mathbf{y}) - \varepsilon^{-1}\Gamma(\varepsilon^{-1}\mathbf{x},\mathbf{O})B\Gamma(\varepsilon^{-1}\mathbf{y},\mathbf{O}) \\ & +\varepsilon\Gamma(\mathbf{x},\mathbf{O})BH(\mathbf{O},\mathbf{y}) + \varepsilon H(\mathbf{x},\mathbf{O})B\Gamma(\mathbf{y},\mathbf{O}) \\ & -\varepsilon H(\mathbf{x},\mathbf{O})BH(\mathbf{O},\mathbf{y}) \\ & +O(\varepsilon^2(|\mathbf{x}||\mathbf{y}|\min\{|\mathbf{x}|,|\mathbf{y}|\})^{-1}) , \end{aligned} \tag{6.270}$$

which is equivalent to

$$\begin{aligned} G_\varepsilon(\mathbf{x},\mathbf{y}) = {} & G(\mathbf{x},\mathbf{y}) - \Gamma(\varepsilon^{-1}\mathbf{x},\mathbf{O})BG(\mathbf{O},\mathbf{y}) \\ & +\varepsilon H(\mathbf{x},\mathbf{O})BG(\mathbf{O},\mathbf{y}) \\ & +O\left(\varepsilon^2(|\mathbf{x}||\mathbf{y}|\min\{|\mathbf{x}|,|\mathbf{y}|\})^{-1}\right) , \end{aligned} \tag{6.271}$$

and from this we obtain (6.265).

(*b*) Since the components of $H(\mathbf{x},\mathbf{y})$ are smooth for $\mathbf{x}, \mathbf{y} \in \Omega$, expanding these components about $(\mathbf{O},\mathbf{O}) \in \Omega \times \Omega$, we may rewrite (6.153) as

$$\begin{aligned} G_\varepsilon(\mathbf{x},\mathbf{y}) = {} & \varepsilon^{-1}g(\varepsilon^{-1}\mathbf{x},\varepsilon^{-1}\mathbf{y}) - H(\mathbf{O},\mathbf{O}) \\ & +(H(\mathbf{O},\mathbf{O}) + O(|\mathbf{x}|))P^T(\varepsilon^{-1}\mathbf{y}) + P(\varepsilon^{-1}\mathbf{x})(H(\mathbf{O},\mathbf{O}) + O(|\mathbf{y}|)) \\ & -P(\varepsilon^{-1}\mathbf{x})H(\mathbf{O},\mathbf{O})P^T(\varepsilon^{-1}\mathbf{y}) + O(\max\{|\mathbf{x}|,|\mathbf{y}|\}) , \end{aligned} \tag{6.272}$$

from which (6.266) follows. □

Next we shall simplify the asymptotic formula given in (6.231) for the case of two dimensions under the same conditions on the points $\mathbf{x}$ and $\mathbf{y}$.

Corollary 6.2. *(a) Let* $\mathbf{x},\mathbf{y} \in \Omega_\varepsilon \subset \mathbb{R}^2$ *such that*

$$\min\{|\mathbf{x}|,|\mathbf{y}|\} > 2\varepsilon . \tag{6.273}$$

Then

$$G_\varepsilon(\mathbf{x},\mathbf{y}) = G(\mathbf{x},\mathbf{y}) - G(\mathbf{x},\mathbf{O})DG(\mathbf{O},\mathbf{y}) + O(\varepsilon(\min\{|\mathbf{x}|,|\mathbf{y}|\})^{-1}) . \tag{6.274}$$

(b) If $\max\{|\mathbf{x}|,|\mathbf{y}|\} < 1/2$, *then*

$$G_\varepsilon(\mathbf{x},\mathbf{y}) = g(\boldsymbol{\xi},\boldsymbol{\eta}) - \zeta(\boldsymbol{\xi})D\zeta^T(\boldsymbol{\eta}) + O(\max\{|\mathbf{x}|,|\mathbf{y}|\}) . \tag{6.275}$$

Both (6.274) and (6.275) are uniform with respect to $(\mathbf{x},\mathbf{y}) \in \Omega_\varepsilon \times \Omega_\varepsilon$.

Proof. (*a*) By Lemma 6.13,

$$h(\boldsymbol{\xi},\boldsymbol{\eta}) = \gamma(\boldsymbol{\xi},\mathbf{O}) - \zeta^T(\boldsymbol{\eta}) + O(|\boldsymbol{\xi}|^{-1}) . \tag{6.276}$$

Also from Lemma 6.14,

$$\zeta(\boldsymbol{\xi}) = -\gamma(\boldsymbol{\xi}, \mathbf{O}) + \zeta^{\infty} + O(|\boldsymbol{\xi}|^{-1}) . \tag{6.277}$$

Substituting (6.277) into (6.204) we obtain

$$P_{\varepsilon}(\mathbf{x}) = \left(G(\mathbf{x}, \mathbf{O}) + O\left(\varepsilon|\mathbf{x}|^{-1}\right)\right) D + O(\varepsilon|\log\varepsilon|^{-1}) . \tag{6.278}$$

Combining (6.276), (6.277) and (6.278) in (6.231), we have

$$\begin{aligned} G_{\varepsilon}(\mathbf{x}, \mathbf{y}) = G(\mathbf{x}, \mathbf{y}) &- (G(\mathbf{x}, \mathbf{O}) + O(\varepsilon|\mathbf{x}|^{-1}))D(G(\mathbf{O}, \mathbf{y}) + O(\varepsilon|\mathbf{y}|^{-1})) \\ &+ O(\varepsilon(\min\{|\mathbf{x}|, |\mathbf{y}|\})^{-1}) , \end{aligned} \tag{6.279}$$

from which we obtain (6.274).

(b) Rewriting formula (6.262) in the form

$$P_{\varepsilon}(\mathbf{x}) = I_2 - (H(\mathbf{O}, \mathbf{O}) - \zeta(\boldsymbol{\xi}) - H(\mathbf{x}, \mathbf{O}))A^{-1} + O(\varepsilon|\log\varepsilon|^{-1}) , \tag{6.280}$$

and substituting this into (6.231) for G_{ε}, we have

$$\begin{aligned} G_{\varepsilon}(\mathbf{x}, \mathbf{y}) = g(\boldsymbol{\xi}, \boldsymbol{\eta}) &- H(\mathbf{x}, \mathbf{y}) \\ &- (H(\mathbf{O}, \mathbf{O}) - \zeta(\boldsymbol{\xi}) - H(\mathbf{x}, \mathbf{O}))D(H(\mathbf{O}, \mathbf{O}) - \zeta^{T}(\boldsymbol{\eta}) - H(\mathbf{O}, \mathbf{y})) \\ &+ H(\mathbf{x}, \mathbf{O}) + H(\mathbf{O}, \mathbf{y}) - H(\mathbf{O}, \mathbf{O}) + O(\varepsilon) . \end{aligned} \tag{6.281}$$

Using the fact that the components of $H(\mathbf{x}, \mathbf{y})$ are smooth for $\mathbf{x}$, $\mathbf{y} \in \Omega$, in the vicinity of the origin we have from (6.281)

$$G(\mathbf{x}, \mathbf{y}) = g(\boldsymbol{\xi}, \boldsymbol{\eta}) - (O(|\mathbf{x}|) - \zeta(\boldsymbol{\xi}))D(O(|\mathbf{y}|) - \zeta^{T}(\boldsymbol{\eta})) + O(\max\{|\mathbf{x}|, |\mathbf{y}|\}) . \tag{6.282}$$

Since from (6.277), $\zeta(\boldsymbol{\xi}) = O\left(|\log(\varepsilon^{-1}|\mathbf{x}|)|\right)$ we have

$$G(\mathbf{x}, \mathbf{y}) = g(\boldsymbol{\xi}, \boldsymbol{\eta}) - \zeta(\boldsymbol{\xi})D\zeta^{T}(\boldsymbol{\eta}) + O(\max\{|\mathbf{x}|, |\mathbf{y}|\}) . \tag{6.283}$$

□

Chapter 7
Green's Tensor in Bodies with Multiple Rigid Inclusions

The results of the previous chapter have been extended here to the case of the elasticity equations in domains with multiple inclusions. Uniform asymptotic approximations have been derived for Green's tensors, taking into account interactions between different small inclusions. Both, three-dimensional and two-dimensional configurations have been considered.

7.1 Estimates for Solutions of the Homogeneous Lamé Equation in a Domain with Multiple Inclusions

In this section, we shall discuss an estimate, analogous to that of Lemma 6.1 of Chap. 6, concerning the solutions of the homogeneous Lamé equation for the Dirichlet problem, in domains with small inclusions. This estimate will aid us in obtaining the uniformity of the remainder estimates for Green's tensors in elastic solids with multiple inclusions.

Let $\mathbf{u}$ be the displacement vector which satisfies the Dirichlet boundary value problem in the domain $\Omega_\varepsilon \subset \mathbb{R}^n$, $n = 2, 3$

$$L\,(\partial_{\mathbf{x}})\,\mathbf{u}(\mathbf{x}) = \mathbf{O}\,, \quad \mathbf{x} \in \Omega_\varepsilon\,, \tag{7.1}$$

$$\mathbf{u}(\mathbf{x}) = \boldsymbol{\psi}(\mathbf{x})\,, \quad \mathbf{x} \in \partial\Omega\,, \tag{7.2}$$

$$\mathbf{u}(\mathbf{x}) = \boldsymbol{\varphi}_\varepsilon^{(j)}(\mathbf{x})\,, \quad \mathbf{x} \in \partial\omega_\varepsilon^{(j)}\,, 1 \le j \le N\,, \tag{7.3}$$

where $\mathbf{O}$ is the zero vector, $\boldsymbol{\varphi}_\varepsilon^{(j)} = \boldsymbol{\varphi}^{(j)}(\varepsilon^{-1}(\mathbf{x}-\mathbf{O}^{(j)}))$ and we assume that $\boldsymbol{\varphi}_\varepsilon^{(j)}$ and $\boldsymbol{\psi}$ are continuous vector functions.

Lemma 7.1. *There exists a unique solution* $\mathbf{u} \in C(\bar{\Omega}_\varepsilon)$ *of problem* (7.1)–(7.3) *which satisfies the estimate*

V. Maz'ya et al., *Green's Kernels and Meso-Scale Approximations in Perforated Domains*, Lecture Notes in Mathematics 2077, DOI 10.1007/978-3-319-00357-3_7,

$$\max_{\bar{\Omega}_\varepsilon} |\mathbf{u}(\mathbf{x})| \le \mathrm{Const}\max\{ \max_{1\le j\le N}\{\|\boldsymbol{\varphi}_\varepsilon^{(j)}\|_{C(\partial\omega_\varepsilon^{(j)})}\}, \|\boldsymbol{\psi}\|_{C(\partial\Omega)}\}. \tag{7.4}$$

In the proof of Lemma 7.1, we use $\Pi^{(j)}$ and Π_Ω that denote the inverse operators of the Dirichlet problem in the domains $C\bar{\omega}^{(j)}$ and in Ω, respectively, similar to those for the case of single inclusion given in Sect. 6.2.3. Also we set the operator $(\Pi_\varepsilon^{(j)}\boldsymbol{\varphi}_\varepsilon^{(j)})(\mathbf{x}) = (\Pi^{(j)}\boldsymbol{\varphi}^{(j)})(\varepsilon^{-1}(\mathbf{x}-\mathbf{O}^{(j)}))$, which corresponds to the inverse kernel of the Dirichlet problem in $C\bar{\omega}_\varepsilon^{(j)}$. Furthermore, let the vector functions $\mathbf{g}_\varepsilon^{(j)}(\mathbf{x}) = \mathbf{g}^{(j)}(\varepsilon^{-1}(\mathbf{x}-\mathbf{O}^{(j)}))$, be defined on $\partial\omega_\varepsilon^{(j)}$, $j = 1, \dots, N$.

We now present a proof for both $n = 2$ and $n = 3$. We note that the two-dimensional case requires the notion of elastic capacitary potentials. Using Fichera's maximum principle (see Lemma 6.2 of Chap. 6), we reduce the proof to the case of when $\boldsymbol{\psi} = \mathbf{O}$ in the boundary condition (7.2).

Proof of Lemma 7.1 for $n = 2$

The matrix $\mathcal{P}_\varepsilon$. We need an auxiliary 2×2 block matrix $\mathcal{P}_\varepsilon = (\mathcal{P}_\varepsilon^{(1)}, \dots, \mathcal{P}_\varepsilon^{(N)})$ whose appearance will become clearer in Sect. 7.2.3, where the elastic capacitary potentials $P_\varepsilon^{(j)}$ are introduced and it is shown that the entries of $\mathcal{P}_\varepsilon$ are the leading order parts in the asymptotic representation of $P_\varepsilon^{(j)}$. We set

$$\mathcal{P}_\varepsilon = \mathcal{F}_\varepsilon(I + \mathcal{G}D_\varepsilon)^{-1}, \tag{7.5}$$

where $I + \mathcal{G}D_\varepsilon$ is the $N \times N$ block matrix, whose (k, j) elements are the 2×2 matrices $\delta_{kj}I_2 + (1-\delta_{kj})G(\mathbf{O}^{(k)}, \mathbf{O}^{(j)})D_\varepsilon^{(j)}$. Here $D_\varepsilon = \mathrm{diag}\,\{D_\varepsilon^{(1)}, \dots, D_\varepsilon^{(N)}\}$, and the 2×2 blocks $D_\varepsilon^{(j)}$ have the entries

$$D_{11}^{(j)} = -(K_1^{(j)})^{-1}(K_2\log\varepsilon - \zeta_{22}^{(\infty,j)} + H_{22}(\mathbf{O}^{(j)}, \mathbf{O}^{(j)})), \tag{7.6}$$

$$D_{12}^{(j)} = -(K_1^{(j)})^{-1}(\zeta_{12}^{(\infty,j)} - H_{12}(\mathbf{O}^{(j)}, \mathbf{O}^{(j)})), \tag{7.7}$$

$$D_{21}^{(j)} = -(K_1^{(j)})^{-1}(\zeta_{21}^{(\infty,j)} - H_{21}(\mathbf{O}^{(j)}, \mathbf{O}^{(j)})), \tag{7.8}$$

$$D_{22}^{(j)} = -(K_1^{(j)})^{-1}(K_2\log\varepsilon - \zeta_{11}^{(\infty,j)} + H_{11}(\mathbf{O}^{(j)}, \mathbf{O}^{(j)})), \tag{7.9}$$

and for $j = 1, \dots, N$

$$\begin{aligned} K_1^{(j)} = &\Big(K_2\log\varepsilon - \zeta_{11}^{(\infty,j)} + H_{11}(\mathbf{O}^{(j)}, \mathbf{O}^{(j)})\Big) \\ &\times \Big(K_2\log\varepsilon - \zeta_{22}^{(\infty,j)} + H_{22}(\mathbf{O}^{(j)}, \mathbf{O}^{(j)})\Big) \\ &-(H_{12}(\mathbf{O}^{(j)}, \mathbf{O}^{(j)}) - \zeta_{12}^{(\infty,j)})(H_{21}(\mathbf{O}^{(j)}, \mathbf{O}^{(j)}) - \zeta_{21}^{(\infty,j)}), \end{aligned} \tag{7.10}$$

with

$$K_2 = \frac{\lambda + 3\mu}{4\pi\mu(\lambda + 2\mu)} \, . \tag{7.11}$$

Here $\zeta_{lk}^{(\infty,j)}$ are the entries of the constant matrix $\zeta^{(\infty,j)}$ analogous to that defined by (6.56) for the individual voids; and H_{lk} are the components of the regular part of Green's tensor in Ω.

The blocks $\mathcal{F}_\varepsilon^{(k)}$ of the matrix $\mathcal{F}_\varepsilon = (\mathcal{F}_\varepsilon^{(1)}, \dots, \mathcal{F}_\varepsilon^{(N)})$ are defined by

$$\begin{aligned} \mathcal{F}_\varepsilon^{(k)}(\mathbf{x}) = (G(\mathbf{x}, \mathbf{O}^{(k)}) - \zeta^{(k)}(\varepsilon^{-1}(\mathbf{x} - \mathbf{O}^{(k)})) \\ -\gamma(\varepsilon^{-1}(\mathbf{x} - \mathbf{O}^{(k)}), \mathbf{O}) + \zeta^{(\infty,k)}) D_\varepsilon^{(k)} \, , \end{aligned} \tag{7.12}$$

where G is Green's matrix in Ω. The matrix functions $\mathcal{P}_\varepsilon^{(j)}$ satisfy the homogeneous Lamé equation. Furthermore,

$$\mathcal{P}_\varepsilon^{(j)} = O(\varepsilon |\log \varepsilon|^{-1}) \, , \text{ for } \mathbf{x} \in \partial\Omega \, , \tag{7.13}$$

and

$$\mathcal{P}_\varepsilon^{(j)} = \delta_{jk} I_2 + O(\varepsilon |\log \varepsilon|^{-1}) \, , \text{ for } \mathbf{x} \in \partial\omega_\varepsilon^{(k)}, k = 1, \dots, N \, . \tag{7.14}$$

Homogeneous boundary condition on $\partial\Omega$. We consider the problem

$$L(\partial_\mathbf{x})\mathbf{u}(\mathbf{x}) = \mathbf{O} \, , \quad \mathbf{x} \in \Omega_\varepsilon \, , \tag{7.15}$$

$$\mathbf{u}(\mathbf{x}) = \boldsymbol{\varphi}^{(j)}(\varepsilon^{-1}(\mathbf{x} - \mathbf{O}^{(j)})) \, , \quad \mathbf{x} \in \partial\omega_\varepsilon^{(j)} \, , \tag{7.16}$$

$$\mathbf{u}(\mathbf{x}) = \mathbf{O} \, , \quad \mathbf{x} \in \partial\Omega \, . \tag{7.17}$$

We are looking for the approximate solution in the form

$$\begin{aligned} \mathbf{u} = \sum_{j=1}^N \Pi_\varepsilon^{(j)}(\mathbf{g}_\varepsilon^{(j)} - \mathcal{A}^{(j)}\mathbf{g}^{(j)}) + \sum_{j=1}^N \mathcal{P}_\varepsilon^{(j)} \mathcal{A}^{(j)} \mathbf{g}^{(j)} \\ -\Pi_\Omega \Big(\sum_{j=1}^N \mathrm{Tr}_{\partial\Omega} \Pi_\varepsilon^{(j)}(\mathbf{g}_\varepsilon^{(j)} - \mathcal{A}^{(j)}\mathbf{g}^{(j)}) \Big) \, , \end{aligned} \tag{7.18}$$

where $\mathbf{g}_\varepsilon^{(j)}(\mathbf{x}) = \mathbf{g}^{(j)}(\varepsilon^{-1}(\mathbf{x} - \mathbf{O}^{(j)}))$, $\mathbf{g}^{(j)}$ are unknown vector functions and the constant vectors $\mathcal{A}^{(j)}\mathbf{g}^{(j)}$ are analogous to that in the formula (6.64) of Sect. 6.2.3.

Evaluating the trace of (7.18) on $\partial\omega_\varepsilon^{(k)}$ we obtain

$$\boldsymbol{\varphi}_\varepsilon^{(k)} = \mathbf{g}_\varepsilon^{(k)} + S_\varepsilon^{(k)}(\mathbf{g}_\varepsilon^{(1)}, \mathbf{g}_\varepsilon^{(2)}, \dots, \mathbf{g}_\varepsilon^{(N)}) \, , \quad k = 1, 2, \dots, N \, , \tag{7.19}$$

where the operators $S_\varepsilon^{(k)}$ are defined by

$$
\begin{aligned}
S_\varepsilon^{(k)}(\mathbf{g}_\varepsilon^{(1)}, \mathbf{g}_\varepsilon^{(2)}, \dots, \mathbf{g}_\varepsilon^{(N)}) = {} & \mathrm{Tr}_{\partial\omega_\varepsilon^{(k)}}(\mathcal{P}_\varepsilon^{(k)} - I_2)\mathcal{A}^{(k)}\mathbf{g}^{(k)} \\
& + \sum_{\substack{j \neq k \\ 1 \le j \le N}} \mathrm{Tr}_{\partial\omega_\varepsilon^{(k)}}(\Pi_\varepsilon^{(j)}(\mathbf{g}_\varepsilon^{(j)} - \mathcal{A}^{(j)}\mathbf{g}^{(j)})) \\
& + \sum_{\substack{j \neq k \\ 1 \le j \le N}} \mathrm{Tr}_{\partial\omega_\varepsilon^{(k)}}(\mathcal{P}_\varepsilon^{(j)}\mathcal{A}^{(j)}\mathbf{g}^{(j)}) \\
& - \mathrm{Tr}_{\partial\omega_\varepsilon^{(k)}}\Pi_\Omega\Big(\sum_{1 \le j \le N} \mathrm{Tr}_{\partial\Omega}\Pi_\varepsilon^{(j)}(\mathbf{g}_\varepsilon^{(j)} - \mathcal{A}^{(j)}\mathbf{g}^{(j)})\Big) .
\end{aligned}
\tag{7.20}
$$

By (7.5)

$$
\|\mathrm{Tr}_{\partial\omega_\varepsilon^{(k)}}(\mathcal{P}_\varepsilon^{(k)} - I_2)\mathcal{A}^{(k)}\mathbf{g}^{(k)}\|_{C(\partial\omega_\varepsilon^{(k)})} \le \text{const}\ \varepsilon|\log\varepsilon|^{-1}\|\mathbf{g}_\varepsilon^{(k)}\|_{C(\partial\omega_\varepsilon^{(k)})} . \tag{7.21}
$$

Lemmas 6.2 and 6.3 imply

$$
\Big\| \sum_{\substack{j \neq k \\ 1 \le j \le N}} \mathrm{Tr}_{\partial\omega_\varepsilon^{(k)}}(\Pi_\varepsilon^{(j)}(\mathbf{g}_\varepsilon^{(j)} - \mathcal{A}^{(j)}\mathbf{g}^{(j)}))\Big\|_{C(\partial\omega_\varepsilon^{(k)})} \le \text{const}\ \varepsilon \max_{1 \le p \le N} \|\mathbf{g}_\varepsilon^{(p)}\|_{C(\partial\omega_\varepsilon^{(p)})} , \tag{7.22}
$$

and

$$
\begin{aligned}
&\Big\|\mathrm{Tr}_{\partial\omega_\varepsilon^{(k)}}\Pi_\Omega\Big(\sum_{1 \le j \le N} \mathrm{Tr}_{\partial\Omega}\Pi_\varepsilon^{(j)}(\mathbf{g}_\varepsilon^{(j)} - \mathcal{A}^{(j)}\mathbf{g}^{(j)})\Big)\Big\|_{C(\partial\omega_\varepsilon^{(k)})} \\
&\quad \le \text{const}\ \varepsilon \max_{1 \le p \le N} \|\mathbf{g}_\varepsilon^{(p)}\|_{C(\partial\omega_\varepsilon^{(p)})} .
\end{aligned}
\tag{7.23}
$$

According to (7.14),

$$
\|\mathrm{Tr}_{\partial\omega_\varepsilon^{(k)}}(\mathcal{P}_\varepsilon^{(j)}\mathcal{A}^{(j)}\mathbf{g}^{(j)})\|_{C(\partial\omega_\varepsilon^{(k)})} \le \text{const}\ \varepsilon|\log\varepsilon|^{-1}\|\mathbf{g}_\varepsilon^{(j)}\|_{C(\partial\omega_\varepsilon^{(j)})} , \tag{7.24}
$$

for $j \neq k$. Combining (7.21)–(7.24) we deduce

$$
\|S_\varepsilon^{(k)}\|_{C(\prod_{j=1}^N \partial\omega_\varepsilon^{(j)}) \to C(\partial\omega_\varepsilon^{(k)})} \le \text{const}\ \varepsilon . \tag{7.25}
$$

It follows from (7.19) that

$$
\mathbf{g}_\varepsilon = (I + S_\varepsilon)^{-1}\boldsymbol{\varphi}_\varepsilon , \tag{7.26}
$$

where $\mathbf{g}_\varepsilon = (\mathbf{g}_\varepsilon^{(1)}, \dots, \mathbf{g}_\varepsilon^{(N)})^T$ and $\boldsymbol{\varphi}_\varepsilon = (\boldsymbol{\varphi}_\varepsilon^{(1)}, \dots, \boldsymbol{\varphi}_\varepsilon^{(N)})^T$, and S_ε is a matrix operator whose rows are $S_\varepsilon^{(1)}, \dots, S_\varepsilon^{(N)}$, then

$$
\|\mathbf{g}_\varepsilon^{(k)}\|_{C(\partial\omega_\varepsilon^{(k)})} \le \text{const} \max_{1 \le p \le N} \|\boldsymbol{\varphi}_\varepsilon^{(p)}\|_{C(\partial\omega_\varepsilon^{(p)})} . \tag{7.27}
$$

Owing to (7.18), (7.27) and Lemmas 6.2, 6.3 we obtain

$$\max_{\bar{\Omega}_\varepsilon} |\mathbf{u}| \le \text{const} \max_{1\le j\le N} \|\boldsymbol{\varphi}_\varepsilon^{(j)}\|_{C(\partial\omega_\varepsilon^{(j)})} \,. \tag{7.28}$$

Proof of Lemma 7.1 for $n = 3$

Let us look for a solution of the problem (7.1)–(7.3) in the form

$$\mathbf{u} = \sum_{j=1}^{N} \Pi_\varepsilon^{(j)} \mathbf{g}_\varepsilon^{(j)} - \Pi_\Omega \Big(\mathrm{Tr}_{\partial\Omega} \sum_{j=1}^{N} \Pi_\varepsilon^{(j)} \mathbf{g}_\varepsilon^{(j)} \Big) \,. \tag{7.29}$$

Evaluating the trace of (7.29) on $\partial\omega_\varepsilon^{(j)}$ we obtain

$$\boldsymbol{\varphi}_\varepsilon^{(j)} = \mathbf{g}_\varepsilon^{(j)} + S_\varepsilon^{(j)}(\mathbf{g}_\varepsilon^{(1)}, \mathbf{g}_\varepsilon^{(2)}, \ldots, \mathbf{g}_\varepsilon^{(N)}) \,, \tag{7.30}$$

where

$$S_\varepsilon^{(j)}(\mathbf{g}_\varepsilon^{(1)}, \mathbf{g}_\varepsilon^{(2)}, \ldots, \mathbf{g}_\varepsilon^{(N)}) = \mathrm{Tr}_{\partial\omega_\varepsilon^{(j)}} \Big(\sum_{\substack{k\neq j \\ 1\le k\le N}} \Pi_\varepsilon^{(k)} \mathbf{g}_\varepsilon^{(k)} \Big)$$
$$- \mathrm{Tr}_{\partial\omega_\varepsilon^{(j)}} \Pi_\Omega \Big(\mathrm{Tr}_{\partial\Omega} \sum_{k=1}^{N} \Pi_\varepsilon^{(k)} \mathbf{g}_\varepsilon^{(k)} \Big) \,. \tag{7.31}$$

By Lemma 6.3

$$\|\mathrm{Tr}_{\partial\omega_\varepsilon^{(j)}} (\Pi_\varepsilon^{(k)} \mathbf{g}_\varepsilon^{(k)})\|_{C(\partial\omega_\varepsilon^{(j)})} \le \text{Const}\, \varepsilon\, \|\mathbf{g}_\varepsilon^{(k)}\|_{C(\partial\omega_\varepsilon^{(k)})} \quad \text{when } k \neq j \,. \tag{7.32}$$

According to Fichera's maximum principle (Lemma 6.2, Chap. 6) and the estimate

$$\|\mathrm{Tr}_{\partial\Omega} \Pi_\varepsilon^{(k)} \mathbf{g}_\varepsilon^{(k)}\|_{C(\partial\Omega)} \le \text{Const}\, \varepsilon\, \|\mathbf{g}_\varepsilon^{(k)}\|_{C(\partial\omega_\varepsilon^{(k)})} \,, \tag{7.33}$$

combined with (7.32), we obtain

$$\|S_\varepsilon^{(j)}\|_{C(\prod_{j=1}^{N} \partial\omega_\varepsilon^{(j)}) \to C(\partial\omega_\varepsilon^{(j)})} \le \text{Const}\, \varepsilon \,. \tag{7.34}$$

Hence

$$\mathbf{g}_\varepsilon = (I + S_\varepsilon)^{-1} \boldsymbol{\varphi}_\varepsilon \,, \tag{7.35}$$

where $\mathbf{g}_\varepsilon = (\mathbf{g}_\varepsilon^{(1)}, \ldots, \mathbf{g}_\varepsilon^{(N)})^T$, $\boldsymbol{\varphi}_\varepsilon = (\boldsymbol{\varphi}_\varepsilon^{(1)}, \ldots, \boldsymbol{\varphi}_\varepsilon^{(N)})^T$ and S_ε is the matrix operator whose rows are $S_\varepsilon^{(1)}, \ldots, S_\varepsilon^{(N)}$, and the estimate

$$\|\mathbf{g}_\varepsilon^{(j)}\|_{C(\partial\omega_\varepsilon^{(j)})} \le \text{Const} \max_{1\le k\le N} \|\boldsymbol{\varphi}_\varepsilon^{(k)}\|_{C(\partial\omega_\varepsilon^{(k)})} \tag{7.36}$$

holds. By (7.36) and Lemmas 6.2, 6.3 we deduce

$$\max_{\bar{\Omega}_\varepsilon} |\mathbf{u}| \le \text{Const} \max_{1\le j\le N} \|\boldsymbol{\varphi}_\varepsilon^{(j)}\|_{C(\partial\omega_\varepsilon^{(j)})} . \tag{7.37}$$

7.2 Green's Tensor for the Lamé Operator in Two-Dimensional Elasticity

In the subsequent sections we study Green's tensor for the Lamé operator in $\Omega_\varepsilon \subset \mathbb{R}^n$, $n = 2, 3$ which will be denoted by G_ε. The tensor G_ε is a solution of

$$\mu\Delta_{\mathbf{x}}G_\varepsilon(\mathbf{x},\mathbf{y})+(\lambda+\mu)\nabla_{\mathbf{x}}(\nabla_{\mathbf{x}}\cdot G_\varepsilon(\mathbf{x},\mathbf{y}))+\delta(\mathbf{x}-\mathbf{y})I_n = 0I_n , \quad \mathbf{x},\mathbf{y}\in\Omega_\varepsilon , \tag{7.38}$$

$$G_\varepsilon(\mathbf{x},\mathbf{y}) = 0I_n , \quad \mathbf{x}\in\partial\Omega_\varepsilon, \mathbf{y}\in\Omega_\varepsilon , \tag{7.39}$$

where I_n is the $n\times n$ identity matrix, and this tensor satisfies the following symmetry relation

$$G_\varepsilon(\mathbf{x},\mathbf{y}) = (G_\varepsilon(\mathbf{y},\mathbf{x}))^T , \quad \mathbf{x},\mathbf{y}\in\Omega_\varepsilon, \mathbf{x}\ne\mathbf{y} . \tag{7.40}$$

We once again use the notation $L(\partial_{\mathbf{x}})$ for the Lamé operator given by (6.11) of Chap. 6.

We present an asymptotic representation for the Green's tensor of the Lamé operator in two dimensions, in this section, and in three dimensions given in Sect. 7.3.

7.2.1 Green's Matrix for a Two-Dimensional Domain with Several Small Inclusions

In this section, we consider the uniform approximation of the tensor $G_\varepsilon(\mathbf{x},\mathbf{y})$ for the case of a planar domain with multiple small inclusions ($n = 2$), defined in Sect. 7.2. We once again introduce model domains and governing equations needed for the study related to this case.

7.2.2 Green's Kernels for Model Domains in Two Dimensions

Let $G(\mathbf{x},\mathbf{y}) = [G^{(1)}(\mathbf{x},\mathbf{y}), G^{(2)}(\mathbf{x},\mathbf{y})]$ and $g^{(j)}(\boldsymbol{\xi}_j,\boldsymbol{\eta}_j) = [g^{(j,1)}(\boldsymbol{\xi}_j,\boldsymbol{\eta}_j), g^{(j,2)}(\boldsymbol{\xi}_j,\boldsymbol{\eta}_j)]$ now denote Green's tensors for the Lamé operator in the domain Ω and

$C\bar{\omega}^{(j)} = \mathbb{R}^2 \backslash \bar{\omega}^{(j)}$, $j = 1, \ldots, N$, respectively. The tensor G is a solution the following problem

$$L(\partial_{\mathbf{x}})G(\mathbf{x},\mathbf{y}) + \delta(\mathbf{x}-\mathbf{y})I_2 = 0I_2 , \quad \mathbf{x},\mathbf{y} \in \Omega , \tag{7.41}$$

$$G(\mathbf{x},\mathbf{y}) = 0I_2 , \quad \mathbf{x} \in \partial\Omega, \mathbf{y} \in \Omega , \tag{7.42}$$

and the tensors $g^{(j)}$ solve

$$L(\partial_{\boldsymbol{\xi}_j})g^{(j)}(\boldsymbol{\xi}_j,\boldsymbol{\eta}_j) + \delta(\boldsymbol{\xi}_j - \boldsymbol{\eta}_j)I_2 = 0I_2 , \quad \boldsymbol{\xi}_j,\boldsymbol{\eta}_j \in C\bar{\omega}^{(j)} , \tag{7.43}$$

$$g^{(j)}(\boldsymbol{\xi}_j,\boldsymbol{\eta}_j) = 0I_2 , \quad \boldsymbol{\xi}_j \in \partial C\bar{\omega}^{(j)}, \boldsymbol{\eta}_j \in C\bar{\omega}^{(j)} , \tag{7.44}$$

$$|g^{(j,k)}(\boldsymbol{\xi}_j,\boldsymbol{\eta}_j)| \text{ is bounded as } |\boldsymbol{\xi}_j| \to \infty , \boldsymbol{\eta}_j \in C\bar{\omega}^{(j)} \text{ for } k = 1,2 . \tag{7.45}$$

From the definition (7.41) and (7.42), we have that G satisfies the symmetry relation (6.176) of Chap. 6, in the domain Ω; and from (7.43)–(7.45), for the tensor $g^{(j)}$, the following relation holds

$$g^{(j)}(\boldsymbol{\xi}_j,\boldsymbol{\eta}_j) = (g^{(j)}(\boldsymbol{\eta}_j,\boldsymbol{\xi}_j))^T, \quad \boldsymbol{\xi}_j,\boldsymbol{\eta}_j \in C\bar{\omega}^{(j)}, \boldsymbol{\xi}_j \neq \boldsymbol{\eta}_j . \tag{7.46}$$

We represent $G(\mathbf{x},\mathbf{y})$ as

$$G(\mathbf{x},\mathbf{y}) = \gamma(\mathbf{x},\mathbf{y}) - H(\mathbf{x},\mathbf{y}) , \tag{7.47}$$

and $g^{(j)}(\boldsymbol{\xi}_j,\boldsymbol{\eta}_j)$ for $j = 1, \ldots, N$ as

$$g^{(j)}(\boldsymbol{\xi}_j,\boldsymbol{\eta}_j) = \gamma(\boldsymbol{\xi}_j,\boldsymbol{\eta}_j) - h^{(j)}(\boldsymbol{\xi}_j,\boldsymbol{\eta}_j) , \tag{7.48}$$

where H and $h^{(j)}$ are the regular parts of G and $g^{(j)}$, respectively, and $\gamma(\mathbf{x},\mathbf{y}) = [\gamma_{ij}(\mathbf{x},\mathbf{y})]_{i,j=1}^2$, is the fundamental solution of the Lamé operator in two dimensions with components

$$\begin{aligned}\gamma_{ij}(\mathbf{x},\mathbf{y}) = {} & K_2(-\log|\mathbf{x}-\mathbf{y}|\delta_{ij} \\ & +(\lambda+\mu)(\lambda+3\mu)^{-1}(x_i-y_i)(x_j-y_j)|\mathbf{x}-\mathbf{y}|^{-2}) ,\end{aligned} \tag{7.49}$$

for $i, j = 1, 2$, where

$$K_2 = \frac{\lambda + 3\mu}{4\pi\mu(\lambda + 2\mu)} . \tag{7.50}$$

We introduce the tensor $\zeta^{(j)}$ as

$$\zeta^{(j)}(\boldsymbol{\eta}_j) = \lim_{|\boldsymbol{\xi}_j|\to\infty} g^{(j)}(\boldsymbol{\eta}_j,\boldsymbol{\xi}_j) , \tag{7.51}$$

and the constant matrix

$$\zeta^{(\infty,j)} = \lim_{|\boldsymbol{\eta}_j|\to\infty} \{\zeta^{(j)}(\boldsymbol{\eta}_j) + \gamma(\boldsymbol{\eta}_j, \mathbf{O})\} , \tag{7.52}$$

for $j = 1, \dots, N$.

In Chap. 6, it was proved that the matrices $\zeta^{(\infty,j)}$, $1 \le j \le N$, were symmetric.

7.2.3 Auxiliary Matrix Functions for Two-Dimensional Elasticity

An Estimate for the Regular Part $h^{(j)}$ of Green's Tensor for the Unbounded Domain

Here we state a result concerning an asymptotic expansion of the regular part $h^{(j)}$ of Green's tensor $g^{(j)}$, which is consequence of Lemma 2 presented in Kondratiev and Oleinik [18], (p. 78).

The proofs of the following Lemmas are identical to those for Lemmas 6.13 and 6.14 of Sect. 6.4, for the case of single inclusion, with some obvious modifications.

Lemma 7.2. *Let $|\boldsymbol{\xi}_j| > 2$. Then the regular part $h^{(j)}(\boldsymbol{\xi}_j, \boldsymbol{\eta}_j)$ of Green's matrix $g^{(j)}(\boldsymbol{\xi}_j, \boldsymbol{\eta}_j)$, in $C\bar{\omega}^{(j)}$ admits the asymptotic representation*

$$h^{(j)}(\boldsymbol{\xi}_j, \boldsymbol{\eta}_j) = \gamma(\boldsymbol{\xi}_j, \mathbf{O}) - \zeta^{(j)T}(\boldsymbol{\eta}_j) + O(|\boldsymbol{\xi}_j|^{-1}) , \tag{7.53}$$

for $j = 1, \dots, N$.

We also have the following asymptotic representation of the matrix function $\zeta^{(j)}$

Lemma 7.3. *For $|\boldsymbol{\xi}_j| > 2$, the asymptotic approximation for $\zeta^{(j)}$ holds*

$$\zeta^{(j)}(\boldsymbol{\xi}_j) = -\gamma(\boldsymbol{\xi}_j, \mathbf{O}) + \zeta^{(\infty,j)} + O(|\boldsymbol{\xi}_j|^{-1}) , \tag{7.54}$$

for $j = 1, \dots, N$.

7.2.3.1 The Elastic Capacitary Potential Matrix

Let $P_\varepsilon^{(j)}(\mathbf{x})$ be the elastic capacitary potential corresponding to the jth inclusion. The matrix $P_\varepsilon^{(j)}(\mathbf{x})$ is defined as a solution of

$$L(\partial_{\mathbf{x}}) P_\varepsilon^{(j)}(\mathbf{x}) = 0 I_2, \quad \mathbf{x} \in \Omega_\varepsilon , \tag{7.55}$$

$$P_\varepsilon^{(j)}(\mathbf{x}) = 0 I_2 , \quad \mathbf{x} \in \partial\Omega , \tag{7.56}$$

$$P_\varepsilon^{(j)}(\mathbf{x}) = I_2\,, \quad \mathbf{x} \in \partial\omega_\varepsilon^{(j)}\,, \tag{7.57}$$

$$P_\varepsilon^{(j)}(\mathbf{x}) = 0I_2\,, \quad \mathbf{x} \in \partial\omega_\varepsilon^{(k)}\,, 1 \le k \le N\,, k \ne j\,. \tag{7.58}$$

Given the above boundary value problem, we now consider the approximation of the matrix $P_\varepsilon^{(j)}(\mathbf{x})$.

Lemma 7.4. *The leading order part $\mathcal{P}_\varepsilon^{(j)}$ of the asymptotic approximation of $P_\varepsilon^{(j)}(\mathbf{x})$ is a solution of the following system of equations*

$$\mathcal{P}_\varepsilon^{(j)}(\mathbf{x}) + \sum_{k=1}^{N} \mathcal{P}_\varepsilon^{(k)}(\mathbf{x})(1-\delta_{jk})G(\mathbf{O}^{(k)}, \mathbf{O}^{(j)})D^{(j)} \tag{7.59}$$
$$= \Big(G(\mathbf{x}, \mathbf{O}^{(j)}) - \zeta^{(j)}(\boldsymbol{\xi}_j) - \gamma(\boldsymbol{\xi}_j, \mathbf{O}) + \zeta^{(\infty,j)}\Big)D^{(j)}\,,$$

where $D^{(j)} = [D_{ik}^{(j)}]_{i,k=1}^2$ has entries given by

$$D_{11}^{(j)} = -(K_1^{(j)})^{-1}(K_2 \log\varepsilon - \zeta_{22}^{(\infty,j)} + H_{22}(\mathbf{O}^{(j)}, \mathbf{O}^{(j)}))\,, \tag{7.60}$$

$$D_{12}^{(j)} = -(K_1^{(j)})^{-1}(\zeta_{12}^{(\infty,j)} - H_{12}(\mathbf{O}^{(j)}, \mathbf{O}^{(j)}))\,, \tag{7.61}$$

$$D_{21}^{(j)} = -(K_1^{(j)})^{-1}(\zeta_{21}^{(\infty,j)} - H_{21}(\mathbf{O}^{(j)}, \mathbf{O}^{(j)}))\,, \tag{7.62}$$

$$D_{22}^{(j)} = -(K_1^{(j)})^{-1}(K_2 \log\varepsilon - \zeta_{11}^{(\infty,j)} + H_{11}(\mathbf{O}^{(j)}, \mathbf{O}^{(j)}))\,, \tag{7.63}$$

and

$$K_1^{(j)} = \Big(K_2 \log\varepsilon - \zeta_{11}^{(\infty,j)} + H_{11}(\mathbf{O}^{(j)}, \mathbf{O}^{(j)})\Big)$$
$$\times \Big(K_2 \log\varepsilon - \zeta_{22}^{(\infty,j)} + H_{22}(\mathbf{O}^{(j)}, \mathbf{O}^{(j)})\Big)$$
$$-(H_{12}(\mathbf{O}^{(j)}, \mathbf{O}^{(j)}) - \zeta_{12}^{(\infty,j)})(H_{21}(\mathbf{O}^{(j)}, \mathbf{O}^{(j)}) - \zeta_{21}^{(\infty,j)})\,, \tag{7.64}$$

for $j = 1, \dots, N$, and K_2 is given by (7.50).

Proof. We represent $P_\varepsilon^{(j)}(\mathbf{x})$ in the form

$$P_\varepsilon^{(j)}(\mathbf{x}) = (G(\mathbf{x}, \mathbf{O}^{(j)}) - \zeta^{(j)}(\boldsymbol{\xi}_j) - \gamma(\boldsymbol{\xi}_j, \mathbf{O}) + \zeta^{(\infty,j)})D^{(j)} + R_\varepsilon^{(j)}(\mathbf{x})\,, \tag{7.65}$$

for $1 \le j \le N$, where the matrix $R_\varepsilon^{(j)}(\mathbf{x})$ satisfies

$$L(\partial_\mathbf{x})R_\varepsilon^{(j)}(\mathbf{x}) = 0I_2, \quad \mathbf{x} \in \Omega_\varepsilon\,, \tag{7.66}$$

$$R_\varepsilon^{(j)}(\mathbf{x}) = (\zeta^{(j)}(\boldsymbol{\xi}_j) + \gamma(\boldsymbol{\xi}_j, \mathbf{O}) - \zeta^{(\infty,j)})D^{(j)}\,, \quad \mathbf{x} \in \partial\Omega\,, \tag{7.67}$$

$$R_\varepsilon^{(j)}(\mathbf{x}) = I_2 - \left(-K_2 \log\varepsilon I_2 - H(\mathbf{x}, \mathbf{O}^{(j)}) + \zeta^{(\infty,j)}\right) D^{(j)}\,, \quad \mathbf{x} \in \partial\omega_\varepsilon^{(j)}\,, \tag{7.68}$$

$$R_\varepsilon^{(j)}(\mathbf{x}) = -(G(\mathbf{x}, \mathbf{O}^{(j)}) - \zeta^{(j)}(\boldsymbol{\xi}_j) - \gamma(\boldsymbol{\xi}_j, \mathbf{O}) + \zeta^{(\infty,j)}) D^{(j)}\,,$$
$$\mathbf{x} \in \partial\omega_\varepsilon^{(k)}\,, 1 \le k \le N\,, k \ne j\,. \tag{7.69}$$

The boundary condition (7.68) is equivalent to

$$R_\varepsilon^{(j)}(\mathbf{x}) = (H(\mathbf{x}, \mathbf{O}^{(j)}) - H(\mathbf{O}^{(j)}, \mathbf{O}^{(j)})) D^{(j)}\,, \quad \mathbf{x} \in \partial\omega_\varepsilon^{(j)}\,, \tag{7.70}$$

where $D^{(j)} = O(|\log\varepsilon|^{-1})$, so $R_\varepsilon^{(j)}(\mathbf{x}) = O(\varepsilon|\log\varepsilon|^{-1})$ for $\mathbf{x} \in \partial\omega_\varepsilon^{(j)}$.

By Lemma 7.3

$$\zeta^{(j)}(\boldsymbol{\xi}_j) + \gamma(\boldsymbol{\xi}_j, \mathbf{O}) - \zeta^{(\infty,j)} = O(\varepsilon), \quad \text{for } \mathbf{x} \in \partial\Omega\,. \tag{7.71}$$

Then in (7.67), we have that $R_\varepsilon^{(j)}(\mathbf{x}) = O(\varepsilon|\log\varepsilon|^{-1})$ for $\mathbf{x} \in \partial\Omega$.

Next, using Lemma 7.3 and the fact that $G(\mathbf{x}, \mathbf{O}^{(j)})$ is smooth for $\mathbf{x} \in \Omega_\varepsilon$, we have in (7.69)

$$R_\varepsilon^{(j)}(\mathbf{x}) = -G(\mathbf{O}^{(k)}, \mathbf{O}^{(j)}) D^{(j)} + O(\varepsilon|\log\varepsilon|^{-1})\,, \tag{7.72}$$

for $\mathbf{x} \in \partial\omega_\varepsilon^{(k)}, 1 \le k \le N, k \ne j$.

Then we may write $R_\varepsilon^{(j)}(\mathbf{x})$, using the elastic capacitary potential for the individual inclusions, as

$$R_\varepsilon^{(j)}(\mathbf{x}) = -\sum_{\substack{k \ne j \\ 1 \le k \le N}} P_\varepsilon^{(k)}(\mathbf{x}) G(\mathbf{O}^{(k)}, \mathbf{O}^{(j)}) D^{(j)} + p_\varepsilon^{(j)}(\mathbf{x})\,. \tag{7.73}$$

Combining (7.65) and (7.73) we arrive at

$$P_\varepsilon^{(j)}(\mathbf{x}) = \Bigg(G(\mathbf{x}, \mathbf{O}^{(j)}) - \zeta^{(j)}(\boldsymbol{\xi}_j) - \gamma(\boldsymbol{\xi}_j, \mathbf{O}) + \zeta^{(\infty,j)}$$
$$- \sum_{\substack{k \ne j \\ 1 \le k \le N}} P_\varepsilon^{(k)}(\mathbf{x}) G(\mathbf{O}^{(k)}, \mathbf{O}^{(j)}) \Bigg) D^{(j)} + p_\varepsilon^{(j)}(\mathbf{x})\,. \tag{7.74}$$

Here $p_\varepsilon^{(j)}(\mathbf{x})$ is a matrix satisfying the homogeneous Lamé equation, and is $O(\varepsilon|\log\varepsilon|^{-1})$ for $\mathbf{x} \in \partial\Omega$ and $\mathbf{x} \in \partial\omega_\varepsilon^{(j)}$, $1 \le j \le N$. Therefore by Lemma 7.1, $p_\varepsilon^{(j)}(\mathbf{x})$ for $1 \le j \le N$ is $O(\varepsilon|\log\varepsilon|^{-1})$ uniformly with respect to $\mathbf{x} \in \Omega_\varepsilon$.

The removal of the remainder term in (7.74), gives the system (7.59). □

7.2.4 A Uniform Asymptotic Formula for Green's Tensor of Dirichlet Problem of Linear Elasticity in a Domain with Multiple Inclusions

Here we consider the approximation of Green's matrix G_ε for a two-dimensional elastic solid with multiple inclusions.

Theorem 7.1. *Green's tensor for the Lamé operator in $\Omega_\varepsilon \subset \mathbb{R}^2$ admits the representation*

$$G_\varepsilon(\mathbf{x},\mathbf{y}) = G(\mathbf{x},\mathbf{y}) + \sum_{j=1}^{N} g^{(j)}(\boldsymbol{\xi}_j, \boldsymbol{\eta}_j) - N\gamma(\varepsilon^{-1}\mathbf{x}, \varepsilon^{-1}\mathbf{y})$$

$$+ \sum_{j=1}^{N} \{P_\varepsilon^{(j)}(\mathbf{x})A^{(j)}P_\varepsilon^{(j)T}(\mathbf{y}) - \zeta^{(j)}(\boldsymbol{\xi}_j) - \zeta^{(j)T}(\boldsymbol{\eta}_j) + \zeta^{(\infty,j)}\}$$

$$- \sum_{j=1}^{N} \sum_{\substack{k\neq j\\ 1\le k\le N}} P_\varepsilon^{(j)}(\mathbf{x})G(\mathbf{O}^{(j)}, \mathbf{O}^{(k)})P_\varepsilon^{(k)T}(\mathbf{y}) + O(\varepsilon) , \tag{7.75}$$

uniformly with respect to $(\mathbf{x},\mathbf{y}) \in \Omega_\varepsilon \times \Omega_\varepsilon$, where

$$A^{(j)} = K_2 \log \varepsilon I_2 + H(\mathbf{O}^{(j)}, \mathbf{O}^{(j)}) - \zeta^{(\infty,j)} , \quad 1 \le j \le N . \tag{7.76}$$

Proof. Let G_ε be sought in the form

$$G_\varepsilon(\mathbf{x},\mathbf{y}) = \gamma(\mathbf{x},\mathbf{y}) - H_\varepsilon(\mathbf{x},\mathbf{y}) - \sum_{j=1}^{N} h_\varepsilon^{(j)}(\mathbf{x},\mathbf{y}) , \tag{7.77}$$

where it suffices to seek the approximation of the tensors $H_\varepsilon(\mathbf{x},\mathbf{y})$ and $h_\varepsilon^{(j)}(\mathbf{x},\mathbf{y})$, which solve the problems

$$L(\partial_\mathbf{x})H_\varepsilon(\mathbf{x},\mathbf{y}) = 0I_2 , \quad \mathbf{x},\mathbf{y} \in \Omega_\varepsilon , \tag{7.78}$$

$$H_\varepsilon(\mathbf{x},\mathbf{y}) = \gamma(\mathbf{x},\mathbf{y}) , \quad \mathbf{x} \in \partial\Omega, \mathbf{y} \in \Omega_\varepsilon , \tag{7.79}$$

$$H_\varepsilon(\mathbf{x},\mathbf{y}) = 0I_2 , \quad \mathbf{x} \in \partial\omega_\varepsilon^{(j)}, \mathbf{y} \in \Omega_\varepsilon, 1 \le j \le N , \tag{7.80}$$

and

$$L(\partial_\mathbf{x})h_\varepsilon^{(j)}(\mathbf{x},\mathbf{y}) = 0I_2 , \quad \mathbf{x},\mathbf{y} \in \Omega_\varepsilon , \tag{7.81}$$

$$h_\varepsilon^{(j)}(\mathbf{x},\mathbf{y}) = 0I_2 , \quad \mathbf{x} \in \partial\Omega, \mathbf{y} \in \Omega_\varepsilon , \tag{7.82}$$

$$h_\varepsilon^{(j)}(\mathbf{x},\mathbf{y}) = \gamma(\mathbf{x},\mathbf{y}) , \quad \mathbf{x} \in \partial\omega_\varepsilon^{(j)}, \mathbf{y} \in \Omega_\varepsilon , \tag{7.83}$$

$$h_\varepsilon^{(j)}(\mathbf{x},\mathbf{y}) = 0I_2 , \quad \mathbf{x} \in \partial\omega_\varepsilon^{(k)}, \mathbf{y} \in \Omega_\varepsilon , 1 \le k \le N , k \ne j . \tag{7.84}$$

The Approximation of $H_\varepsilon(\mathbf{x},\mathbf{y})$

Let $H_\varepsilon(\mathbf{x},\mathbf{y})$ be given by

$$H_\varepsilon(\mathbf{x},\mathbf{y}) = -P_\varepsilon^{(j)}(\mathbf{x})H(\mathbf{O}^{(j)},\mathbf{y}) + H(\mathbf{x},\mathbf{y}) + V_\varepsilon(\mathbf{x},\mathbf{y}) , \tag{7.85}$$

where the index j is fixed and $V(\mathbf{x},\mathbf{y})$ satisfies

$$L(\partial_\mathbf{x})V_\varepsilon(\mathbf{x},\mathbf{y}) = 0I_2 , \quad \mathbf{x},\mathbf{y} \in \Omega_\varepsilon , \tag{7.86}$$

$$V_\varepsilon(\mathbf{x},\mathbf{y}) = 0I_2 , \quad \mathbf{x} \in \partial\Omega, \mathbf{y} \in \Omega_\varepsilon , \tag{7.87}$$

$$V_\varepsilon(\mathbf{x},\mathbf{y}) = H(\mathbf{O}^{(j)},\mathbf{y}) - H(\mathbf{x},\mathbf{y}) , \quad \mathbf{x} \in \partial\omega_\varepsilon^{(j)}, \mathbf{y} \in \Omega_\varepsilon , \tag{7.88}$$

$$V_\varepsilon(\mathbf{x},\mathbf{y}) = -H(\mathbf{x},\mathbf{y}) , \quad \mathbf{x} \in \partial\omega_\varepsilon^{(k)}, \mathbf{y} \in \Omega_\varepsilon , k \ne j , 1 \le k \le N . \tag{7.89}$$

Since $\omega_\varepsilon^{(j)}$, $1 \le j \le N$, are small inclusions and H is a smooth tensor in Ω we may expand H about their centres. Namely, for the boundary condition (7.88) we have

$$V_\varepsilon(\mathbf{x},\mathbf{y}) = H(\mathbf{O}^{(j)},\mathbf{y}) - H(\mathbf{x},\mathbf{y}) = O(\varepsilon) , \quad \mathbf{x} \in \partial\omega_\varepsilon^{(j)}, \mathbf{y} \in \Omega_\varepsilon , \tag{7.90}$$

and from (7.89)

$$V_\varepsilon(\mathbf{x},\mathbf{y}) = -H(\mathbf{x},\mathbf{y}) = -H(\mathbf{O}^{(k)},\mathbf{y}) + O(\varepsilon) , \\ \mathbf{x} \in \partial\omega_\varepsilon^{(k)}, \mathbf{y} \in \Omega_\varepsilon , k \ne j , 1 \le k \le N . \tag{7.91}$$

Therefore, using the elastic capacitary potential of the individual inclusions, we represent the tensor $V_\varepsilon(\mathbf{x},\mathbf{y})$ as

$$V_\varepsilon(\mathbf{x},\mathbf{y}) = - \sum_{\substack{k \ne j \\ 1 \le k \le N}} P_\varepsilon^{(k)}(\mathbf{x})H(\mathbf{O}^{(k)},\mathbf{y}) + \mathfrak{H}_\varepsilon(\mathbf{x},\mathbf{y}) . \tag{7.92}$$

Substituting (7.92) into (7.85) we have

$$H_\varepsilon(\mathbf{x},\mathbf{y}) = -\sum_{j=1}^{N} P_\varepsilon^{(j)}(\mathbf{x})H(\mathbf{O}^{(j)},\mathbf{y}) + H(\mathbf{x},\mathbf{y}) + \mathfrak{H}_\varepsilon(\mathbf{x},\mathbf{y}) , \tag{7.93}$$

where $\mathfrak{H}_\varepsilon(\mathbf{x}, \mathbf{y})$ is the remainder term satisfying

$$L(\partial_{\mathbf{x}})\mathfrak{H}_\varepsilon(\mathbf{x}, \mathbf{y}) = 0I_2 , \quad \mathbf{x}, \mathbf{y} \in \Omega_\varepsilon , \tag{7.94}$$

$$\mathfrak{H}_\varepsilon(\mathbf{x}, \mathbf{y}) = 0I_2 , \quad \mathbf{x} \in \partial\Omega, \mathbf{y} \in \Omega_\varepsilon , \tag{7.95}$$

$$\begin{aligned} \mathfrak{H}_\varepsilon(\mathbf{x}, \mathbf{y}) &= H(\mathbf{O}^{(j)}, \mathbf{y}) - H(\mathbf{x}, \mathbf{y}) \\ &= O(\varepsilon) , \quad \mathbf{x} \in \partial\omega_\varepsilon^{(j)}, \mathbf{y} \in \Omega_\varepsilon, 1 \le j \le N . \end{aligned} \tag{7.96}$$

Therefore, by Lemma 7.1, we have $\mathfrak{H}_\varepsilon(\mathbf{x}, \mathbf{y}) = O(\varepsilon)$ uniformly with respect to $\mathbf{x}$ and $\mathbf{y}$ in Ω_ε.

The Approximation of $h_\varepsilon^{(j)}(\mathbf{x}, \mathbf{y})$

We begin by writing the boundary condition (7.83) on $\partial\omega_\varepsilon^{(j)}$ as

$$h_\varepsilon^{(j)}(\mathbf{x}, \mathbf{y}) = -K_2 \log \varepsilon I_2 + \gamma(\boldsymbol{\xi}_j, \boldsymbol{\eta}_j) , \quad \mathbf{x} \in \partial\omega_\varepsilon^{(j)} , \mathbf{y} \in \Omega_\varepsilon . \tag{7.97}$$

Thus we seek $h_\varepsilon^{(j)}(\mathbf{x}, \mathbf{y})$ in the form

$$h_\varepsilon^{(j)}(\mathbf{x}, \mathbf{y}) = -K_2 \log \varepsilon I_2 + h^{(j)}(\boldsymbol{\xi}_j, \boldsymbol{\eta}_j) + \chi_\varepsilon^{(j)}(\mathbf{x}, \mathbf{y}) , \tag{7.98}$$

for $\mathbf{x}, \mathbf{y} \in \Omega_\varepsilon$, where the remainder $\chi_\varepsilon^{(j)}$ satisfies

$$L(\partial_{\mathbf{x}})\chi_\varepsilon^{(j)}(\mathbf{x}, \mathbf{y}) = 0I_2 , \quad \mathbf{x}, \mathbf{y} \in \Omega_\varepsilon , \tag{7.99}$$

$$\chi_\varepsilon^{(j)}(\mathbf{x}, \mathbf{y}) = K_2 \log \varepsilon I_2 - h^{(j)}(\boldsymbol{\xi}_j, \boldsymbol{\eta}_j) , \quad \mathbf{x} \in \partial\Omega, \mathbf{y} \in \Omega_\varepsilon , \tag{7.100}$$

$$\chi_\varepsilon^{(j)}(\mathbf{x}, \mathbf{y}) = 0I_2 , \quad \mathbf{x} \in \partial\omega_\varepsilon^{(j)}, \mathbf{y} \in \Omega_\varepsilon , \tag{7.101}$$

$$\begin{aligned} \chi_\varepsilon^{(j)}(\mathbf{x}, \mathbf{y}) = {}& K_2 \log \varepsilon I_2 - h^{(j)}(\boldsymbol{\xi}_j, \boldsymbol{\eta}_j) , \quad \mathbf{x} \in \partial\omega_\varepsilon^{(k)}, \mathbf{y} \in \Omega_\varepsilon , \\ & 1 \le k \le N , k \ne j . \end{aligned} \tag{7.102}$$

Using Lemma 7.2, we rewrite boundary conditions (7.100) and (7.102) as

$$\chi_\varepsilon^{(j)}(\mathbf{x}, \mathbf{y}) = -\gamma(\mathbf{x}, \mathbf{O}^{(j)}) + \zeta^{(j)T}(\boldsymbol{\eta}_j) + O(\varepsilon) , \quad \mathbf{x} \in \partial\Omega, \mathbf{y} \in \Omega_\varepsilon , \tag{7.103}$$

and

$$\chi_\varepsilon^{(j)}(\mathbf{x}, \mathbf{y}) = -\gamma(\mathbf{x}, \mathbf{O}^{(j)}) + \zeta^{(j)T}(\boldsymbol{\eta}_j) + O(\varepsilon) , \tag{7.104}$$

for $\mathbf{x} \in \partial\omega_\varepsilon^{(k)}, \mathbf{y} \in \Omega_\varepsilon$, $1 \le k \le N$, $k \ne j$. Then, using the elastic capacitary potential, $\chi_\varepsilon^{(j)}$ is sought in the form

$$\chi_\varepsilon^{(j)}(\mathbf{x},\mathbf{y}) = -H(\mathbf{x},\mathbf{O}^{(j)}) + (I_2 - P_\varepsilon^{(j)}(\mathbf{x}))\zeta^{(j)T}(\boldsymbol{\eta}_j) + \mathfrak{h}_\varepsilon^{(j)}(\mathbf{x},\mathbf{y}) , \tag{7.105}$$

where the matrix $\mathfrak{h}_\varepsilon^{(j)}(\mathbf{x},\mathbf{y})$ satisfies

$$L(\partial_{\mathbf{x}})\mathfrak{h}_\varepsilon^{(j)}(\mathbf{x},\mathbf{y}) = 0I_2 , \quad \mathbf{x},\mathbf{y} \in \Omega_\varepsilon , \tag{7.106}$$

$$\mathfrak{h}_\varepsilon^{(j)}(\mathbf{x},\mathbf{y}) = O(\varepsilon) , \quad \mathbf{x} \in \partial\Omega, \mathbf{y} \in \Omega_\varepsilon , \tag{7.107}$$

$$\mathfrak{h}_\varepsilon^{(j)}(\mathbf{x},\mathbf{y}) = H(\mathbf{x},\mathbf{O}^{(j)}) , \quad \mathbf{x} \in \partial\omega_\varepsilon^{(j)}, \mathbf{y} \in \Omega_\varepsilon , \tag{7.108}$$

$$\mathfrak{h}_\varepsilon^{(j)}(\mathbf{x},\mathbf{y}) = -G(\mathbf{x},\mathbf{O}^{(j)}) + O(\varepsilon) , \quad \mathbf{x} \in \partial\omega_\varepsilon^{(k)}, \mathbf{y} \in \Omega_\varepsilon , 1 \le k \le N , k \ne j . \tag{7.109}$$

From the fact that $G(\mathbf{x},\mathbf{O}^{(j)})$ and its regular part are smooth in Ω_ε, in the vicinity of the small inclusions we expand these matrices about the centres of these inclusions, in such a way that boundary conditions (7.108) and (7.109) become

$$\mathfrak{h}_\varepsilon^{(j)}(\mathbf{x},\mathbf{y}) = H(\mathbf{O}^{(j)},\mathbf{O}^{(j)}) + O(\varepsilon) , \quad \mathbf{x} \in \partial\omega_\varepsilon^{(j)}, \mathbf{y} \in \Omega_\varepsilon , \tag{7.110}$$

$$\mathfrak{h}_\varepsilon^{(j)}(\mathbf{x},\mathbf{y}) = -G(\mathbf{O}^{(k)},\mathbf{O}^{(j)}) + O(\varepsilon) , \quad \mathbf{x} \in \partial\omega_\varepsilon^{(k)}, \mathbf{y} \in \Omega_\varepsilon , 1 \le k \le N , k \ne j . \tag{7.111}$$

Then, using the elastic capacitary potential, we represent $\mathfrak{h}_\varepsilon^{(j)}(\mathbf{x},\mathbf{y})$ as

$$\mathfrak{h}_\varepsilon^{(j)}(\mathbf{x},\mathbf{y}) = P_\varepsilon^{(j)}(\mathbf{x})H(\mathbf{O}^{(j)},\mathbf{O}^{(j)}) - \sum_{\substack{k \ne j \\ 1 \le k \le N}} P_\varepsilon^{(k)}(\mathbf{x})G(\mathbf{O}^{(k)},\mathbf{O}^{(j)}) + O(\varepsilon) , \tag{7.112}$$

which is uniform for $\mathbf{x},\mathbf{y} \in \Omega_\varepsilon$, by Lemma 7.1.

Placing (7.105) and (7.112) into (7.98), we obtain the approximation of $h_\varepsilon^{(j)}(\mathbf{x},\mathbf{y})$ in the form

$$\begin{aligned} h_\varepsilon^{(j)}(\mathbf{x},\mathbf{y}) = & -K_2 \log \varepsilon I_2 + h^{(j)}(\boldsymbol{\xi}_j,\boldsymbol{\eta}_j) - H(\mathbf{x},\mathbf{O}^{(j)}) \\ & + (I_2 - P_\varepsilon^{(j)}(\mathbf{x}))\zeta^{(j)T}(\boldsymbol{\eta}_j) + P_\varepsilon^{(j)}(\mathbf{x})H(\mathbf{O}^{(j)},\mathbf{O}^{(j)}) \\ & - \sum_{\substack{k \ne j \\ 1 \le k \le N}} P_\varepsilon^{(k)}(\mathbf{x})G(\mathbf{O}^{(k)},\mathbf{O}^{(j)}) + O(\varepsilon) . \end{aligned} \tag{7.113}$$

Combined Formula

Now substituting (7.93), (7.113) into (7.77) we obtain

$$G_\varepsilon(\mathbf{x},\mathbf{y}) = G(\mathbf{x},\mathbf{y}) + \sum_{j=1}^{N} g^{(j)}(\boldsymbol{\xi}_j, \boldsymbol{\eta}_j) - N\gamma(\mathbf{x},\mathbf{y})$$

$$+\sum_{j=1}^{N}(I_2 - P_\varepsilon^{(j)}(\mathbf{x}))(H(\mathbf{O}^{(j)},\mathbf{O}^{(j)}) - \zeta^{(j)T}(\boldsymbol{\eta}_j) - H(\mathbf{O}^{(j)},\mathbf{y}))$$

$$+\sum_{j=1}^{N}(H(\mathbf{x},\mathbf{O}^{(j)}) + H(\mathbf{O}^{(j)},\mathbf{y}) - H(\mathbf{O}^{(j)},\mathbf{O}^{(j)}))$$

$$+\sum_{j=1}^{N}\sum_{\substack{k\neq j\\1\le k\le N}} P_\varepsilon^{(k)}(\mathbf{x})G(\mathbf{O}^{(k)},\mathbf{O}^{(j)}) + O(\varepsilon)\,. \tag{7.114}$$

Using the following relation obtained from the approximation of $P_\varepsilon^{(j)}(\mathbf{x})$, (see (7.74))

$$(A^{(j)})^{-1}(H(\mathbf{O}^{(j)},\mathbf{O}^{(j)}) - \zeta^{(j)T}(\boldsymbol{\eta}_j) - H(\mathbf{O}^{(j)},\mathbf{y}))$$

$$= I_2 - P_\varepsilon^{(j)T}(\mathbf{y}) + \sum_{\substack{k\neq j\\1\le k\le N}} (A^{(j)})^{-1}G(\mathbf{O}^{(j)},\mathbf{O}^{(k)})P_\varepsilon^{(k)T}(\mathbf{y})$$

$$+O(\varepsilon|\log\varepsilon|^{-1})\,, \tag{7.115}$$

where $A^{(j)} = -(D^{(j)})^{-1}$, and substituting in (7.114) we have

$$G_\varepsilon(\mathbf{x},\mathbf{y}) = G(\mathbf{x},\mathbf{y}) + \sum_{j=1}^{N} g^{(j)}(\boldsymbol{\xi}_j, \boldsymbol{\eta}_j) - N\gamma(\mathbf{x},\mathbf{y})$$

$$+\sum_{j=1}^{N}(I_2 - P_\varepsilon^{(j)}(\mathbf{x}))A^{(j)}(I_2 - P_\varepsilon^{(j)T}(\mathbf{y}))$$

$$+\sum_{j=1}^{N}(H(\mathbf{x},\mathbf{O}^{(j)}) + H(\mathbf{O}^{(j)},\mathbf{y}) - H(\mathbf{O}^{(j)},\mathbf{O}^{(j)}))$$

$$+\sum_{j=1}^{N}\sum_{\substack{k\neq j\\1\le k\le N}} P_\varepsilon^{(k)}(\mathbf{x})G(\mathbf{O}^{(k)},\mathbf{O}^{(j)})$$

$$+\sum_{j=1}^{N}\sum_{\substack{k\neq j\\1\le k\le N}} (I_2 - P_\varepsilon^{(j)}(\mathbf{x}))G(\mathbf{O}^{(j)},\mathbf{O}^{(k)})P_\varepsilon^{(k)T}(\mathbf{y})$$

$$+O(\varepsilon)\,. \tag{7.116}$$

Then, using the approximation of the elastic capacitary potential to simplify the second sum in the right-hand side gives

$$\sum_{j=1}^{N}(I_2 - P_\varepsilon^{(j)}(\mathbf{x}))A^{(j)}(I_2 - P_\varepsilon^{(j)T}(\mathbf{y}))$$

$$= -\sum_{j=1}^{N}(H(\mathbf{x}, \mathbf{O}^{(j)}) + H(\mathbf{O}^{(j)}, \mathbf{y}) - H(\mathbf{O}^{(j)}, \mathbf{O}^{(j)}))$$

$$-\sum_{j=1}^{N}(\zeta^{(j)}(\boldsymbol{\xi}_j) + \zeta^{(j)T}(\boldsymbol{\eta}_j) - \zeta^{(\infty,j)}) - NK_2 \log \varepsilon I_2$$

$$-\sum_{j=1}^{N}\sum_{\substack{k\neq j\\1\leq k\leq N}}\{G(\mathbf{O}^{(j)}, \mathbf{O}^{(k)})P_\varepsilon^{(k)T}(\mathbf{y}) + P_\varepsilon^{(k)}(\mathbf{x})G(\mathbf{O}^{(k)}, \mathbf{O}^{(j)})\}$$

$$+\sum_{j=1}^{N}P_\varepsilon^{(j)}(\mathbf{x})A^{(j)}P_\varepsilon^{(j)T}(\mathbf{y}) + O(\varepsilon)\,. \tag{7.117}$$

Substitution of (7.117) in (7.116) yields the formula (7.75). The proof is complete. □

7.3 Green's Matrix for a Three-Dimensional Domain with Several Small Rigid Inclusions

Now that the study of the approximation of Green's kernel for the situation of plane strain of elasticity have been considered, we now produce an approximation of Green's matrix for the system of elasticity in a three-dimensional domain with multiple inclusions.

7.3.1 Green's Tensors for Model Domains in Three Dimensions

Let $G(\mathbf{x}, \mathbf{y}) = [G^{(1)}(\mathbf{x}, \mathbf{y}), G^{(2)}(\mathbf{x}, \mathbf{y}), G^{(3)}(\mathbf{x}, \mathbf{y})]$ and $g^{(j)}(\boldsymbol{\xi}_j, \boldsymbol{\eta}_j) = [g^{(j,1)}(\boldsymbol{\xi}_j, \boldsymbol{\eta}_j), g^{(j,2)}(\boldsymbol{\xi}_j, \boldsymbol{\eta}_j), g^{(j,3)}(\boldsymbol{\xi}_j, \boldsymbol{\eta}_j)]$ denote Green's tensors in the sets Ω and $C\bar{\omega}^{(j)} = \mathbb{R}^3\backslash\bar{\omega}^{(j)}$, $j = 1, \dots, N$, respectively, for the Lamé operator $L(\partial_{\mathbf{x}})$, for the case of three dimensions (see (6.11) of Chap. 6).

The tensor G solves the following problem

$$L(\partial_{\mathbf{x}})G(\mathbf{x},\mathbf{y})+\delta(\mathbf{x}-\mathbf{y})I_3=0I_3\ ,\quad \mathbf{x},\mathbf{y}\in\Omega\ , \tag{7.118}$$

$$G(\mathbf{x},\mathbf{y})=0I_3\ ,\quad \mathbf{x}\in\partial\Omega,\mathbf{y}\in\Omega\ , \tag{7.119}$$

and the tensors $g^{(j)}$ are solutions of

$$L(\partial_{\boldsymbol{\xi}_j})g^{(j)}(\boldsymbol{\xi}_j,\boldsymbol{\eta}_j)+\delta(\boldsymbol{\xi}_j-\boldsymbol{\eta}_j)I_3=0I_3\ ,\quad \boldsymbol{\xi}_j,\boldsymbol{\eta}_j\in C\bar{\omega}^{(j)}\ , \tag{7.120}$$

$$g^{(j)}(\boldsymbol{\xi}_j,\boldsymbol{\eta}_j)=0I_3\ ,\quad \boldsymbol{\xi}_j\in\partial C\bar{\omega}^{(j)},\boldsymbol{\eta}_j\in C\bar{\omega}^{(j)}\ , \tag{7.121}$$

$$g^{(j)}(\boldsymbol{\xi}_j,\boldsymbol{\eta}_j)\to 0I_3\quad \text{as}\quad |\boldsymbol{\xi}_j|\to\infty,\boldsymbol{\eta}_j\in C\bar{\omega}^{(j)}\ , \tag{7.122}$$

for $j=1,\dots,N$.

We represent $G(\mathbf{x},\mathbf{y})$ and $g^{(j)}(\boldsymbol{\xi}_j,\boldsymbol{\eta}_j)$ as

$$G(\mathbf{x},\mathbf{y})=\Gamma(\mathbf{x},\mathbf{y})-H(\mathbf{x},\mathbf{y})\ , \tag{7.123}$$

and

$$g^{(j)}(\boldsymbol{\xi}_j,\boldsymbol{\eta}_j)=\Gamma(\boldsymbol{\xi}_j,\boldsymbol{\eta}_j)-h^{(j)}(\boldsymbol{\xi}_j,\boldsymbol{\eta}_j)\ , \tag{7.124}$$

where $\Gamma(\mathbf{x},\mathbf{y})=[\Gamma_{mn}(\mathbf{x},\mathbf{y})]_{m,n=1}^3$, is the fundamental solution of the Lamé operator in three dimensions (see (6.87)), and H, $h^{(j)}$ are the regular parts of G, $g^{(j)}$, $j=1,\dots,N$, respectively.

7.3.2 *Elastic Capacitary Potential in Three Dimensions*

The Elastic Capacitary Potential Matrix

We denote by $P^{(j)}(\boldsymbol{\xi}_j)=[P^{(j,1)}(\boldsymbol{\xi}_j),P^{(j,2)}(\boldsymbol{\xi}_j),P^{(j,3)}(\boldsymbol{\xi}_j)]$ the elastic capacitary potential matrix of the set $\omega^{(j)}$, which is defined as a solution of

$$L(\partial_{\boldsymbol{\xi}_j})P^{(j)}(\boldsymbol{\xi}_j)=0I_3\ ,\quad \boldsymbol{\xi}_j\in C\bar{\omega}^{(j)}\ , \tag{7.125}$$

$$P^{(j)}(\boldsymbol{\xi}_j)=I_3\ ,\quad \boldsymbol{\xi}_j\in\partial\omega^{(j)}\ , \tag{7.126}$$

$$P^{(j)}(\boldsymbol{\xi}_j)\to 0I_3\quad \text{as}\quad |\boldsymbol{\xi}_j|\to\infty\ , \tag{7.127}$$

for $j=1,\dots,N$.

Let $B^{(j)}=[B^{(j,i)}]_{i=1}^3$ be the elastic capacity matrix for the set $\omega^{(j)}$, for $j=1,\dots,N$. This matrix was introduced and its properties where studied in Chap. 6. In particular, it was shown that this matrix is symmetric.

For the proof of the following Lemma, we refer to Sect. 6.3.2, Lemmas 6.4 and 6.6.

Lemma 7.5. *(i) If* $|\boldsymbol{\xi}_j| > 2$*, then for* $P^{(j,i)}$, $i = 1, 2, 3$*, the estimate holds:*

$$|P^{(j,i)}(\boldsymbol{\xi}_j) - \Gamma(\boldsymbol{\xi}_j, \mathbf{O})B^{(j,i)}| \le \text{Const}\, |\boldsymbol{\xi}_j|^{-2}\,, \tag{7.128}$$

where $B^{(j,i)}$ *are the columns of the symmetric elastic capacity matrix* $B^{(j)}$ *of the set* $\omega^{(j)}$.

(ii) The columns $P^{(j,i)}$, $i = 1, 2$ *or* 3*, of the elastic capacitary potential of* $\omega^{(j)}$, $j = 1, \dots, N$ *satisfy the inequality*

$$\sup_{\boldsymbol{\xi}_j \in C\bar{\omega}^{(j)}} \{|\boldsymbol{\xi}_j||P^{(j,i)}(\boldsymbol{\xi}_j)|\} \le \text{Const}\,, \quad j = 1, \dots, N\,. \tag{7.129}$$

An Estimate for the Regular Part $h^{(j)}$ of Green's Tensor in the Unbounded Domain

Now we present an asymptotic expansion for the regular part $h^{(j)}$ of Green's tensor $g^{(j)}$, whose proof is found in Sect. 6.3, Lemma 6.11.

Lemma 7.6. *For all* $\boldsymbol{\eta}_j \in C\bar{\omega}^{(j)}$ *and* $\boldsymbol{\xi}_j$ *with* $|\boldsymbol{\xi}_j| > 2$*, the following estimate for the columns* $h^{(j,i)}$, $i = 1, 2,$ *or* 3*, of the regular part of* $g^{(j,i)}$ *holds*

$$|h^{(j,i)}(\boldsymbol{\xi}_j, \boldsymbol{\eta}_j) - \Gamma(\boldsymbol{\xi}_j, \mathbf{O})(P^{(j)T})^{(i)}(\boldsymbol{\eta}_j)| \le \text{Const}\, |\boldsymbol{\xi}_j|^{-2}|\boldsymbol{\eta}_j|^{-1}\,, \tag{7.130}$$

where $j = 1, \dots, N$*, and* $(P^{(j)T})^{(i)}$ *is the i-th column of* $P^{(j)T}$.

7.3.3 A Uniform Asymptotic Formula for Green's Tensor in a Three-Dimensional Domain with Several Inclusions

Now we present the main result concerning the approximation of the matrix G_ε, for a three-dimensional domain with multiple inclusions.

Theorem 7.2. *Green's tensor* G_ε *for the Lamé operator in the domain* $\Omega_\varepsilon \subset \mathbb{R}^3$ *admits the representation*

$$G_\varepsilon(\mathbf{x}, \mathbf{y}) = G(\mathbf{x}, \mathbf{y}) + \varepsilon^{-1} \sum_{j=1}^{N} g^{(j)}(\boldsymbol{\xi}_j, \boldsymbol{\eta}_j) - N\Gamma(\mathbf{x}, \mathbf{y})$$

$$+ \sum_{j=1}^{N} \{P^{(j)}(\boldsymbol{\xi}_j)H(\mathbf{O}^{(j)}, \mathbf{y}) + H(\mathbf{x}, \mathbf{O}^{(j)})P^{(j)T}(\boldsymbol{\eta}_j)$$

$$-P^{(j)}(\boldsymbol{\xi}_j)H(\mathbf{O}^{(j)}, \mathbf{O}^{(j)})P^{(j)T}(\boldsymbol{\eta}_j) - \varepsilon H(\mathbf{x}, \mathbf{O}^{(j)})B^{(j)}H(\mathbf{O}^{(j)}, \mathbf{y})\}$$

$$+\sum_{j=1}^{N} \sum_{\substack{k \neq j \\ 1 \leq k \leq N}} P^{(k)}(\boldsymbol{\xi}_k) G(\mathbf{O}^{(k)}, \mathbf{O}^{(j)}) P^{(j)T}(\boldsymbol{\eta}_j)$$

$$+O\left(\sum_{j=1}^{N} \varepsilon^2 (\min\{|\mathbf{x} - \mathbf{O}^{(j)}|\,, |\mathbf{y} - \mathbf{O}^{(j)}|\})^{-1}\right), \tag{7.131}$$

uniformly with respect to $(\mathbf{x}, \mathbf{y}) \in \Omega_\varepsilon \times \Omega_\varepsilon$.

Proof. For the proof of Theorem 7.2, we first present a formal argument of how to obtain the leading order part of (7.131), after which we will give a rigorous proof of the remainder estimate.

Formal Argument

As in Chap. 3, we seek G_ε in the form

$$G_\varepsilon(\mathbf{x}, \mathbf{y}) = \Gamma(\mathbf{x}, \mathbf{y}) - H_\varepsilon(\mathbf{x}, \mathbf{y}) - \sum_{j=1}^{N} h_\varepsilon^{(j)}(\mathbf{x}, \mathbf{y}), \tag{7.132}$$

where the tensors $H_\varepsilon(\mathbf{x}, \mathbf{y})$ and $h_\varepsilon^{(j)}(\mathbf{x}, \mathbf{y})$ are solutions of the problems

$$L(\partial_{\mathbf{x}}) H_\varepsilon(\mathbf{x}, \mathbf{y}) = 0I_3\,, \quad \mathbf{x}, \mathbf{y} \in \Omega_\varepsilon\,, \tag{7.133}$$

$$H_\varepsilon(\mathbf{x}, \mathbf{y}) = \Gamma(\mathbf{x}, \mathbf{y})\,, \quad \mathbf{x} \in \partial\Omega, \mathbf{y} \in \Omega_\varepsilon\,, \tag{7.134}$$

$$H_\varepsilon(\mathbf{x}, \mathbf{y}) = 0I_3\,, \quad \mathbf{x} \in \partial\omega_\varepsilon^{(j)}, \mathbf{y} \in \Omega_\varepsilon, 1 \leq j \leq N\,, \tag{7.135}$$

and

$$L(\partial_{\mathbf{x}}) h_\varepsilon^{(j)}(\mathbf{x}, \mathbf{y}) = 0I_3\,, \quad \mathbf{x}, \mathbf{y} \in \Omega_\varepsilon\,, \tag{7.136}$$

$$h_\varepsilon^{(j)}(\mathbf{x}, \mathbf{y}) = 0I_3\,, \quad \mathbf{x} \in \partial\Omega, \mathbf{y} \in \Omega_\varepsilon\,, \tag{7.137}$$

$$h_\varepsilon^{(j)}(\mathbf{x}, \mathbf{y}) = \Gamma(\mathbf{x}, \mathbf{y})\,, \quad \mathbf{x} \in \partial\omega_\varepsilon^{(j)}, \mathbf{y} \in \Omega_\varepsilon\,, \tag{7.138}$$

$$h_\varepsilon^{(j)}(\mathbf{x}, \mathbf{y}) = 0I_3\,, \quad \mathbf{x} \in \partial\omega_\varepsilon^{(k)}, \mathbf{y} \in \Omega_\varepsilon\,, 1 \leq k \leq N\,, k \neq j\,. \tag{7.139}$$

The Approximation of $H_\varepsilon(\mathbf{x}, \mathbf{y})$

Consider the tensor $H_\varepsilon(\mathbf{x}, \mathbf{y}) - H(\mathbf{x}, \mathbf{y})$. This satisfies the homogeneous Lamé equation and has zero boundary data for $\mathbf{x} \in \partial\Omega, \mathbf{y} \in \Omega_\varepsilon$. For $\mathbf{x} \in \partial\omega_\varepsilon^{(j)}, \mathbf{y} \in \Omega_\varepsilon$, $1 \leq j \leq N$, this matrix is equal to $-H(\mathbf{x}, \mathbf{y})$, whose leading order part is $-H(\mathbf{O}^{(j)}, \mathbf{y})$. Then we may approximate H_ε, using the elastic capacitary potential, by

$$H_\varepsilon(\mathbf{x},\mathbf{y}) - H(\mathbf{x},\mathbf{y}) = -\sum_{j=1}^{N} P^{(j)}(\boldsymbol{\xi}_j) H(\mathbf{O}^{(j)},\mathbf{y}) + \mathfrak{S}_\varepsilon(\mathbf{x},\mathbf{y}) , \tag{7.140}$$

where the remainder term $\mathfrak{S}_\varepsilon$ on the right is a solution of the homogeneous Lamé equation, is $O(\varepsilon)$ for $\mathbf{x} \in \partial\omega_\varepsilon^{(j)}, \mathbf{y} \in \Omega_\varepsilon$, $1 \le j \le N$ and by Lemma 7.5 i) the leading order part of $\mathfrak{S}_\varepsilon$ is

$$\sum_{j=1}^{N} \varepsilon \Gamma(\mathbf{x},\mathbf{O}^{(j)}) B^{(j)} H(\mathbf{O}^{(j)},\mathbf{y}) \text{ for } \mathbf{x} \in \partial\Omega, \mathbf{y} \in \Omega_\varepsilon . \tag{7.141}$$

Then the approximation of $\mathfrak{S}_\varepsilon(\mathbf{x},\mathbf{y})$ may be given by

$$\mathfrak{S}_\varepsilon(\mathbf{x},\mathbf{y}) = \sum_{j=1}^{N} \varepsilon H(\mathbf{x},\mathbf{O}^{(j)}) B^{(j)} H(\mathbf{O}^{(j)},\mathbf{y}) + \mathfrak{H}_\varepsilon(\mathbf{x},\mathbf{y}) , \tag{7.142}$$

then upon substitution of this into (7.140) we obtain the following approximation for H_ε

$$\begin{aligned} H_\varepsilon(\mathbf{x},\mathbf{y}) = H(\mathbf{x},\mathbf{y}) - \sum_{j=1}^{N} \{ & P^{(j)}(\boldsymbol{\xi}_j) H(\mathbf{O}^{(j)},\mathbf{y}) \\ & - \varepsilon H(\mathbf{x},\mathbf{O}^{(j)}) B^{(j)} H(\mathbf{O}^{(j)},\mathbf{y}) \} + \mathfrak{H}_\varepsilon(\mathbf{x},\mathbf{y}) , \end{aligned} \tag{7.143}$$

where $\mathfrak{H}_\varepsilon(\mathbf{x},\mathbf{y})$ represents the remainder given by this approximation.

The Approximation of $h_\varepsilon^{(j)}(\mathbf{x},\mathbf{y})$

The matrix

$$W^{(j)}(\mathbf{x},\mathbf{y}) = h_\varepsilon^{(j)}(\mathbf{x},\mathbf{y}) - \varepsilon^{-1} h^{(j)}(\boldsymbol{\xi}_j, \boldsymbol{\eta}_j) , \tag{7.144}$$

satisfies the homogeneous Lamé equation, is equal to $0 I_3$ on the boundary of the inclusion $\partial\omega_\varepsilon^{(j)}$ and

$$W^{(j)}(\mathbf{x},\mathbf{y}) = -\varepsilon^{-1} h^{(j)}(\boldsymbol{\xi}_j, \boldsymbol{\eta}_j) , \quad \mathbf{x} \in \partial\Omega , \mathbf{y} \in \Omega_\varepsilon , \tag{7.145}$$

$$W^{(j)}(\mathbf{x},\mathbf{y}) = -\varepsilon^{-1} h^{(j)}(\boldsymbol{\xi}_j, \boldsymbol{\eta}_j) , \quad \mathbf{x} \in \partial\omega_\varepsilon^{(k)} , \mathbf{y} \in \Omega_\varepsilon , k \neq j , 1 \le k \le N . \tag{7.146}$$

By Lemma 7.6, the boundary conditions (7.145), (7.146) are equivalent to

$$W^{(j)}(\mathbf{x},\mathbf{y}) = -\Gamma(\mathbf{x},\mathbf{O}^{(j)})P^{(j)T}(\boldsymbol{\eta}_j) + O(\varepsilon^2|\mathbf{y}-\mathbf{O}^{(j)}|^{-1}) , \quad \mathbf{x} \in \partial\Omega , \mathbf{y} \in \Omega_\varepsilon , \tag{7.147}$$

$$W^{(j)}(\mathbf{x},\mathbf{y}) = -\Gamma(\mathbf{x},\mathbf{O}^{(j)})P^{(j)T}(\boldsymbol{\eta}_j) + O(\varepsilon^2|\mathbf{y}-\mathbf{O}^{(j)}|^{-1}) , \tag{7.148}$$

for $\mathbf{x} \in \partial\omega_\varepsilon^{(k)}, \mathbf{y} \in \Omega_\varepsilon, k \neq j, 1 \leq k \leq N$.

Then the matrix $W^{(j)}(\mathbf{x},\mathbf{y})$ is sought in the form

$$W^{(j)}(\mathbf{x},\mathbf{y}) = -H(\mathbf{x},\mathbf{O}^{(j)})P^{(j)T}(\boldsymbol{\eta}_j) + \chi_\varepsilon^{(j)}(\mathbf{x},\mathbf{y}) , \tag{7.149}$$

where the matrix $\chi_\varepsilon^{(j)}(\mathbf{x},\mathbf{y})$ is a solution of the boundary value problem

$$L(\partial_{\mathbf{x}})\chi_\varepsilon^{(j)}(\mathbf{x},\mathbf{y}) = 0I_3 , \quad \mathbf{x},\mathbf{y} \in \Omega_\varepsilon , \tag{7.150}$$

$$\chi_\varepsilon^{(j)}(\mathbf{x},\mathbf{y}) = O(\varepsilon^2|\mathbf{y}-\mathbf{O}^{(j)}|^{-1}) , \quad \mathbf{x} \in \partial\Omega, \mathbf{y} \in \Omega_\varepsilon , \tag{7.151}$$

$$\chi_\varepsilon^{(j)}(\mathbf{x},\mathbf{y}) = H(\mathbf{x},\mathbf{O}^{(j)})P^{(j)T}(\boldsymbol{\eta}_j) , \quad \mathbf{x} \in \partial\omega_\varepsilon^{(j)}, \mathbf{y} \in \Omega_\varepsilon , \tag{7.152}$$

$$\chi_\varepsilon^{(j)}(\mathbf{x},\mathbf{y}) = -G(\mathbf{x},\mathbf{O}^{(j)})P^{(j)T}(\boldsymbol{\eta}_j) + O(\varepsilon^2|\mathbf{y}-\mathbf{O}^{(j)}|^{-1}) , \quad \mathbf{x} \in \partial\omega_\varepsilon^{(k)}, \mathbf{y} \in \Omega_\varepsilon , 1 \leq k \leq N , k \neq j . \tag{7.153}$$

Since the tensor $G(\mathbf{x},\mathbf{O}^{(j)})$ and the regular part $H(\mathbf{x},\mathbf{y})$ of Green's tensor for the domain Ω, have smooth components for $\mathbf{x}$, $\mathbf{y} \in \Omega_\varepsilon$, then on $\partial\omega_\varepsilon^{(j)}$ we may expand these tensors about the centres of $\omega_\varepsilon^{(j)}$ $(1 \leq j \leq N)$. Thus from (7.152), (7.153) we obtain

$$\chi_\varepsilon^{(j)}(\mathbf{x},\mathbf{y}) = H(\mathbf{O}^{(j)},\mathbf{O}^{(j)})P^{(j)T}(\boldsymbol{\eta}_j) + O(\varepsilon^2|\mathbf{y}-\mathbf{O}^{(j)}|^{-1}) , \tag{7.154}$$

for $\mathbf{x} \in \partial\omega_\varepsilon^{(j)}, \mathbf{y} \in \Omega_\varepsilon$, and

$$\chi_\varepsilon^{(j)}(\mathbf{x},\mathbf{y}) = -G(\mathbf{O}^{(k)},\mathbf{O}^{(j)})P^{(j)T}(\boldsymbol{\eta}_j) + O(\varepsilon^2|\mathbf{y}-\mathbf{O}^{(j)}|^{-1}) , \tag{7.155}$$

for $\mathbf{x} \in \partial\omega_\varepsilon^{(k)}, \mathbf{y} \in \Omega_\varepsilon, 1 \leq k \leq N, k \neq j$.

However, (7.154) and (7.155) are not small on the exterior boundary $\partial\Omega$. Therefore, using the elastic capacitary potential we represent $\chi_\varepsilon^{(j)}(\mathbf{x},\mathbf{y})$ as

$$\begin{aligned}\chi_\varepsilon^{(j)}(\mathbf{x},\mathbf{y}) = {} & P^{(j)}(\boldsymbol{\xi}_j)H(\mathbf{O}^{(j)},\mathbf{O}^{(j)})P^{(j)T}(\boldsymbol{\eta}_j) \\ & - \sum_{\substack{k\neq j \\ 1\leq k\leq N}} P^{(k)}(\boldsymbol{\xi}_k)G(\mathbf{O}^{(k)},\mathbf{O}^{(j)})P^{(j)T}(\boldsymbol{\eta}_j) \\ & + \mathfrak{h}_\varepsilon^{(j)}(\mathbf{x},\mathbf{y}) ,\end{aligned} \tag{7.156}$$

where the matrix $\mathfrak{h}_\varepsilon^{(j)}(\mathbf{x},\mathbf{y})$ is the remainder term.

Collecting (7.149) and (7.156) in (7.144), we have the following approximation for the tensor $h_\varepsilon^{(j)}$

$$\begin{aligned} h_\varepsilon^{(j)}(\mathbf{x},\mathbf{y}) = {}& \varepsilon^{-1}h^{(j)}(\boldsymbol{\xi}_j,\boldsymbol{\eta}_j) - H(\mathbf{x},\mathbf{O}^{(j)})P^{(j)T}(\boldsymbol{\eta}_j) \\ & + P^{(j)}(\boldsymbol{\xi}_j)H(\mathbf{O}^{(j)},\mathbf{O}^{(j)})P^{(j)T}(\boldsymbol{\eta}_j) \\ & - \sum_{\substack{k\neq j\\ 1\le k\le N}} P^{(k)}(\boldsymbol{\xi}_k)G(\mathbf{O}^{(k)},\mathbf{O}^{(j)})P^{(j)T}(\boldsymbol{\eta}_j) \\ & + \mathfrak{h}_\varepsilon^{(j)}(\mathbf{x},\mathbf{y})\,. \end{aligned} \tag{7.157}$$

Combined Formula

Substituting (7.143) and (7.157) in (7.132) we obtain

$$\begin{aligned} G_\varepsilon(\mathbf{x},\mathbf{y}) = {}& G(\mathbf{x},\mathbf{y}) + \varepsilon^{-1}\sum_{j=1}^{N} g^{(j)}(\boldsymbol{\xi}_j,\boldsymbol{\eta}_j) - N\Gamma(\mathbf{x},\mathbf{y}) \\ & + \sum_{j=1}^{N}\{P^{(j)}(\boldsymbol{\xi}_j)H(\mathbf{O}^{(j)},\mathbf{y}) + H(\mathbf{x},\mathbf{O}^{(j)})P^{(j)T}(\boldsymbol{\eta}_j) \\ & - P^{(j)}(\boldsymbol{\xi}_j)H(\mathbf{O}^{(j)},\mathbf{O}^{(j)})P^{(j)T}(\boldsymbol{\eta}_j) - \varepsilon H(\mathbf{x},\mathbf{O}^{(j)})B^{(j)}H(\mathbf{O}^{(j)},\mathbf{y})\} \\ & + \sum_{j=1}^{N}\sum_{\substack{k\neq j\\ 1\le k\le N}} P^{(k)}(\boldsymbol{\xi}_k)G(\mathbf{O}^{(k)},\mathbf{O}^{(j)})P^{(j)T}(\boldsymbol{\eta}_j) + R_\varepsilon(\mathbf{x},\mathbf{y})\,, \end{aligned} \tag{7.158}$$

where the matrix R_ε represents the combination of the remainder terms $\mathfrak{H}_\varepsilon$ and $\mathfrak{h}_\varepsilon^{(j)}$, $j = 1,\dots,N$, given in the approximations (7.143) and (7.157), respectively.

We now give a proof of (7.131), including the remainder estimate.

Proof of Theorem 7.2

From (7.158), the columns $R_\varepsilon^{(k)}(\mathbf{x},\mathbf{y})$, $k = 1,2,3$ of the remainder, satisfy the boundary value problem

$$L(\partial_{\mathbf{x}})R_\varepsilon^{(k)}(\mathbf{x},\mathbf{y}) = \mathbf{O}\,, \quad \mathbf{x},\mathbf{y}\in\Omega_\varepsilon\,, \tag{7.159}$$

$$R_\varepsilon^{(k)}(\mathbf{x},\mathbf{y}) = \varepsilon^{-1}\sum_{j=1}^{N} h^{(j,k)}(\boldsymbol{\xi}_j,\boldsymbol{\eta}_j) - \sum_{j=1}^{N}\{P^{(j)}(\boldsymbol{\xi}_j)H^{(k)}(\mathbf{O}^{(j)},\mathbf{y})$$
$$+H(\mathbf{x},\mathbf{O}^{(j)})(P^{(j)T})^{(k)}(\boldsymbol{\eta}_j) - \varepsilon H(\mathbf{x},\mathbf{O}^{(j)})B^{(j)}H^{(k)}(\mathbf{O}^{(j)},\mathbf{y})$$
$$-P^{(j)}(\boldsymbol{\xi}_j)H(\mathbf{O}^{(j)},\mathbf{O}^{(j)})(P^{(j)T})^{(k)}(\boldsymbol{\eta}_j)\}$$
$$-\sum_{j=1}^{N}\sum_{\substack{l\neq j\\ 1\leq l\leq N}} P^{(l)}(\boldsymbol{\xi}_l)G(\mathbf{O}^{(l)},\mathbf{O}^{(j)})(P^{(j)T})^{(k)}(\boldsymbol{\eta}_j)\,,$$
$$\text{for } \mathbf{x}\in\partial\Omega,\ \mathbf{y}\in\Omega_\varepsilon\,, \tag{7.160}$$

$$R_\varepsilon^{(k)}(\mathbf{x},\mathbf{y}) = H^{(k)}(\mathbf{x},\mathbf{y}) - H^{(k)}(\mathbf{O}^{(m)},\mathbf{y}) + \varepsilon^{-1}\sum_{\substack{j\neq m\\ 1\leq j\leq N}} h^{(j,k)}(\boldsymbol{\xi}_j,\boldsymbol{\eta}_j)$$
$$-\{H(\mathbf{x},\mathbf{O}^{(m)}) - H(\mathbf{O}^{(m)},\mathbf{O}^{(m)})\}(P^{(m)T})^{(k)}(\boldsymbol{\eta}_m)$$
$$-\sum_{\substack{j\neq m\\ 1\leq j\leq N}}\{P^{(j)}(\boldsymbol{\xi}_j)H^{(k)}(\mathbf{O}^{(j)},\mathbf{y}) + H(\mathbf{x},\mathbf{O}^{(j)})(P^{(j)T})^{(k)}(\boldsymbol{\eta}_j)$$
$$-P^{(j)}(\boldsymbol{\xi}_j)H(\mathbf{O}^{(j)},\mathbf{O}^{(j)})(P^{(j)T})^{(k)}(\boldsymbol{\eta}_j)\}$$
$$+\sum_{j=1}^{N}\varepsilon H(\mathbf{x},\mathbf{O}^{(j)})B^{(j)}H^{(k)}(\mathbf{O}^{(j)},\mathbf{y})$$
$$-\sum_{\substack{j\neq m\\ 1\leq j\leq N}} G(\mathbf{O}^{(m)},\mathbf{O}^{(j)})(P^{(j)T})^{(k)}(\boldsymbol{\eta}_j)$$
$$-\sum_{j=1}^{N}\sum_{\substack{l\neq j\\ l\neq m\\ 1\leq l\leq N}} P^{(l)}(\boldsymbol{\xi}_l)G(\mathbf{O}^{(l)},\mathbf{O}^{(j)})(P^{(j)T})^{(k)}(\boldsymbol{\eta}_j)$$
$$\text{for } \mathbf{x}\in\partial\omega_\varepsilon^{(m)},\ \mathbf{y}\in\Omega_\varepsilon\,,\ 1\leq m\leq N. \tag{7.161}$$

The components of $H^{(k)}(\mathbf{x},\mathbf{O}^{(j)})$ and $H^{(k)}(\mathbf{O}^{(j)},\mathbf{y})$ are bounded in Ω and the components of $H^{(k)}(\mathbf{x},\mathbf{O}^{(j)})$ are bounded on $\partial\Omega$. They are also bounded for $\mathbf{x}\in\partial\omega_\varepsilon^{(m)}$, $\mathbf{y}\in\Omega_\varepsilon$, $1\leq m\leq N$. Therefore, the norms of the terms

$$\sum_{j=1}^{N}\varepsilon H(\mathbf{x},\mathbf{O}^{(j)})B^{(j)}H^{(k)}(\mathbf{O}^{(k)},\mathbf{y})\,,$$

are bounded by Const ε in (7.161).

By Lemma 7.5 ii), since the entries of $P^{(j)}(\boldsymbol{\eta}_j)$ are bounded, we have

$$|H^{(k)}(\mathbf{x},\mathbf{y}) - H^{(k)}(\mathbf{O}^{(m)},\mathbf{y}) - (H(\mathbf{x},\mathbf{O}^{(m)}) - H(\mathbf{O}^{(m)},\mathbf{O}^{(m)}))(P^{(m)T})^{(k)}(\boldsymbol{\eta}_m)|$$

$$\leq \text{Const}\,\varepsilon\,, \quad \text{for } \mathbf{x} \in \partial\omega_\varepsilon^{(m)}, \mathbf{y} \in \Omega_\varepsilon\,, 1 \leq m \leq N\,. \tag{7.162}$$

Then using the estimate given in Lemma 7.6 for the columns of $h^{(j)}$, $j \neq m$, we have

$$\Big| \sum_{\substack{j \neq m \\ 1 \leq j \leq N}} \{\varepsilon^{-1} h^{(j,k)}(\boldsymbol{\xi}_j, \boldsymbol{\eta}_j) - H(\mathbf{x},\mathbf{O}^{(j)})(P^{(j)T})^{(k)}(\boldsymbol{\eta}_j) - G(\mathbf{O}^{(m)},\mathbf{O}^{(j)})(P^{(j)T})^{(k)}(\boldsymbol{\eta}_j)\} \Big|$$

$$\leq \Big| \sum_{\substack{j \neq m \\ 1 \leq j \leq N}} \{G(\mathbf{x},\mathbf{O}^{(j)}) - G(\mathbf{O}^{(m)},\mathbf{O}^{(j)})\}(P^{(j)T})^{(k)}(\boldsymbol{\eta}_j) \Big|$$

$$+\text{Const} \sum_{\substack{j \neq m \\ 1 \leq j \leq N}} \varepsilon^2 |\mathbf{y} - \mathbf{O}^{(j)}|^{-1} \leq \text{Const} \sum_{\substack{j \neq m \\ 1 \leq j \leq N}} \varepsilon^2 |\mathbf{y} - \mathbf{O}^{(j)}|^{-1}\,, \tag{7.163}$$

for $\mathbf{x} \in \partial\omega_\varepsilon^{(m)}$, $\mathbf{y} \in \Omega_\varepsilon$.

Finally, using the estimate for $P^{(j)}$ of Lemma 7.5 *ii)* for $j \neq m$ and also the fact that the components of H and $G(\mathbf{O}^{(l)},\mathbf{O}^{(j)})$, $j \neq l$ are bounded in Ω, we obtain

$$\sum_{\substack{j \neq m \\ 1 \leq j \leq N}} \{P^{(j)}(\boldsymbol{\xi}_j) H^{(k)}(\mathbf{O}^{(j)},\mathbf{y}) - P^{(j)}(\boldsymbol{\xi}_j) H(\mathbf{O}^{(j)},\mathbf{O}^{(j)})(P^{(j)T})^{(k)}(\boldsymbol{\eta}_j)\} = O(\varepsilon)\,, \tag{7.164}$$

and

$$\sum_{j=1}^{N} \sum_{\substack{l \neq j \\ l \neq m \\ 1 \leq l \leq N}} P^{(l)}(\boldsymbol{\xi}_l) G(\mathbf{O}^{(l)},\mathbf{O}^{(j)})(P^{(j)T})^{(k)}(\boldsymbol{\eta}_j) = O\Big(\sum_{j=1}^{N} \varepsilon^2 |\mathbf{y} - \mathbf{O}^{(j)}|^{-1}\Big)\,, \tag{7.165}$$

for $\mathbf{x} \in \partial\omega_\varepsilon^{(m)}$, $\mathbf{y} \in \Omega_\varepsilon$.

Thus combining the estimates (7.162)–(7.165) in (7.161), we have

$$|R_\varepsilon^{(k)}(\mathbf{x},\mathbf{y})| \leq \text{Const}\,\varepsilon\,, \tag{7.166}$$

for $\mathbf{x} \in \partial\omega_\varepsilon^{(m)}$, $\mathbf{y} \in \Omega_\varepsilon$, $1 \leq m \leq N$.

Now we estimate the right-hand side of the boundary condition (7.160).

Using Lemma 7.5 i), we obtain

$$\left|\sum_{j=1}^{N}\{P^{(j)}(\boldsymbol{\xi}_j)H^{(k)}(\mathbf{O}^{(j)},\mathbf{y})-\varepsilon H(\mathbf{x},\mathbf{O}^{(j)})B^{(j)}H^{(k)}(\mathbf{O}^{(j)},\mathbf{y})\}\right|$$

$$=\left|\sum_{j=1}^{N}\{(P^{(j)}(\boldsymbol{\xi}_j)-\Gamma(\boldsymbol{\xi}_j,\mathbf{O})B^{(j)})H^{(k)}(\mathbf{O}^{(j)},\mathbf{y})\}\right|$$

$$\leq \text{Const}\sum_{j=1}^{N}\varepsilon^2|\mathbf{x}-\mathbf{O}^{(j)}|^{-2}\leq \text{Const}\,\varepsilon^2\,,\mathbf{x}\in\partial\Omega\,,\mathbf{y}\in\Omega_\varepsilon\,, \tag{7.167}$$

where we have used the fact that for $\mathbf{x}\in\partial\Omega$, $1\leq|\mathbf{x}-\mathbf{O}^{(j)}|$, $1\leq j\leq N$.

From Lemma 7.5 *ii*), we also have

$$|P^{(j,k)}(\boldsymbol{\xi}^{(j)})|\leq \text{Const}\,\varepsilon|\mathbf{x}-\mathbf{O}^{(j)}|^{-1}\,. \tag{7.168}$$

Owing to Lemma 7.6 we have

$$\left|\varepsilon^{-1}\sum_{j=1}^{N}\{h^{(j,k)}(\boldsymbol{\xi}_j,\boldsymbol{\eta}_j)-H(\mathbf{x},\mathbf{O}^{(j)})(P^{(j)T})^{(k)}(\boldsymbol{\eta}_j)\}\right|$$

$$=\varepsilon^{-1}\left|\sum_{j=1}^{N}\{h^{(j,k)}(\boldsymbol{\xi}_j,\boldsymbol{\eta}_j)-\Gamma(\boldsymbol{\xi}_j,\mathbf{O})(P^{(j)T})^{(k)}(\boldsymbol{\eta}_j)\}\right|$$

$$\leq \text{Const}\sum_{j=1}^{N}\varepsilon^2|\mathbf{x}-\mathbf{O}^{(j)}|^{-2}|\mathbf{y}-\mathbf{O}^{(j)}|^{-1}$$

$$\leq \text{Const}\sum_{j=1}^{N}\varepsilon^2|\mathbf{y}-\mathbf{O}^{(j)}|^{-1}\,,\ \text{for } \mathbf{x}\in\partial\Omega\,,\mathbf{y}\in\Omega_\varepsilon\,. \tag{7.169}$$

Then, by (7.168) and the definition of G and its regular part H, the estimates

$$|P^{(j)}(\boldsymbol{\xi}_j)H(\mathbf{O}^{(j)},\mathbf{O}^{(j)})(P^{(j)T})^{(k)}(\boldsymbol{\eta}_j)|\leq \text{Const}\,\varepsilon^2|\mathbf{y}-\mathbf{O}^{(j)}|^{-1}\,, \tag{7.170}$$

and

$$|P^{(l)}(\boldsymbol{\xi}_l)G(\mathbf{O}^{(l)},\mathbf{O}^{(j)})(P^{(j)T})^{(k)}(\boldsymbol{\eta}_j)|\leq \text{Const}\,\varepsilon^2|\mathbf{y}-\mathbf{O}^{(j)}|^{-1}\,,\quad \text{for } l\neq j\,, \tag{7.171}$$

holds for $\mathbf{x}\in\partial\Omega$, $\mathbf{y}\in\Omega_\varepsilon$.

Therefore, combining the estimates (7.167), (7.169)–(7.171) we have

$$|R_\varepsilon^{(k)}(\mathbf{x},\mathbf{y})| \le \text{Const} \sum_{j=1}^{N} \varepsilon^2 |\mathbf{y} - \mathbf{O}^{(j)}|^{-1} , \tag{7.172}$$

for $\mathbf{x} \in \partial\Omega$, $\mathbf{y} \in \Omega_\varepsilon$.

Then (7.166), (7.172) and Lemma 7.1 imply

$$\begin{aligned} |R_\varepsilon^{(k)}(\mathbf{x},\mathbf{y})| &\le \text{Const} \max\left\{ \sum_{j=1}^{N} \varepsilon^2 |\mathbf{x} - \mathbf{O}^{(j)}|^{-1} , \sum_{j=1}^{N} \varepsilon^2 |\mathbf{y} - \mathbf{O}^{(j)}|^{-1} \right\} \\ &\le \text{Const} \sum_{j=1}^{N} \varepsilon^2 (\min\{|\mathbf{x} - \mathbf{O}^{(j)}| , |\mathbf{y} - \mathbf{O}^{(j)}|\})^{-1} . \end{aligned} \tag{7.173}$$

The proof is complete. □

7.4 Simplified Asymptotic Formulae for the Case of a Three-Dimensional Elastic Solid with Several Small Inclusions

Here we show how the asymptotic formula (7.131) simplifies under certain constraints on the independent variables.

Corollary 7.1. *(a) Let* $\mathbf{x}, \mathbf{y} \in \Omega_\varepsilon \subset \mathbb{R}^3$ *such that*

$$\min\{|\mathbf{x} - \mathbf{O}^{(j)}|, |\mathbf{y} - \mathbf{O}^{(j)}|\} > 2\,\varepsilon \textit{ for all } j = 1, \dots, N. \tag{7.174}$$

Then

$$\begin{aligned} G_\varepsilon(\mathbf{x},\mathbf{y}) &= G(\mathbf{x},\mathbf{y}) - \varepsilon \sum_{j=1}^{N} G(\mathbf{x}, \mathbf{O}^{(j)}) B^{(j)} G(\mathbf{O}^{(j)}, \mathbf{y}) \\ &+ O\left(\sum_{j=1}^{N} \varepsilon^2 (|\mathbf{x} - \mathbf{O}^{(j)}||\mathbf{y} - \mathbf{O}^{(j)}| \min\{|\mathbf{x} - \mathbf{O}^{(j)}|, |\mathbf{y} - \mathbf{O}^{(j)}|\})^{-1} \right) . \end{aligned} \tag{7.175}$$

(b) If $\max\{|\mathbf{x} - \mathbf{O}^{(m)}|, |\mathbf{y} - \mathbf{O}^{(m)}|\} < 1/2$, *then*

$$\begin{aligned} G_\varepsilon(\mathbf{x},\mathbf{y}) &= \varepsilon^{-1} g^{(m)}(\boldsymbol{\xi}_m, \boldsymbol{\eta}_m) \\ &\quad - (I_3 - P^{(m)}(\boldsymbol{\xi}_m)) H(\mathbf{O}^{(m)}, \mathbf{O}^{(m)}) (I_3 - P^{(m)T}(\boldsymbol{\eta}_m)) \\ &\quad + O(\max\{|\mathbf{x} - \mathbf{O}^{(m)}| , |\mathbf{y} - \mathbf{O}^{(m)}|\}) . \end{aligned} \tag{7.176}$$

Both (7.175) *and* (7.176) *are uniform with respect to* $\mathbf{x}, \mathbf{y} \in \Omega_\varepsilon$.

We note that the formula (7.175) presented in part a) of the above Corollary is similar to that presented in the paper by Ozawa [36] (p. 215), for the approximate Green's function of the eigenvalue problem for the Laplacian in a bounded domain in $\mathbb{R}^3$ containing several spherical inclusions, which makes use of the Green's function in the unperturbed domain.

Proof. (*a*) From (7.131), G_ε can be rewritten as

$$\begin{aligned}G_\varepsilon(\mathbf{x},\mathbf{y}) &= G(\mathbf{x},\mathbf{y}) - \varepsilon^{-1}\sum_{j=1}^{N} h^{(j)}(\boldsymbol{\xi}_j, \boldsymbol{\eta}_j)\\ &+ \sum_{j=1}^{N}\{P^{(j)}(\boldsymbol{\xi}_j)H(\mathbf{O}^{(j)},\mathbf{y}) + H(\mathbf{x},\mathbf{O}^{(j)})P^{(j)T}(\boldsymbol{\eta}_j)\\ &- P^{(j)}(\boldsymbol{\xi}_j)H(\mathbf{O}^{(j)},\mathbf{O}^{(j)})P^{(j)T}(\boldsymbol{\eta}_j) - \varepsilon H(\mathbf{x},\mathbf{O}^{(j)})B^{(j)}H(\mathbf{O}^{(j)},\mathbf{y})\}\\ &+ \sum_{j=1}^{N}\sum_{\substack{k\neq j\\ 1\le k\le N}} P^{(k)}(\boldsymbol{\xi}_k)G(\mathbf{O}^{(k)},\mathbf{O}^{(j)})P^{(j)T}(\boldsymbol{\eta}_j)\\ &+ O\left(\sum_{j=1}^{N}\varepsilon^2(\min\{|\mathbf{x}-\mathbf{O}^{(j)}|\,,|\mathbf{y}-\mathbf{O}^{(j)}|\})^{-1}\right)\,. \end{aligned} \tag{7.177}$$

By Lemma 7.5 *i*), we have the following estimate for the elastic capacitary potential

$$P^{(j)}(\boldsymbol{\xi}_j) = \varepsilon\Gamma(\mathbf{x},\mathbf{O}^{(j)})B^{(j)} + O\left(\varepsilon^2|\mathbf{x}-\mathbf{O}^{(j)}|^{-2}\right)\,, \tag{7.178}$$

and from Lemma 7.6 we also have the approximation

$$\begin{aligned}\varepsilon^{-1}h^{(j)}(\boldsymbol{\xi}_j,\boldsymbol{\eta}_j) &= \Gamma(\mathbf{x},\mathbf{O}^{(j)})P^{(j)T}(\boldsymbol{\eta}_j) + O\left(\varepsilon^2(|\mathbf{x}-\mathbf{O}^{(j)}|^2|\mathbf{y}-\mathbf{O}^{(j)}|)^{-1}\right)\\ &= \varepsilon\Gamma(\mathbf{x},\mathbf{O}^{(j)})B^{(j)}\Gamma(\mathbf{y},\mathbf{O}^{(j)})\\ &+ O\left(\varepsilon^2(|\mathbf{x}-\mathbf{O}^{(j)}||\mathbf{y}-\mathbf{O}^{(j)}|\min\{|\mathbf{x}-\mathbf{O}^{(j)}|,|\mathbf{y}-\mathbf{O}^{(j)}|\})^{-1}\right)\,,\end{aligned} \tag{7.179}$$

where in (7.179) we have combined both of the above mentioned results.

In (7.177), using the (7.178) and (7.179), we have

$$G_\varepsilon(\mathbf{x},\mathbf{y}) = G(\mathbf{x},\mathbf{y}) - \varepsilon\sum_{j=1}^{N}\Gamma(\mathbf{x},\mathbf{O}^{(j)})B^{(j)}\Gamma(\mathbf{y},\mathbf{O}^{(j)})$$

$$+\sum_{j=1}^{N}\{\varepsilon\Gamma(\mathbf{x},\mathbf{O}^{(j)})B^{(j)}H(\mathbf{O}^{(j)},\mathbf{y}) + \varepsilon H(\mathbf{x},\mathbf{O}^{(j)})B^{(j)}\Gamma(\mathbf{y},\mathbf{O}^{(j)})$$

$$-\varepsilon H(\mathbf{x},\mathbf{O}^{(j)})B^{(j)}H(\mathbf{O}^{(j)},\mathbf{y})\}$$

$$+O\left(\sum_{j=1}^{N}\varepsilon^2(|\mathbf{x}-\mathbf{O}^{(j)}||\mathbf{y}-\mathbf{O}^{(j)}|\min\{|\mathbf{x}-\mathbf{O}^{(j)}|,|\mathbf{y}-\mathbf{O}^{(j)}|\})^{-1}\right). \quad (7.180)$$

Using the definition of the matrix function G given in (7.123), we may rewrite the preceding formula as

$$G_\varepsilon(\mathbf{x},\mathbf{y}) = G(\mathbf{x},\mathbf{y}) - \varepsilon\sum_{j=1}^{N}G(\mathbf{x},\mathbf{O}^{(j)})B^{(j)}\Gamma(\mathbf{y},\mathbf{O}^{(j)})$$

$$+\varepsilon\sum_{j=1}^{N}G(\mathbf{x},\mathbf{O}^{(j)})B^{(j)}H(\mathbf{O}^{(j)},\mathbf{y})$$

$$+O\left(\sum_{j=1}^{N}\varepsilon^2(|\mathbf{x}-\mathbf{O}^{(j)}||\mathbf{y}-\mathbf{O}^{(j)}|\min\{|\mathbf{x}-\mathbf{O}^{(j)}|,|\mathbf{y}-\mathbf{O}^{(j)}|\})^{-1}\right), \quad (7.181)$$

from which (7.175) follows.

(*b*) Due to the condition $\max\{|\mathbf{x}-\mathbf{O}^{(m)}|,|\mathbf{y}-\mathbf{O}^{(m)}|\} < 1/2$, and since $H(\mathbf{x},\mathbf{y})$ has smooth components for $\mathbf{x},\mathbf{y}\in\Omega$, in the vicinity of $(\mathbf{O}^{(m)},\mathbf{O}^{(m)})$ we have from (7.131)

$$G_\varepsilon(\mathbf{x},\mathbf{y}) = -H(\mathbf{O}^{(m)},\mathbf{O}^{(m)}) + \varepsilon^{-1}\sum_{j=1}^{N}g^{(j)}(\boldsymbol{\xi}_j,\boldsymbol{\eta}_j) - (N-1)\Gamma(\mathbf{x},\mathbf{y})$$

$$+P^{(m)}(\boldsymbol{\xi}_m)(H(\mathbf{O}^{(m)},\mathbf{O}^{(m)}) + O(|\mathbf{y}-\mathbf{O}^{(m)}|))$$

$$+(H(\mathbf{O}^{(m)},\mathbf{O}^{(m)}) + O(|\mathbf{x}-\mathbf{O}^{(m)}|))P^{(m)T}(\boldsymbol{\eta}_m)$$

$$-P^{(m)}(\boldsymbol{\xi}_m)H(\mathbf{O}^{(m)},\mathbf{O}^{(m)})P^{(m)T}(\boldsymbol{\eta}_m)$$

$$+\sum_{\substack{j\neq m\\1\le j\le N}}\{P^{(j)}(\boldsymbol{\xi}_j)(H(\mathbf{O}^{(j)},\mathbf{O}^{(m)}) + O(|\mathbf{y}-\mathbf{O}^{(m)}|))$$

$$+(H(\mathbf{O}^{(m)},\mathbf{O}^{(j)}) + O(|\mathbf{x}-\mathbf{O}^{(m)}|))P^{(j)T}(\boldsymbol{\eta}_j)$$

$$
-P^{(j)}(\boldsymbol{\xi}_j)H(\mathbf{O}^{(j)}, \mathbf{O}^{(j)})P^{(j)T}(\boldsymbol{\eta}_j)\}
$$

$$
+\sum_{j=1}^{N} \sum_{\substack{k \neq j \\ 1 \leq k \leq N}} P^{(k)}(\boldsymbol{\xi}_k)G(\mathbf{O}^{(k)}, \mathbf{O}^{(j)})P^{(j)T}(\boldsymbol{\eta}_j)
$$

$$
+O(\max\{|\mathbf{x} - \mathbf{O}^{(m)}|, |\mathbf{y} - \mathbf{O}^{(m)}|\}). \tag{7.182}
$$

Now using the estimate for the regular part $h^{(j)}$ given in (7.179), and that for the elastic capacitary potential (7.178) for $j \neq m$ we arrive at (7.176). □

Chapter 8
Green's Tensor for the Mixed Boundary Value Problem in a Domain with a Small Hole

In this chapter, we derive and justify the asymptotic approximation of the Green's tensor for the Lamé system in the situation when the traction boundary condition is prescribed on the small hole and the displacement condition is set on the exterior boundary.

Naturally, as a result of considering the traction condition on the boundary of the hole, we would expect new features to appear, when dealing with the approximation of the Green's tensor. One important model field discussed here will be the Neumann tensor for the unbounded domain corresponding to the exterior of the void. We will also see that in comparison to the Dirichlet problem for the Lamé system, where we used the notion of the elastic capacitary potential of the small holes in order to construct the approximation of Green's tensor, we will need other auxiliary fields defined in the unbounded domain corresponding to the hole, which are known as the *dipole fields*.

Following the main definitions outlined in Sect. 8.1, we state and prove an estimate related to solutions of the homogeneous Lamé equation for the Neumann problem in the unbounded domain, as described in Sect. 8.2. This result will then be used, in Sect. 8.3, to prove an estimate for solutions of the mixed problem for the Lamé equation in a domain with a single void. We introduce, in Sect. 8.4, the dipole fields and their properties. This section also contains an asymptotic estimate for the regular part of the Neumann tensor in the unbounded domain at infinity. We give the main result of this chapter, concerning the uniform asymptotic approximation of Green's tensor for the mixed boundary value problem, in Sect. 8.5. Once we obtain this approximation, we then aim to simplify this under assumptions on the independent spatial variables, and these results are given in Sect. 8.6.

V. Maz'ya et al., *Green's Kernels and Meso-Scale Approximations in Perforated Domains*, Lecture Notes in Mathematics 2077, DOI 10.1007/978-3-319-00357-3_8,

8.1 Definition of Green's Tensor in a Domain with a Single Void

The main object of our study in this chapter is Green's tensor for the mixed boundary value problem for the Lamé operator in the domain Ω_ε defined as in Chap. 6, Sect. 6.1.1. We will denote this tensor by G_ε and use the operator notations of Sect. 6.1

$$L(\partial_{\mathbf{x}})G_\varepsilon(\mathbf{x},\mathbf{y}) = \mathbf{D}(\partial_{\mathbf{x}})\mathcal{C}\mathbf{D}(\partial_{\mathbf{x}})^T G_\varepsilon(\mathbf{x},\mathbf{y}) = -\delta(\mathbf{x}-\mathbf{y})I_2\,, \quad \mathbf{x},\mathbf{y} \in \Omega_\varepsilon\,, \tag{8.1}$$

$$G_\varepsilon(\mathbf{x},\mathbf{y}) = 0I_2, \quad \mathbf{x} \in \partial\Omega, \mathbf{y} \in \Omega_\varepsilon\,, \tag{8.2}$$

$$T_n(\partial_{\mathbf{x}})G_\varepsilon(\mathbf{x},\mathbf{y}) = \mathbf{D}(\mathbf{n})\mathcal{C}\mathbf{D}(\partial_{\mathbf{x}})^T G_\varepsilon(\mathbf{x},\mathbf{y}) = 0I_2\,, \quad \mathbf{x} \in \partial\omega_\varepsilon, \mathbf{y} \in \Omega_\varepsilon\,, \tag{8.3}$$

where Ω_ε is the domain with the small void. Here in the boundary condition (8.3), $T_n(\partial_{\mathbf{x}})$ is the differential operator of tractions in two dimensions (cf. Chap. 6, (6.8), (6.9)).

8.2 An Estimate for Solutions of the Exterior Neumann Problem for the Homogeneous Lamé Equation

Now we formulate and prove a result concerning the estimate for the solution of the Neumann problem for the homogeneous Lamé operator in the unbounded domain $C\bar{\omega}$. This result will be shown to be useful when constructing asymptotic estimates for the model fields defined in $C\bar{\omega}$ involved in the algorithm.

Lemma 8.1. *Let* $\mathbf{u}$ *be a solution, which decays at infinity, of the exterior Neumann problem*

$$L(\partial_{\boldsymbol{\xi}})\mathbf{u}(\boldsymbol{\xi}) = \mathbf{O}\,, \quad \boldsymbol{\xi} \in C\bar{\omega}\,,$$

$$T_n(\partial_{\boldsymbol{\xi}})\mathbf{u}(\boldsymbol{\xi}) = \boldsymbol{\varphi}(\boldsymbol{\xi})\,, \quad \boldsymbol{\xi} \in \partial\omega\,,$$

where $\boldsymbol{\varphi} \in L_\infty(\partial\omega)$, *such that*

$$\int_{\partial\omega} \mathbf{c}\cdot\boldsymbol{\varphi}(\boldsymbol{\xi})\,dS_{\boldsymbol{\xi}} = 0\,, \tag{8.4}$$

where $\mathbf{c}$ *is an arbitrary constant vector.*

Then there exists a constant C, *depending on* ω, *such that*

$$\sup_{\boldsymbol{\xi}\in C\bar{\omega}} \{|\boldsymbol{\xi}||\mathbf{u}(\boldsymbol{\xi})|\} \le C\, D_\omega^2\, \|\boldsymbol{\varphi}\|_{L_\infty(\partial\omega)}\,,$$

where D_ω *is the diameter of* ω.

Proof. By dilation, we may assume without loss of generality that $D_\omega = 1$. We note that (8.4) implies the asymptotic behaviour for $\mathbf{u}(\boldsymbol{\xi})$ and its derivatives

$$\mathscr{D}^\alpha \mathbf{u}(\boldsymbol{\xi}) = O(|\boldsymbol{\xi}|^{-1-|\alpha|}) , \quad \text{as} \quad |\boldsymbol{\xi}| \to \infty ,$$

where $\alpha = (\alpha_1, \alpha_2)$ is any multi-index.

Using this and Betti's formula, we obtain the classical identity

$$\mathbf{u}(\boldsymbol{\xi}) = (\mathbf{V}\, T_n(\partial_{\boldsymbol{\xi}})\mathbf{u})(\boldsymbol{\xi}) - (\mathbf{W}\mathbf{u})(\boldsymbol{\xi}) , \quad \boldsymbol{\xi} \in C\bar{\omega} , \tag{8.5}$$

where $\mathbf{V}$ and $\mathbf{W}$ are single and double layer elastic potentials, respectively, with densities on $\partial\omega$. By the continuity of the single layer potential and the jump relation for the double layer potential, one arrives at the integral equation

$$(2^{-1} I_2 + \mathbf{W})\mathbf{u}(\boldsymbol{\xi}) = (\mathbf{V}\boldsymbol{\varphi})(\boldsymbol{\xi}) , \quad \boldsymbol{\xi} \in \partial\omega . \tag{8.6}$$

(This is the so-called direct method of boundary integral equations.)

Let us consider an auxiliary exterior Dirichlet problem

$$L(\partial_{\boldsymbol{\xi}})\mathbf{v}(\boldsymbol{\xi}) = \mathbf{O} , \quad \text{in } C\bar{\omega} ,$$

$$\mathbf{v}(\boldsymbol{\xi}) = \boldsymbol{\psi}(\boldsymbol{\xi}) , \quad \text{on } \partial\omega ,$$

$$|\mathbf{v}| \quad \text{is bounded as} \quad |\boldsymbol{\xi}| \to \infty .$$

It is standard that representing $\mathbf{v}$ as the double layer potential $\mathbf{W}\sigma$, one arrives at the singular integral equation

$$(2^{-1} I_2 + \mathbf{W})\sigma(\boldsymbol{\xi}) = \boldsymbol{\psi}(\boldsymbol{\xi}) \quad \text{on } \partial\omega ,$$

which is uniquely solvable. Moreover, the inverse operator $(2^{-1} I_2 + \mathbf{W})^{-1}$ is bounded in $W_p^{1-1/p}(\partial\omega)$. Therefore, from (8.6), we obtain the estimate

$$\|\mathbf{u}\|_{W_p^{1-1/p}(\partial\omega)} \leq C\, \|\mathbf{V}\boldsymbol{\varphi}\|_{W_p^{1-1/p}(\partial\omega)} . \tag{8.7}$$

Since the kernel of the integral operator $\mathbf{V}$ has only the logarithmic singularity, the estimate

$$\|\mathbf{V}\boldsymbol{\varphi}\|_{W_p^{1-1/p}(\partial\omega)} \leq C\, \|\boldsymbol{\varphi}\|_{L_p(\partial\omega)} , \tag{8.8}$$

holds. In particular, by (8.7) and (8.8) we arrive at

$$\|\mathbf{u}\|_{W_p^{1-1/p}(\partial\omega)} \leq C\, \|\boldsymbol{\varphi}\|_{L_p(\partial\omega)} , \tag{8.9}$$

which implies

$$\|\mathbf{u}\|_{L_\infty(\partial\omega)} \le C\, \|\boldsymbol{\varphi}\|_{L_p(\partial\omega)}\,. \tag{8.10}$$

By (8.4)

$$|\mathbf{V}\boldsymbol{\varphi}(\boldsymbol{\xi})| \le C|\boldsymbol{\xi}|^{-1}\|\boldsymbol{\varphi}\|_{L_1(\partial\omega)}\,, \quad \text{for } |\boldsymbol{\xi}| \ge 2\,,$$

which in combination with (8.5) and (8.9) gives for $|\boldsymbol{\xi}| \ge 2$

$$\begin{aligned}|\mathbf{u}(\boldsymbol{\xi})| &\le C|\boldsymbol{\xi}|^{-1}(\|\boldsymbol{\varphi}\|_{L_1(\partial\omega)} + \|\mathbf{u}\|_{L_1(\partial\omega)})\\ &\le C|\boldsymbol{\xi}|^{-1}\|\boldsymbol{\varphi}\|_{L_p(\partial\omega)}\end{aligned} \tag{8.11}$$

where $p \in (1,\infty)$.

Now, using the inequalities (8.10) and (8.11), by Lemma 6.1, we have

$$\begin{aligned}\|\mathbf{u}\|_{L_\infty(B_3\setminus\bar{\omega})} &\le C(\|\mathbf{u}\|_{L_\infty(\partial B_3)} + \|\mathbf{u}\|_{L_\infty(\partial\omega)})\\ &\le C\,\|\boldsymbol{\varphi}\|_{L_p(\partial\omega)}\end{aligned}$$

which gives the final result owing to (8.11). □

8.3 An Estimate for Solutions to the Mixed Problem for the Lamé Equation in the Perforated Domain Ω_ε

The following result is a consequence of Lemma 8.1 and Fichera's maximum principle (Lemma 6.2, Chap. 6).

Lemma 8.2. *Let* $\mathbf{u}$ *be a vector function in* Ω_ε *such that* $\nabla\mathbf{u}$ *is square integrable in a neighborhood of* $\partial\omega_\varepsilon$ *and let* $\mathbf{u}$ *be a variational solution of the mixed boundary value problem*

$$L(\partial_{\mathbf{x}})\mathbf{u}(\mathbf{x}) = \mathbf{O}\,, \quad \mathbf{x} \in \Omega_\varepsilon\,, \tag{8.12}$$

$$T_n(\partial_{\mathbf{x}})\mathbf{u}(\mathbf{x}) = \boldsymbol{\varphi}_\varepsilon(\mathbf{x})\,, \quad \mathbf{x} \in \partial\omega_\varepsilon\,, \tag{8.13}$$

$$\mathbf{u}(\mathbf{x}) = \boldsymbol{\psi}(\mathbf{x})\,, \quad \mathbf{x} \in \partial\Omega\,, \tag{8.14}$$

where $\boldsymbol{\psi} \in L_\infty(\partial\Omega)$, $\boldsymbol{\varphi}_\varepsilon \in L_\infty(\partial\omega_\varepsilon)$, *and*

$$\int_{\partial\omega_\varepsilon} \boldsymbol{\varphi}_\varepsilon(\mathbf{x})\, dS_{\mathbf{x}} = \mathbf{O}\,. \tag{8.15}$$

Then there exists a positive constant A such that

$$\|\mathbf{u}\|_{L_\infty(\Omega_\varepsilon)} \le A\{\|\boldsymbol{\psi}\|_{L_\infty(\partial\Omega)} + \varepsilon\|\boldsymbol{\varphi}_\varepsilon\|_{L_\infty(\partial\omega_\varepsilon)}\} . \tag{8.16}$$

Proof. We introduce the inverse operators

$$\Pi : \boldsymbol{\psi} \to \mathbf{w} \quad \text{and} \quad N : \boldsymbol{\varphi} \to \mathbf{v} ,$$

for the boundary value problems

$$L(\partial_{\mathbf{x}})\mathbf{w}(\mathbf{x}) = \mathbf{O} , \quad \mathbf{x} \in \Omega ,$$

$$\mathbf{w}(\mathbf{x}) = \boldsymbol{\psi}(\mathbf{x}) , \quad \mathbf{x} \in \partial\Omega ,$$

and

$$\left.\begin{aligned} &L(\partial_{\boldsymbol{\xi}})\mathbf{v}(\boldsymbol{\xi}) = \mathbf{O} , \quad \boldsymbol{\xi} \in C\bar{\omega} , \\ &T_n(\partial_{\boldsymbol{\xi}})\mathbf{v}(\boldsymbol{\xi}) = \boldsymbol{\varphi}(\boldsymbol{\xi}) , \quad \boldsymbol{\xi} \in \partial\omega , \\ &\mathbf{v}(\boldsymbol{\xi}) \to \mathbf{O} \quad \text{as} \quad |\boldsymbol{\xi}| \to \infty , \end{aligned}\right\} \tag{8.17}$$

where $\boldsymbol{\psi} \in L_\infty(\partial\Omega)$ and $\boldsymbol{\varphi} \in L_\infty(\partial\omega)$.

Note that problem (8.17) is solvable if and only if

$$\int_{\partial\omega} \boldsymbol{\varphi}(\boldsymbol{\xi})\, dS_{\boldsymbol{\xi}} = \mathbf{O} .$$

We also need the operator N_ε given by

$$(N_\varepsilon \boldsymbol{\varphi}_\varepsilon)(\mathbf{x}) = (N\boldsymbol{\varphi})(\boldsymbol{\xi})$$

where $\boldsymbol{\varphi}_\varepsilon(\mathbf{x}) = \varepsilon^{-1}\boldsymbol{\varphi}(\varepsilon^{-1}\mathbf{x})$.

The case of the homogeneous displacement condition on $\partial\Omega$. We start by assuming zero boundary condition (8.14) on $\partial\Omega$.

Let us look for a solution of the problem (8.12)–(8.15) in the form

$$\mathbf{u} = N_\varepsilon \mathbf{g}_\varepsilon - \Pi(\mathrm{Tr}_{\partial\Omega} N_\varepsilon \mathbf{g}_\varepsilon) \tag{8.18}$$

with the unknown vector function $\mathbf{g}_\varepsilon$ defined on $\partial\omega_\varepsilon$ such that

$$\int_{\partial\omega} \mathbf{g}(\boldsymbol{\xi})\, d\boldsymbol{\xi} = \mathbf{O} ,$$

where we use the notation $\mathbf{g}_\varepsilon(\mathbf{x}) = \varepsilon^{-1}\mathbf{g}(\varepsilon^{-1}\mathbf{x})$.

Obviously, $\mathrm{Tr}_{\partial\Omega}\mathbf{u} = \mathbf{O}$. Furthermore, when $\mathbf{x} \in \partial\omega_\varepsilon$, we have

$$\boldsymbol{\varphi}_\varepsilon = \mathbf{g}_\varepsilon + S_\varepsilon \mathbf{g}_\varepsilon ,$$

where

$$S_\varepsilon \mathbf{g}_\varepsilon = -\mathrm{Tr}_{\partial\omega_\varepsilon} T_n(\partial_{\mathbf{x}})(\Pi(\mathrm{Tr}_{\partial\Omega} N_\varepsilon \mathbf{g}_\varepsilon)) .$$

Let B be a disk centered at the origin containing $\partial\omega_\varepsilon$, which doesn't intersect $\partial\Omega$.

By local regularity of solutions to the homogeneous Lamé system and Fichera's maximum principle (Lemma 6.2, Chap. 6), we have

$$\begin{aligned}\|T_n(\partial_{\mathbf{x}})(\Pi(\mathrm{Tr}_{\partial\Omega} N_\varepsilon \mathbf{g}_\varepsilon))\|_{L_\infty(\partial\omega_\varepsilon)} &\le \mathrm{Const}\ \|\Pi(\mathrm{Tr}_{\partial\Omega} N_\varepsilon \mathbf{g}_\varepsilon)\|_{L_\infty(B)} \\ &\le \mathrm{Const}\ \|N_\varepsilon \mathbf{g}_\varepsilon\|_{L_\infty(\partial\Omega)} ,\end{aligned}$$

and from this, by Lemma 8.1, we have

$$\|T_n(\partial_{\mathbf{x}})(\Pi(\mathrm{Tr}_{\partial\Omega} N_\varepsilon \mathbf{g}_\varepsilon))\|_{C(\partial\omega_\varepsilon)} \le \mathrm{Const}\ \varepsilon^2 \|\mathbf{g}_\varepsilon\|_{C(\partial\omega_\varepsilon)} .$$

Hence

$$\|S_\varepsilon\|_{C(\partial\omega_\varepsilon)\to C(\partial\omega_\varepsilon)} \le \mathrm{Const}\ \varepsilon^2 ,$$

thus the smallness of S_ε enables one to write

$$\mathbf{g}_\varepsilon = (I + S_\varepsilon)^{-1} \boldsymbol{\varphi}_\varepsilon$$

and

$$\|\mathbf{g}_\varepsilon\|_{L_\infty(\partial\omega_\varepsilon)} \le \mathrm{Const}\ \|\boldsymbol{\varphi}_\varepsilon\|_{L_\infty(\partial\omega_\varepsilon)} . \tag{8.19}$$

It follows from (8.18), using Lemmas 6.2, 8.1 and (8.19)

$$\sup_{\Omega_\varepsilon} |\mathbf{u}| \le \mathrm{Const}\ \varepsilon \|\mathbf{g}_\varepsilon\|_{C(\partial\omega_\varepsilon)} \le \mathrm{Const}\ \varepsilon \|\boldsymbol{\varphi}_\varepsilon\|_{C(\partial\omega_\varepsilon)} . \tag{8.20}$$

The case of the homogeneous traction condition on $\partial\omega_\varepsilon$. The solution of problem (8.12)–(8.15), is written in the form

$$\mathbf{u} = \Pi \boldsymbol{\psi} + \mathbf{v} ,$$

where the second term $\mathbf{v}$ is a solution of (8.12)–(8.15) with the homogeneous boundary condition on $\partial\Omega$ in (8.14) and the condition (8.13) is replaced by

$$T_n(\partial_{\mathbf{x}})\mathbf{v}(\mathbf{x}) = -T_n(\partial_{\mathbf{x}})(\Pi\boldsymbol{\phi})(\mathbf{x}) , \quad \mathbf{x} \in \partial\omega_\varepsilon .$$

According to the result (8.20) of the first part of the proof,

$$\sup_{\Omega_\varepsilon} |\mathbf{v}| \le \text{Const}\, \varepsilon \| T_n(\partial_x) \Pi \boldsymbol{\psi} \|_{L_\infty(\partial\omega_\varepsilon)} \,.$$

Then, using the local regularity of solutions to the homogeneous Lamé system and Lemma 6.2 (Fichera's maximum principle) we have

$$\sup_{\Omega_\varepsilon} |\mathbf{v}| \le \text{Const}\, \varepsilon \| \boldsymbol{\psi} \|_{L_\infty(\partial\Omega)} \,.$$

Thus

$$\sup_{\Omega_\varepsilon} |\mathbf{u}| \le \text{Const}\, \| \boldsymbol{\psi} \|_{L_\infty(\partial\Omega)} \,. \tag{8.21}$$

Combining (8.20) and (8.21) we complete the proof of (8.16). □

The aim of the next two sections is to obtain a uniform asymptotic formula for G_ε, defined as a solution of (8.1)–(8.3). In the first section we introduce the model tensors necessary for the representation of G_ε, the second section gives the main result and develops the asymptotic algorithm related to the current problem.

8.4 Model Boundary Value Problems

1. *The regular part H of Green's tensor in Ω.* Let $H(\mathbf{x},\mathbf{y}) = [H_{ij}(\mathbf{x},\mathbf{y})]_{i,j=1}^2$ denote the regular part of Green's tensor in the domain Ω, which solves

$$L(\partial_{\mathbf{x}}) H(\mathbf{x},\mathbf{y}) = 0 I_2 \,, \quad \mathbf{x},\mathbf{y} \in \Omega \,,$$

$$H(\mathbf{x},\mathbf{y}) = \gamma(\mathbf{x},\mathbf{y}) \,, \quad \mathbf{x} \in \partial\Omega, \mathbf{y} \in \Omega \,. \tag{8.22}$$

Here, $\gamma(\mathbf{x},\mathbf{y}) = [\gamma_{ij}(\mathbf{x},\mathbf{y})]_{i,j=1}^2$ is the fundamental solution of the Lamé operator in two dimensions, with entries given by

$$\begin{aligned}\gamma_{ij}(\mathbf{x},\mathbf{y}) = K_2 (&-\log|\mathbf{x}-\mathbf{y}|\delta_{ij} \\ &+(\lambda+\mu)(\lambda+3\mu)^{-1}(x_i-y_i)(x_j-y_j)|\mathbf{x}-\mathbf{y}|^{-2}) \,,\end{aligned}$$

for $i, j = 1, 2$, where K_2 is given by (7.50) of Chap. 7. The tensor G is related to H by

$$G(\mathbf{x},\mathbf{y}) = \gamma(\mathbf{x},\mathbf{y}) - H(\mathbf{x},\mathbf{y}) \,, \tag{8.23}$$

where as discussed in Chap. 6, G satisfies the symmetry relation (6.176).

2. *The Neumann tensor in* $C\bar{\omega}$. We also make use of the Neumann tensor $\mathcal{N}(\boldsymbol{\xi},\boldsymbol{\eta}) = [\mathcal{N}_{ij}(\boldsymbol{\xi},\boldsymbol{\eta})]_{i,j=1}^2$ in the domain $C\bar{\omega}$, and this solves the problem

$$L(\partial_{\boldsymbol{\xi}})\mathcal{N}(\boldsymbol{\xi},\boldsymbol{\eta}) = -\delta(\boldsymbol{\xi}-\boldsymbol{\eta})I_2\,, \quad \boldsymbol{\xi} \in C\overline{\omega},$$

$$T_n(\partial_{\boldsymbol{\xi}})\mathcal{N}(\boldsymbol{\xi},\boldsymbol{\eta}) = 0I_2\,, \quad \boldsymbol{\xi} \in \partial\omega,$$

$$\mathcal{N}(\boldsymbol{\xi},\boldsymbol{\eta}) \sim \gamma(\boldsymbol{\xi},\boldsymbol{\eta}) \quad \text{as} \quad |\boldsymbol{\xi}| \to \infty,$$

where $\boldsymbol{\eta} \in C\bar{\omega}$. From the above definitions it follows that the Neumann tensor satisfies the symmetry relation

$$\mathcal{N}(\boldsymbol{\xi},\boldsymbol{\eta}) = (\mathcal{N}(\boldsymbol{\eta},\boldsymbol{\xi}))^T, \quad \boldsymbol{\xi},\boldsymbol{\eta} \in C\bar{\omega}, \boldsymbol{\xi} \neq \boldsymbol{\eta}\,.$$

Similarly to G, $\mathcal{N}$ is written as

$$\mathcal{N}(\boldsymbol{\xi},\boldsymbol{\eta}) = \gamma(\boldsymbol{\xi},\boldsymbol{\eta}) - h(\boldsymbol{\xi},\boldsymbol{\eta})\,, \tag{8.24}$$

where h is the regular part of $\mathcal{N}$.

8.4.1 *The Dipole Fields*

By $\mathcal{W}^{(p)} = \{\mathcal{W}_{ip}(\boldsymbol{\xi})\}_{i=1}^2$, $p = 1, 2, 3$ we mean the dipole fields for the void ω. These vectors comprise the columns of the 2×3 matrix $\mathcal{W}$, which solves

$$L(\partial_{\boldsymbol{\xi}})\mathcal{W}(\boldsymbol{\xi}) = \mathbf{0}_{2\times 3}\,, \quad \boldsymbol{\xi} \in C\bar{\omega}\,, \tag{8.25}$$

$$T_n(\partial_{\boldsymbol{\xi}})\mathcal{W}(\boldsymbol{\xi}) = T_n(\partial_{\boldsymbol{\xi}})\mathbf{D}(\boldsymbol{\xi})\,, \quad \boldsymbol{\xi} \in \partial\omega\,, \tag{8.26}$$

$$\mathcal{W}(\boldsymbol{\xi}) \to \mathbf{0}_{2\times 3}\,, \quad \text{as} \quad |\boldsymbol{\xi}| \to \infty\,, \tag{8.27}$$

where $\mathbf{0}_{2\times 3}$, is the 2×3 null matrix, and $\mathbf{D}(\boldsymbol{\xi})$ is a 2×3 matrix given by

$$\mathbf{D}(\boldsymbol{\xi}) = \begin{pmatrix} \xi_1 & 0 & 2^{-1/2}\xi_2 \\ 0 & \xi_2 & 2^{-1/2}\xi_1 \end{pmatrix}\,. \tag{8.28}$$

Therefore, the right-hand side of (8.26) is equal to

$$T_n(\partial_{\boldsymbol{\xi}})\mathbf{D}(\boldsymbol{\xi}) = \mathbf{D}(\mathbf{n})\mathcal{C} = \begin{pmatrix} (\lambda+2\mu)n_1 & \lambda n_1 & 2^{1/2}\mu n_2 \\ \lambda n_2 & (\lambda+2\mu)n_2 & 2^{1/2}\mu n_1 \end{pmatrix}, \mathbf{x} \in \partial\omega\,, \tag{8.29}$$

where $\mathbf{n} = (n_1, n_2)^T$ is the unit outward normal to ω

We note that from the problem (8.25)–(8.29), it can be shown that the columns of the boundary condition (8.26) are self-balanced i.e. we recall that for the resultant vector of forces on the boundary we have

$$\int_{\partial\omega} T_n(\partial_{\boldsymbol{\xi}})\mathcal{W}^{(p)}(\boldsymbol{\xi})\, dS_{\boldsymbol{\xi}} = \mathbf{O}\,, \tag{8.30}$$

and for the resultant moment

$$\int_{\partial\omega} \{\xi_1 t_2(\mathcal{W}^{(p)}(\boldsymbol{\xi})) - \xi_2 t_1(\mathcal{W}^{(p)}(\boldsymbol{\xi}))\} dS_{\boldsymbol{\xi}} = 0\,, \tag{8.31}$$

where $p = 1, 2, 3$, and $t_i, i = 1, 2$, are components of tractions.

8.4.1.1 An Estimate for the Columns of $\mathcal{W}$

The next result contains an estimate for the columns of $\mathcal{W}$:

Lemma 8.3. *For* $\mathcal{W}^{(p)}$, $p = 1, 2, 3$, *the estimate*

$$\sup_{\boldsymbol{\xi}\in C\bar{\omega}} \{|\boldsymbol{\xi}||\mathcal{W}^{(p)}(\boldsymbol{\xi})|\} \leq \text{Const} \tag{8.32}$$

holds.

Proof. Since the columns of traction boundary condition on $\partial\omega$ for the matrix $\mathcal{W}$ (see (8.26)) are self-balanced, the above estimate (8.32) follows from Lemma 8.1. □

8.4.2 *The Elastic Dipole Matrix*

The asymptotics of the dipole fields in the neighborhood of infinity can be described using the dipole matrix M for the void ω, which is a 3×3 symmetric matrix and is an integral characteristic for the void (see Movchan, Movchan and Poulton [34]).

In Chap. 6, we showed for the case of three-dimensional elasticity that the symmetric elastic capacity matrix B was a Cartesian tensor of rank 2 (cf. Lemma 6.8), and the quantity $2^{-1}B$ represents the elastic energy matrix for the capacitary potential.

We can use a Cartesian tensor of rank 4, say $\mathcal{M}_{ijkl}$, to represent the dipole matrix. This matrix characterizes the energy increment of a field when a void is introduced.

For let $\mathbf{u}_0$ be an unperturbed field in the infinite plane before a void is introduced at the origin, and consider its vector of strain

$$S(\mathbf{u}_0) = (e_{11}(\mathbf{u}_0), e_{22}(\mathbf{u}_0), \sqrt{2}\, e_{12}(\mathbf{u}_0))^T ,$$

where $e_{ij}(\mathbf{u}_0)$ are the components of the strain tensor for the vector $\mathbf{u}_0$.

Then, when a void is placed within the plane, we have that the increment in the elastic energy $\delta\mathscr{E}$ is characterized by

$$2\delta\mathscr{E} = e_{ij}(\mathbf{u}_0)\Big|_{\mathbf{x}=\mathbf{O}} \mathcal{M}_{ijkl} e_{kl}(\mathbf{u}_0)\Big|_{\mathbf{x}=\mathbf{O}} = S^T(\mathbf{u}_0)\Big|_{\mathbf{x}=\mathbf{O}} M\, S(\mathbf{u}_0)\Big|_{\mathbf{x}=\mathbf{O}} .$$

In the asymptotic representation of the elastic capacitary potential matrix at infinity, the elastic capacity B is the coefficient near the fundamental solution in three-dimensional elasticity (cf. Lemma 6.6, Chap. 6). The dipole matrix is also present in the asymptotic behaviour of the dipole fields $\mathcal{W}^{(p)}$, $p = 1, 2, 3$, in the neighborhood of infinity. The latter information is contained in the next subsection.

8.4.3 The Asymptotics of the Matrix $\mathcal{W}$ at Infinity

In order to construct an asymptotic approximation for the dipole fields $\mathcal{W}^{(p)}$, $p = 1, 2, 3$ we need the following lemma, which is particular case of that considered in Kondratiev and Oleinik [18].

Lemma 8.4. *Suppose the columns $u^{(j)}(\xi)$ of the matrix $u(\xi)$ are solutions of*

$$\mu\Delta u^{(j)}(\xi) + (\lambda+\mu)\nabla(\nabla\cdot u^{(j)}(\xi)) = \mathbf{O} , \quad \text{in } C\bar{\omega} ,$$

and that $|u^{(j)}(\xi)| \le \text{Const}\,(1+|\xi|)^k$, $k \ge 0$, for $j = 1, 2$.

Then for $|\xi| > 2$

$$u^{(j)}(\xi) = \mathscr{P}_k^{(j)}(\xi) + \sum_{0\le|\alpha|\le 1} \mathscr{D}_\xi^\alpha \gamma(\xi, \mathbf{O}) C^{(j,\alpha)} + O(|\xi|^{-2}) , \tag{8.33}$$

where $\mathscr{P}_k^{(j)}(\xi) = \{\mathscr{P}_i^{(j,k)}(\xi)\}_{i=1}^2$, $\mathscr{P}_i^{(j,k)}(\xi)$ are polynomials of order not greater than k, $\alpha = (\alpha_1, \alpha_2)^T$ is a multi-index, $\mathscr{D}_\xi^\alpha = \partial^{|\alpha|}/(\partial\xi_1^{\alpha_1}\partial\xi_2^{\alpha_2})$, $C^{(j,\alpha)} = \{C_i^{(j,\alpha)}\}_{i=1}^2$, where $C_i^{(j,\alpha)}$ are constants.

The next lemma will be used when we address the simplification of the uniform asymptotics of Green's tensor under constraints on the spatial variables.

Lemma 8.5. *For $|\xi| > 2$, the matrix $\mathcal{W}(\xi)$ admits the representation*

$$\mathcal{W}(\xi) = -(\mathbf{D}(\partial_\xi)^T \gamma(\xi, \mathbf{O}))^T M + O(|\xi|^{-2}) .$$

Proof. Since the columns $\mathcal{W}^{(p)}$, $p = 1,2,3$ are a solutions of the Lamé equation, by Lemma 8.4 in $C\bar{\omega}$, they admit the following the representation

$$\mathcal{W}^{(p)}(\boldsymbol{\xi}) = \mathscr{P}_k^{(p)}(\boldsymbol{\xi}) + \sum_{0\le|\alpha|\le 1} \mathscr{D}_{\boldsymbol{\xi}}^{\alpha}\gamma(\boldsymbol{\xi},\mathbf{O})C^{(p,\alpha)} + O(|\boldsymbol{\xi}|^{-2}) , \tag{8.34}$$

where all items on the right-hand side of the preceding equation are as in the formulation of the previous lemma.

Next consulting the Lemma 8.3 we can assume $\mathcal{W}^{(p)}(\boldsymbol{\xi}) = O(|\boldsymbol{\xi}|^{-1})$, $p = 1,2,3$ for $\boldsymbol{\xi} \in C\bar{\omega}$. Thus the terms $\mathscr{P}_k^{(p)}(\boldsymbol{\xi})$ and the coefficient near γ are equal to the zero vector.

Therefore we are left with the approximation

$$\mathcal{W}^{(p)}(\boldsymbol{\xi}) = \sum_{|\alpha|=1} \mathscr{D}_{\boldsymbol{\xi}}^{\alpha}\gamma(\boldsymbol{\xi},\mathbf{O})C^{(p,\alpha)} + O(|\boldsymbol{\xi}|^{-2}) , \tag{8.35}$$

for $p = 1,2,3$, where the leading order term here may be rewritten in the form given in that of (8.5). □

8.4.4 The Matrix Function $\boldsymbol{\Upsilon}$

In the following, it is convenient to introduce the notation

$$\boldsymbol{\Upsilon}(\boldsymbol{\xi}) = \mathbf{D}(\boldsymbol{\xi}) - \mathcal{W}(\boldsymbol{\xi}) . \tag{8.36}$$

Therefore, the tensor $\boldsymbol{\Upsilon}$ solves

$$L(\partial_{\boldsymbol{\xi}})\boldsymbol{\Upsilon}(\boldsymbol{\xi}) = \mathbf{0}_{2\times 3} , \quad \boldsymbol{\xi} \in C\bar{\omega} , \tag{8.37}$$

$$T_n(\partial_{\boldsymbol{\xi}})\boldsymbol{\Upsilon}(\boldsymbol{\xi}) = \mathbf{0}_{2\times 3} , \quad \boldsymbol{\xi} \in \partial\omega , \tag{8.38}$$

$$\boldsymbol{\Upsilon}(\boldsymbol{\xi}) \sim \mathbf{D}(\boldsymbol{\xi}) \quad \text{as} \quad |\boldsymbol{\xi}| \to \infty , \tag{8.39}$$

which is consistent with (8.25)–(8.29).

8.4.5 An Estimate for the Regular Part of the Neumann Tensor in the Unbounded Domain

We also obtain an approximation of the regular part of the Neumann tensor which is contained in the following lemma

Lemma 8.6. *For $|\xi| > 2$ and $\eta \in C\bar{\omega}$, the regular part h of the Neumann tensor in $C\bar{\omega}$ admits the representation*

$$h(\xi,\eta) = -(\mathrm{D}(\partial_\xi)^T \gamma(\xi,\mathbf{O}))^T \mathcal{W}^T(\eta) + O(|\xi|^{-2}|\eta|^{-1}). \qquad (8.40)$$

Proof. Let $h^{(l)}(\xi,\eta)$, $l = 1,2$, be a column of the regular part h of the Neumann tensor, and $\boldsymbol{\Upsilon}^{(k)}(\xi)$, $k = 1,2,3$ a column of the matrix function $\boldsymbol{\Upsilon}(\xi)$ (see (8.36)–(8.39)).

Take $B_R(\mathbf{O}) = \{\xi : |\xi| < R\}$ to be a disk with sufficiently large radius R. We begin by applying Betti's formula to the vectors $h^{(l)}(\xi,\eta)$ and $\boldsymbol{\Upsilon}^{(k)}(\xi)$ in the domain $B_R\backslash\bar{\omega}$ to obtain

$$\begin{aligned}0 = &\int_{\partial B_R} \{h^{(l)}(\xi,\eta) \cdot T_n(\partial_\xi)\boldsymbol{\Upsilon}^{(k)}(\xi) - \boldsymbol{\Upsilon}^{(k)}(\xi) \cdot T_n(\partial_\xi) h^{(l)}(\xi,\eta)\, dS_\xi \\ &- \int_{\partial\omega} \boldsymbol{\Upsilon}^{(k)}(\xi) \cdot T_n(\partial_\xi) h^{(l)}(\xi,\eta)\, dS_\xi\,, \end{aligned} \qquad (8.41)$$

where we have used that $\boldsymbol{\Upsilon}^{(k)}$ and $h^{(l)}$ are solutions of the homogeneous Lamé equation and the boundary condition (8.38). Dealing with the last integral in (8.41), we have by the definition of h and $\boldsymbol{\Upsilon}$, this integral is equal to

$$\begin{aligned}&-\int_{\partial\omega} \boldsymbol{\Upsilon}^{(k)}(\xi) \cdot T_n(\partial_\xi)\gamma^{(l)}(\xi,\eta)\, dS_\xi \\ &= \int_{\partial\omega} \{\mathcal{W}^{(k)}(\xi) \cdot T_n(\partial_\xi)\gamma^{(l)}(\xi,\eta) - \mathrm{D}^{(k)}(\xi) \cdot T_n(\partial_\xi)\gamma^{(l)}(\xi,\eta)\}\, dS_\xi \\ &= \int_{\partial\omega} \{\mathcal{W}^{(k)}(\xi) \cdot T_n(\partial_\xi)\gamma^{(l)}(\xi,\eta) - \gamma^{(l)}(\xi,\eta) \cdot T_n(\partial_\xi)\mathrm{D}^{(k)}(\xi)\}\, dS_\xi, \end{aligned} \qquad (8.42)$$

where $\mathrm{D}^{(k)} = \{D_{ik}\}_{i=1}^2$ is k^{th} column of the matrix D. In moving from the second line in (8.42) to the last, we applied Betti's formula to the vectors $\gamma^{(l)}(\xi,\eta)$ and $\mathbf{D}^{(k)}(\xi)$ in the domain ω.

Now, by applying Betti's formula to the vectors $\mathcal{W}^{(k)}$ and $\gamma^{(l)}$ in the domain $B_R\backslash\bar{\omega}$, the integral on the right-hand side of (8.42) is equivalent to

$$\begin{aligned}&\int_{\partial\omega} \{\mathcal{W}^{(k)}(\xi) \cdot T_n(\partial_\xi)\gamma^{(l)}(\xi,\eta) - \gamma^{(l)}(\xi,\eta) \cdot T_n(\partial_\xi)\mathbf{D}^{(k)}(\xi)\}\, dS_\xi \\ &= -\mathcal{W}_{lk}(\eta) - \int_{\partial B_R} \{\mathcal{W}^{(k)}(\xi) \cdot T_n(\partial_\xi)\gamma^{(l)}(\xi,\eta) \\ &\quad -\gamma^{(l)}(\xi,\eta) \cdot T_n(\partial_\xi)\mathcal{W}^{(k)}(\xi)\}\, dS_\xi\,. \end{aligned} \qquad (8.43)$$

The last identity holds for all sufficiently large R and taking the limit as R tends to infinity, the integral on the right-hand side of (8.43) by Lemma 8.5 tends to zero.

Thus we have shown

$$\int_{\partial\omega} \boldsymbol{\Upsilon}^{(k)}(\boldsymbol{\xi}) \cdot T_n(\partial_{\boldsymbol{\xi}}) h^{(l)}(\boldsymbol{\xi}, \boldsymbol{\eta})\, dS_{\boldsymbol{\xi}} = \mathcal{W}_{lk}(\boldsymbol{\eta}) \,. \tag{8.44}$$

Combining (8.44) with (8.41) we have

$$\mathcal{W}_{lk}(\boldsymbol{\eta}) = \int_{\partial B_R} \{h^{(l)}(\boldsymbol{\xi}, \boldsymbol{\eta}) \cdot T_n(\partial_{\boldsymbol{\xi}}) \boldsymbol{\Upsilon}^{(k)}(\boldsymbol{\xi}) - \boldsymbol{\Upsilon}^{(k)}(\boldsymbol{\xi}) \cdot T_n(\partial_{\boldsymbol{\xi}}) h^{(l)}(\boldsymbol{\xi}, \boldsymbol{\eta})\}\, dS_{\boldsymbol{\xi}}, \tag{8.45}$$

which once again holds for all sufficiently large R.

From the definition of h (see (8.24)), the columns of this matrix function, owing to Lemma 8.4 and in a similar way to the proof of Lemma 8.5, for $|\boldsymbol{\xi}| > 2$ admits an estimate of the form

$$h^{(l)}(\boldsymbol{\xi}, \boldsymbol{\eta}) = C_{sl}(\boldsymbol{\eta}) D_{vs}(\partial_{\boldsymbol{\xi}}) \gamma^{(v)}(\boldsymbol{\xi}, \mathbf{O}) + r^{(l)}(\boldsymbol{\xi}, \boldsymbol{\eta}) \,, \tag{8.46}$$

where $r^{(l)}(\boldsymbol{\xi}, \boldsymbol{\eta})$ are columns of the remainder such that its behaviour in $\boldsymbol{\xi}$ is estimated by $O(|\boldsymbol{\xi}|^{-2})$, and the constant in this estimate can depend on $\boldsymbol{\eta}$.

Then, returning to (8.45) and passing to the limit as $R \to \infty$, and using (8.46) we obtain

$$\begin{aligned} \mathcal{W}_{lk}(\boldsymbol{\eta}) = \lim_{R\to\infty} \int_{\partial B_R} &\{(C_{sl}(\boldsymbol{\eta}) D_{vs}(\partial_{\boldsymbol{\xi}}) \gamma^{(v)}(\boldsymbol{\xi}, \mathbf{O})) \cdot T_n(\partial_{\boldsymbol{\xi}}) \mathbf{D}^{(k)}(\boldsymbol{\xi}) \\ &- \mathbf{D}^{(k)}(\boldsymbol{\xi}) \cdot (C_{sl}(\boldsymbol{\eta}) D_{vs}(\partial_{\boldsymbol{\xi}}) T_n(\partial_{\boldsymbol{\xi}}) \gamma^{(v)}(\boldsymbol{\xi}, \mathbf{O}))\}\, dS_{\boldsymbol{\xi}} \,. \end{aligned} \tag{8.47}$$

One more application of Betti's formula to the vectors $C_{sl}(\boldsymbol{\eta}) D_{vs}(\partial_{\boldsymbol{\xi}}) \gamma^{(v)}(\boldsymbol{\xi}, \mathbf{O})$ and $\mathbf{D}^{(k)}(\boldsymbol{\xi})$ in B_R yields the relation

$$\mathcal{W}_{lk}(\boldsymbol{\eta}) = \int_{B_R} D_{vk}(\boldsymbol{\xi}) C_{sl}(\boldsymbol{\eta}) D_{vs}(\partial_{\boldsymbol{\xi}}) \delta(\boldsymbol{\xi})\, d\boldsymbol{\xi} \,,$$

and computing the right-hand side of this gives

$$\mathcal{W}_{lk}(\boldsymbol{\eta}) = -C_{sl}(\boldsymbol{\eta}) (D_{vs}(\partial_{\boldsymbol{\xi}}) D_{vk}(\xi))|_{\boldsymbol{\xi}=0} \,. \tag{8.48}$$

Then using (8.28) in (8.48) we obtain

$$C_{kl}(\boldsymbol{\eta}) = -\mathcal{W}_{lk}(\boldsymbol{\eta}) \,. \tag{8.49}$$

Estimation of the Remainder. Now we investigate the estimate of the remainder produced by this approximation for $h^{(j)}$, $j = 1, 2$.

Using (8.46), (8.49) we have the representation for the matrix h for $|\boldsymbol{\xi}| > 2$

$$h(\boldsymbol{\xi}, \boldsymbol{\eta}) = -(\mathrm{D}(\partial_{\boldsymbol{\xi}})^T \gamma(\boldsymbol{\xi}, \mathbf{O}))^T \mathcal{W}^T(\boldsymbol{\eta}) + r(\boldsymbol{\xi}, \boldsymbol{\eta}) , \tag{8.50}$$

where r is a matrix whose components are $O(|\boldsymbol{\xi}|^{-2})$.

Consider the matrix $h(\boldsymbol{\eta}, \boldsymbol{\xi})$, which satisfies

$$L(\partial_{\boldsymbol{\eta}})h(\boldsymbol{\eta}, \boldsymbol{\xi}) = 0I_2 , \quad \boldsymbol{\eta}, \boldsymbol{\xi} \in C\bar{\omega} , \tag{8.51}$$

$$T_n(\partial_{\boldsymbol{\eta}})h(\boldsymbol{\eta}, \boldsymbol{\xi}) = T_n(\partial_{\boldsymbol{\eta}})\gamma(\boldsymbol{\eta}, \boldsymbol{\xi}) , \quad \boldsymbol{\eta} \in \partial\omega, \boldsymbol{\xi} \in C\bar{\omega} , \tag{8.52}$$

$$h(\boldsymbol{\eta}, \boldsymbol{\xi}) \to 0I_2 , \quad \text{as} \quad |\boldsymbol{\eta}| \to \infty, \boldsymbol{\xi} \in C\bar{\omega} , \tag{8.53}$$

where the columns of the boundary condition (8.52) are self-balanced.

We recall from the symmetry relation of the Neumann tensor, that $(h(\boldsymbol{\xi}, \boldsymbol{\eta}))^T = h(\boldsymbol{\eta}, \boldsymbol{\xi})$, and set $r^T(\boldsymbol{\xi}, \boldsymbol{\eta}) = h(\boldsymbol{\eta}, \boldsymbol{\xi}) + \mathcal{W}(\boldsymbol{\eta})\mathbf{D}(\partial_{\boldsymbol{\xi}})^T \gamma(\boldsymbol{\xi}, \mathbf{O})$. The problem for r^T is then

$$L(\partial_{\boldsymbol{\eta}})r^T(\boldsymbol{\xi}, \boldsymbol{\eta}) = 0I_2 , \quad \boldsymbol{\eta}, \boldsymbol{\xi} \in C\bar{\omega} , \tag{8.54}$$

$$T_n(\partial_{\boldsymbol{\eta}})r^T(\boldsymbol{\xi}, \boldsymbol{\eta}) = T_n(\partial_{\boldsymbol{\eta}})\{\gamma(\boldsymbol{\eta}, \boldsymbol{\xi}) + \mathcal{W}(\boldsymbol{\eta})D(\partial_{\boldsymbol{\xi}})^T \gamma(\boldsymbol{\xi}, \mathbf{O})\}, \ \text{for } \boldsymbol{\eta} \in \partial\omega, \boldsymbol{\xi} \in C\bar{\omega}, \tag{8.55}$$

$$r^T(\boldsymbol{\xi}, \boldsymbol{\eta}) \to 0I_2 , \quad \text{as} \quad |\boldsymbol{\eta}| \to \infty, \boldsymbol{\xi} \in C\bar{\omega} , \tag{8.56}$$

where the right-hand side of condition (8.55) is also self-balanced.

Now we note that

$$\begin{aligned} T_n(\partial_{\boldsymbol{\eta}})\mathcal{W}(\boldsymbol{\eta})\mathbf{D}(\partial_{\boldsymbol{\xi}})^T \gamma(\boldsymbol{\xi}, \mathbf{O}) &= -\lim_{\mathbf{z}\to\mathbf{O}} T_n(\partial_{\boldsymbol{\eta}})\mathcal{W}(\boldsymbol{\eta})\mathbf{D}(\partial_{\mathbf{z}})^T \gamma(\boldsymbol{\xi}, \mathbf{z}) \\ &= -\lim_{\mathbf{z}\to\mathbf{O}} T_n(\partial_{\mathbf{z}})\gamma(\boldsymbol{\xi}, \mathbf{z}) , \end{aligned} \tag{8.57}$$

for $\boldsymbol{\eta} \in \partial\omega, \boldsymbol{\xi} \in C\bar{\omega}$.

Let $|\boldsymbol{\xi}| > 2$, $\boldsymbol{\eta} \in \partial\omega$, and consider condition (8.55). Using (8.57), we estimate the matrix norm of (8.55) as follows

$$\begin{aligned} ||T_n(\partial_{\boldsymbol{\eta}})r^T(\boldsymbol{\xi}, \boldsymbol{\eta})|| &= ||T_n(\partial_{\boldsymbol{\eta}})\gamma(\boldsymbol{\eta}, \boldsymbol{\xi}) - \lim_{\mathbf{z}\to\mathbf{O}} T_n(\partial_{\mathbf{z}})\gamma(\boldsymbol{\xi}, \mathbf{z})|| \\ &\leq \text{Const } |\boldsymbol{\eta}||\boldsymbol{\xi}|^{-2} \leq \text{Const } |\boldsymbol{\xi}|^{-2} , \end{aligned} \tag{8.58}$$

where it has been used for $\boldsymbol{\eta} \in \partial\omega$, $|\boldsymbol{\eta}| \leq 1$. Then, by Lemma 8.1, we obtain that $r(\boldsymbol{\xi}, \boldsymbol{\eta}) = O(|\boldsymbol{\xi}|^{-2}|\boldsymbol{\eta}|^{-1})$. □

8.5 A Uniform Asymptotic Formula for G_ε of the Mixed Problem in a Domain with a Void

Now we have described the model fields and associated asymptotic estimates for the algorithm, we will obtain a uniform asymptotic approximation of G_ε for the mixed problem. We have the theorem

Theorem 8.1. *Green's tensor for the mixed boundary value problem of the Lamé operator in $\Omega_\varepsilon \subset \mathbb{R}^2$ admits the representation*

$$G_\varepsilon(\mathbf{x},\mathbf{y}) = G(\mathbf{x},\mathbf{y}) + \mathcal{N}(\boldsymbol{\xi},\boldsymbol{\eta}) - \gamma(\boldsymbol{\xi},\boldsymbol{\eta})$$
$$+\varepsilon\mathcal{W}(\boldsymbol{\xi})\mathbf{D}(\partial_{\mathbf{x}})^T H(\mathbf{O},\mathbf{y}) + \varepsilon(\mathbf{D}(\partial_{\mathbf{y}})^T H(\mathbf{O},\mathbf{x}))^T \mathcal{W}^T(\boldsymbol{\eta}) + O(\varepsilon^2) \tag{8.59}$$

which is uniform with respect to $\mathbf{x},\mathbf{y} \in \Omega_\varepsilon$.

Proof. We deal with the proof in two parts. First we present a formal argument which will enable one to obtain the leading order term in (8.59). Second we give a rigorous proof of the remainder in (8.59).

Formal Argument

Let G_ε have the representation

$$G_\varepsilon(\mathbf{x},\mathbf{y}) = \gamma(\mathbf{x},\mathbf{y}) - \mathfrak{M}_\varepsilon(\mathbf{x},\mathbf{y}) , \tag{8.60}$$

where it suffices to seek the approximation of the tensor $\mathfrak{M}_\varepsilon(\mathbf{x},\mathbf{y})$, which is a solution of the problem

$$L(\partial_{\mathbf{x}})\mathfrak{M}_\varepsilon(\mathbf{x},\mathbf{y}) = 0I_2 , \quad \mathbf{x},\mathbf{y} \in \Omega_\varepsilon ,$$

$$\mathfrak{M}_\varepsilon(\mathbf{x},\mathbf{y}) = \gamma(\mathbf{x},\mathbf{y}) , \quad \mathbf{x} \in \partial\Omega, \mathbf{y} \in \Omega_\varepsilon , \tag{8.61}$$

$$T_n(\partial_{\mathbf{x}})\mathfrak{M}_\varepsilon(\mathbf{x},\mathbf{y}) = T_n(\partial_{\mathbf{x}})\gamma(\mathbf{x},\mathbf{y}) , \quad \mathbf{x} \in \partial\omega_\varepsilon, \mathbf{y} \in \Omega_\varepsilon . \tag{8.62}$$

The Approximation of $\mathfrak{M}_\varepsilon$

Using the scaled coordinates in the boundary condition (8.62), we have

$$T_n(\partial_{\mathbf{x}})\mathfrak{M}_\varepsilon(\mathbf{x},\mathbf{y}) = T_n(\partial_{\mathbf{x}})\gamma(\boldsymbol{\xi},\boldsymbol{\eta}) , \quad \mathbf{x} \in \partial\omega_\varepsilon, \mathbf{y} \in \Omega_\varepsilon . \tag{8.63}$$

In view of the boundary conditions (8.61), (8.63), we write $\mathfrak{M}_\varepsilon$ in the form

$$\mathfrak{M}_\varepsilon(\mathbf{x}, \mathbf{y}) = H(\mathbf{x}, \mathbf{y}) + h(\boldsymbol{\xi}, \boldsymbol{\eta}) + R_\varepsilon^{(1)}(\mathbf{x}, \mathbf{y}) . \tag{8.64}$$

Here $R_\varepsilon^{(1)}$ is a solution of the homogeneous Lamé equation for $\mathbf{x}, \mathbf{y} \in \Omega_\varepsilon$. The displacement condition for $R_\varepsilon^{(1)}$ is given by

$$R_\varepsilon^{(1)}(\mathbf{x}, \mathbf{y}) = -h(\boldsymbol{\xi}, \boldsymbol{\eta}) \ , \ \text{for } \mathbf{x} \in \partial\Omega, \mathbf{y} \in \Omega_\varepsilon \ ,$$

where the asymptotics of h in Lemma 8.6 allows one to replace this condition by

$$\begin{aligned} R_\varepsilon^{(1)}(\mathbf{x}, \mathbf{y}) = & -\varepsilon \lim_{\mathbf{z} \to \mathbf{O}} (\mathbf{D}(\partial_\mathbf{z})^T \gamma(\mathbf{x}, \mathbf{z}))^T \mathcal{W}^T(\boldsymbol{\eta}) \\ & + O(\varepsilon^3 |\mathbf{y}|^{-1}) \text{ for } \mathbf{x} \in \partial\Omega, \mathbf{y} \in \Omega_\varepsilon . \end{aligned} \tag{8.65}$$

The boundary condition for $R_\varepsilon^{(1)}$ on the interior contour $\partial\omega_\varepsilon$ takes the form

$$T_n(\partial_\mathbf{x}) R_\varepsilon^{(1)}(\mathbf{x}, \mathbf{y}) = -T_n(\partial_\mathbf{x}) H(\mathbf{x}, \mathbf{y}) \ , \quad \mathbf{x} \in \partial\omega_\varepsilon, \mathbf{y} \in \Omega_\varepsilon \ .$$

Then using the Taylor expansion of H about $\mathbf{x} = \mathbf{O}$, this boundary condition is equivalent to

$$\begin{aligned} T_n(\partial_\mathbf{x}) R_\varepsilon^{(1)}(\mathbf{x}, \mathbf{y}) = & -\mathbf{D}(\mathbf{n}) \mathcal{C} \mathbf{D}(\partial_\mathbf{x})^T H(\mathbf{O}, \mathbf{y}) \\ & + O(\varepsilon) \ , \quad \mathbf{x} \in \partial\omega_\varepsilon, \mathbf{y} \in \Omega_\varepsilon \ . \end{aligned} \tag{8.66}$$

In order to correct for the discrepancies present in (8.65) and (8.66), we consult the boundary conditions for the regular part H in (8.22) and that for the matrix $\mathcal{W}$ in (8.29), and construct $R_\varepsilon^{(1)}$ to leading order in the form

$$R_\varepsilon^{(1)}(\mathbf{x}, \mathbf{y}) \sim -\varepsilon \mathcal{W}(\boldsymbol{\xi}) \mathbf{D}(\partial_\mathbf{x})^T H(\mathbf{O}, \mathbf{y}) - \varepsilon (\mathbf{D}(\partial_\mathbf{y})^T H(\mathbf{O}, \mathbf{x}))^T \mathcal{W}^T(\boldsymbol{\eta}) . \tag{8.67}$$

Combined Formula

Substituting (8.64) and (8.67) into (8.60) we have the following representation for G_ε

$$\begin{aligned} G_\varepsilon(\mathbf{x}, \mathbf{y}) = & \gamma(\mathbf{x}, \mathbf{y}) - H(\mathbf{x}, \mathbf{y}) - h(\boldsymbol{\xi}, \boldsymbol{\eta}) \\ & + \varepsilon \mathcal{W}(\boldsymbol{\xi}) \mathbf{D}(\partial_\mathbf{x})^T H(\mathbf{O}, \mathbf{y}) + \varepsilon (\mathbf{D}(\partial_\mathbf{y})^T H(\mathbf{O}, \mathbf{x}))^T \mathcal{W}^T(\boldsymbol{\eta}) + R_\varepsilon(\mathbf{x}, \mathbf{y}) \end{aligned} \tag{8.68}$$

where R_ε is the remainder. Finally, from the definition of G and $\mathcal{N}$ we obtain the leading order part of (8.59).

Now we give a rigorous proof of Theorem 8.1.

The Remainder Estimate

The remainder R_ε, present in (8.68), is a solution of the problem

$$L(\partial_{\mathbf{x}})R_\varepsilon(\mathbf{x},\mathbf{y}) = \mathbf{O}\,, \quad \mathbf{x},\mathbf{y} \in \Omega_\varepsilon\,,$$

$$\begin{aligned} R_\varepsilon(\mathbf{x},\mathbf{y}) &= h(\boldsymbol{\xi},\boldsymbol{\eta}) - \varepsilon\mathcal{W}(\boldsymbol{\xi})\mathbf{D}(\partial_{\mathbf{x}})^T H(\mathbf{O},\mathbf{y}) \\ &-\varepsilon(\mathbf{D}(\partial_{\mathbf{y}})^T H(\mathbf{O},\mathbf{x}))^T \mathcal{W}^T(\boldsymbol{\eta})\,, \quad \mathbf{x} \in \partial\Omega, \mathbf{y} \in \Omega_\varepsilon\,, \end{aligned} \tag{8.69}$$

$$\begin{aligned} T_n(\partial_{\mathbf{x}})R_\varepsilon(\mathbf{x},\mathbf{y}) = &\; T_n(\partial_{\mathbf{x}})H(\mathbf{x},\mathbf{y}) - \varepsilon T_n(\partial_{\mathbf{x}})\mathcal{W}(\boldsymbol{\xi})\mathbf{D}(\partial_{\mathbf{x}})^T H(\mathbf{O},\mathbf{y}) \\ &-\varepsilon T_n(\partial_{\mathbf{x}})(\mathbf{D}(\partial_{\mathbf{y}})^T H(\mathbf{O},\mathbf{x}))^T \mathcal{W}^T(\boldsymbol{\eta}) \\ &\text{for } \mathbf{x} \in \partial\omega_\varepsilon, \mathbf{y} \in \Omega_\varepsilon\,, \end{aligned} \tag{8.70}$$

where the boundary condition (8.70) is self-balanced.

Estimate for $R_\varepsilon(\mathbf{x},\mathbf{y})$ *on* $\partial\Omega$. Since the derivatives of the components of H are bounded for $\mathbf{x} \in \partial\Omega$, $\mathbf{y} \in \Omega_\varepsilon$, by Lemma 8.3

$$\begin{aligned} |\varepsilon\mathcal{W}(\boldsymbol{\xi})\mathbf{D}(\partial_{\mathbf{x}})^T H(\mathbf{O},\mathbf{y})| &\le \text{Const}\,\varepsilon^2|\mathbf{x}|^{-1} \\ &\le \text{Const}\,\varepsilon^2 \quad \mathbf{x} \in \partial\Omega, \mathbf{y} \in \Omega_\varepsilon\,, \end{aligned} \tag{8.71}$$

where we have used for $\mathbf{x} \in \partial\Omega$, $|\mathbf{x}| \ge 1$.

Owing to Lemma 8.6 and the boundary condition (8.22) for H, one obtains

$$\begin{aligned} &|h(\boldsymbol{\xi},\boldsymbol{\eta}) - \varepsilon(\mathbf{D}(\partial_{\mathbf{y}})^T H(\mathbf{O},\mathbf{x}))^T \mathcal{W}^T(\boldsymbol{\eta})| \\ &= |h(\boldsymbol{\xi},\boldsymbol{\eta}) - \varepsilon \lim_{\mathbf{z}\to\mathbf{O}} (\mathbf{D}(\partial_{\mathbf{z}})^T \gamma(\mathbf{x},\mathbf{z}))^T \mathcal{W}^T(\boldsymbol{\eta})| \\ &\le \text{Const}\,\varepsilon^3|\mathbf{x}|^{-2}|\mathbf{y}|^{-1} \le \text{Const}\,\varepsilon^3|\mathbf{y}|^{-1}\,, \mathbf{x} \in \partial\Omega, \mathbf{y} \in \Omega_\varepsilon\,. \end{aligned} \tag{8.72}$$

Thus estimates (8.71), (8.72) lead to

$$|R_\varepsilon(\mathbf{x},\mathbf{y})| \le \text{Const}\,\varepsilon^2\,, \quad \mathbf{x} \in \partial\Omega, \mathbf{y} \in \Omega_\varepsilon\,. \tag{8.73}$$

Estimate for $R_\varepsilon(\mathbf{x},\mathbf{y})$ *on* $\partial\omega_\varepsilon$. The boundary condition (8.29) for the matrix $\mathcal{W}$ implies

$$\begin{aligned} &|T_n(\partial_{\mathbf{x}})H(\mathbf{x},\mathbf{y}) - \varepsilon T_n(\partial_{\mathbf{x}})\mathcal{W}(\boldsymbol{\xi})\mathbf{D}(\partial_{\mathbf{x}})^T H(\mathbf{O},\mathbf{y})| \\ &= |\mathbf{D}(\mathbf{n})\mathcal{C}\mathbf{D}(\partial_{\mathbf{x}})^T H(\mathbf{x},\mathbf{y}) - \mathbf{D}(\mathbf{n})\mathcal{C}\mathbf{D}(\partial_{\mathbf{x}})^T H(\mathbf{O},\mathbf{y})|\,. \end{aligned} \tag{8.74}$$

Next, using the Taylor expansion we expand H about $\mathbf{x} = \mathbf{O}$ to derive the inequality

$$
\begin{aligned}
&|T_n(\partial_{\mathbf{x}})H(\mathbf{x},\mathbf{y}) - \varepsilon T_n(\partial_{\mathbf{x}})\mathcal{W}(\boldsymbol{\xi})\mathbf{D}(\partial_{\mathbf{x}})^T H(\mathbf{O},\mathbf{y})| \\
&\leq \text{Const}\,\varepsilon\,, \quad \mathbf{x} \in \partial\omega_\varepsilon, \mathbf{y} \in \Omega_\varepsilon\,. \qquad (8.75)
\end{aligned}
$$

Lemma 8.3 then gives

$$
|\varepsilon T_n(\partial_{\mathbf{x}})(\mathbf{D}(\partial_{\mathbf{y}})^T H(\mathbf{O},\mathbf{x}))^T \mathcal{W}^T(\boldsymbol{\eta})| \leq \text{Const}\,\varepsilon^2|\mathbf{y}|^{-1}\,. \qquad (8.76)
$$

Then, (8.75) and (8.76) yield

$$
|T_n(\partial_{\mathbf{x}})R_\varepsilon(\mathbf{x},\mathbf{y})| \leq \text{Const}\,\varepsilon\,, \quad \mathbf{x} \in \partial\omega_\varepsilon, \mathbf{y} \in \Omega_\varepsilon\,. \qquad (8.77)
$$

By Lemma 8.2, (8.73), (8.77) and the fact (8.70) is self-balanced, we have $R_\varepsilon(\mathbf{x},\mathbf{y})$ is $O(\varepsilon^2)$. □

8.6 Simplified Asymptotic Formulae for G_ε Under Constraints on the Independent Spatial Variables for a Domain with a Small Hole

Now that the uniform asymptotic formulae have been obtained for the entries of G_ε for the mixed boundary value problem, we now show how these formulae simplify under constraints on the points $\mathbf{x}$ and $\mathbf{y}$.

Corollary 8.1. *a) Let* $\mathbf{x}$ *and* $\mathbf{y}$ *be points of* $\Omega_\varepsilon \subset \mathbb{R}^2$ *such that*

$$
\min\{|\mathbf{x}|, |\mathbf{y}|\} > 2\varepsilon\,. \qquad (8.78)
$$

Then

$$
\begin{aligned}
G_\varepsilon(\mathbf{x},\mathbf{y}) = {}& G(\mathbf{x},\mathbf{y}) - \varepsilon^2(\mathbf{D}(\partial_{\mathbf{x}})^T\gamma(\mathbf{x},\mathbf{O}))^T M(\mathbf{D}(\partial_{\mathbf{y}})^T\gamma(\mathbf{O},\mathbf{y})) \\
&- \varepsilon^2(\mathbf{D}(\partial_{\mathbf{x}})^T\gamma(\mathbf{x},\mathbf{O}))^T M\mathbf{D}(\partial_{\mathbf{x}})^T H(\mathbf{O},\mathbf{y}) \qquad (8.79) \\
&- \varepsilon^2(\mathbf{D}(\partial_{\mathbf{y}})^T H(\mathbf{O},\mathbf{x}))^T M\mathbf{D}(\partial_{\mathbf{y}})^T\gamma(\mathbf{O},\mathbf{y}) + O(\varepsilon^2(|\mathbf{x}|^{-2} + |\mathbf{y}|^{-2}))\,.
\end{aligned}
$$

b) If $\max\{|\mathbf{x}|, |\mathbf{y}|\} < 1/2$, *then*

$$
\begin{aligned}
G_\varepsilon(\mathbf{x},\mathbf{y}) = {}& \mathcal{N}(\boldsymbol{\xi},\boldsymbol{\eta}) - H(\mathbf{O},\mathbf{O}) - (\mathbf{r}(\partial_{\mathbf{y}})^T H(\mathbf{O},\mathbf{x}))^T \otimes \mathbf{r}(\mathbf{y})^T \\
&- \mathbf{r}(\mathbf{x}) \otimes \mathbf{r}(\partial_{\mathbf{x}})^T H(\mathbf{O},\mathbf{y}) - \varepsilon\boldsymbol{\Upsilon}(\boldsymbol{\xi})\mathbf{D}(\partial_{\mathbf{x}})^T H(\mathbf{O},\mathbf{y}) \\
&- \varepsilon(\mathbf{D}(\partial_{\mathbf{y}})^T H(\mathbf{O},\mathbf{x}))^T \boldsymbol{\Upsilon}^T(\boldsymbol{\eta}) + O(\varepsilon^2 + |\mathbf{x}|^2 + |\mathbf{y}|^2)\,. \qquad (8.80)
\end{aligned}
$$

Both (8.79) and (8.80) are uniform with respect to $\mathbf{x}$ *and* $\mathbf{y}$ *of* Ω_ε.

Proof. We write (8.59) as

$$G_\varepsilon(\mathbf{x},\mathbf{y}) = G(\mathbf{x},\mathbf{y}) - h(\boldsymbol{\xi},\boldsymbol{\eta}) + \varepsilon \mathcal{W}(\boldsymbol{\xi})\mathbf{D}(\partial_{\mathbf{x}})^T H(\mathbf{O},\mathbf{y})$$
$$+\varepsilon(\mathbf{D}(\partial_{\mathbf{y}})^T H(\mathbf{O},\mathbf{x}))^T \mathcal{W}^T(\boldsymbol{\eta}) + O(\varepsilon^2)\,. \tag{8.81}$$

Due to the constraint (8.78), from Lemmas 8.5 and 8.6, we have the estimate

$$h(\boldsymbol{\xi},\boldsymbol{\eta}) = \varepsilon^2(\mathbf{D}(\partial_{\mathbf{x}})^T \gamma(\mathbf{x},\mathbf{O}))^T M\mathbf{D}(\partial_{\mathbf{y}})^T\gamma(\mathbf{O},\mathbf{y}) + O\Big(\frac{\varepsilon^3(|\mathbf{x}|+|\mathbf{y}|)}{|\mathbf{x}|^2|\mathbf{y}|^2}\Big)\,.$$

Substitution of this into (8.1) and again using Lemma 8.5, leads to

$$G(\mathbf{x},\mathbf{y}) = G(\mathbf{x},\mathbf{y}) - \varepsilon^2(\mathbf{D}(\partial_{\mathbf{x}})^T\gamma(\mathbf{x},\mathbf{O}))^T M\mathbf{D}(\partial_{\mathbf{y}})^T\gamma(\mathbf{O},\mathbf{y})$$

$$-\varepsilon^2(\mathbf{D}(\partial_{\mathbf{x}})^T\gamma(\mathbf{x},\mathbf{O}))^T M\mathbf{D}(\partial_{\mathbf{x}})^T H(\mathbf{O},\mathbf{y}) + O\Big(\frac{\varepsilon^2}{|\mathbf{x}|^2}\Big)$$

$$-\varepsilon^2(\mathbf{D}(\partial_{\mathbf{y}})^T H(\mathbf{O},\mathbf{x}))^T M\mathbf{D}(\partial_{\mathbf{y}})^T\gamma(\mathbf{O},\mathbf{y}) + O\Big(\frac{\varepsilon^2}{|\mathbf{y}|^2}\Big)$$

$$+O\Big(\frac{\varepsilon^3(|\mathbf{x}|+|\mathbf{y}|)}{|\mathbf{x}|^2|\mathbf{y}|^2}\Big)\,.$$

Noting that the remainder term in the above formula can be written as $O(\varepsilon^2(|\mathbf{x}|^{-2} + |\mathbf{y}|^{-2}))$, we obtain (8.79).

b) The function G_ε in (8.59) also takes the form

$$G_\varepsilon(\mathbf{x},\mathbf{y}) = -H(\mathbf{x},\mathbf{y}) + \mathcal{N}(\boldsymbol{\xi},\boldsymbol{\eta}) + \varepsilon \mathcal{W}(\boldsymbol{\xi})\mathbf{D}(\partial_{\mathbf{x}})^T H(\mathbf{O},\mathbf{y})$$
$$+\varepsilon(\mathbf{D}(\partial_{\mathbf{y}})^T H(\mathbf{O},\mathbf{x}))^T \mathcal{W}^T(\boldsymbol{\eta}) + O(\varepsilon^2)\,. \tag{8.82}$$

The components of the matrix H and their derivatives are smooth for $\mathbf{x},\mathbf{y}\in\Omega_\varepsilon$ and we can use the linear expansion of this matrix in the vicinity of the origin in the above formula. According to (6.17) the Taylor expansion for H in the vicinity of $\mathbf{x}=\mathbf{O}$ takes the form

$$H(\mathbf{x},\mathbf{y}) = H(\mathbf{y},\mathbf{O})^T + \mathbf{r}(\mathbf{x})\otimes\mathbf{r}(\partial_{\mathbf{x}})^T H(\mathbf{O},\mathbf{y})$$
$$+\mathbf{D}(\mathbf{x})\mathbf{D}(\partial_{\mathbf{x}})^T H(\mathbf{O},\mathbf{y}) + O(|\mathbf{x}|^2)\,.$$

Next using the symmetry condition $H(\mathbf{x},\mathbf{y}) = (H(\mathbf{y},\mathbf{x}))^T$, and the expansion (6.17) for H about $\mathbf{y}=\mathbf{O}$ gives

$$H(\mathbf{x},\mathbf{y}) = H(\mathbf{O},\mathbf{O}) + (\mathbf{r}(\partial_{\mathbf{y}})^T H(\mathbf{O},\mathbf{x}))^T \otimes \mathbf{r}(\mathbf{y})^T + (\mathbf{D}(\partial_{\mathbf{y}})^T H(\mathbf{O},\mathbf{x}))^T \mathbf{D}(\mathbf{y})^T$$
$$+\mathbf{r}(\mathbf{x}) \otimes \mathbf{r}(\partial_{\mathbf{x}})^T H(\mathbf{O},\mathbf{y}) + \mathbf{D}(\mathbf{x})\mathbf{D}(\partial_{\mathbf{x}})^T H(\mathbf{O},\mathbf{y}) + O(|\mathbf{x}|^2 + |\mathbf{y}|^2) .$$

Inserting this into (8.82) and recalling the definition of $\boldsymbol{\Upsilon}$ in (8.36) we arrive at (8.80).

Part III
Meso-scale Approximations: Asymptotic Treatment of Perforated Domains Without Homogenization

Chapter 9
Meso-scale Approximations for Solutions of Dirichlet Problems

In this chapter, we address the Dirichlet problem for the Poisson equation $-\Delta u = f$ in a multiply perforated domain.

The asymptotic approximations constructed here are efficient for certain geometries, intermediate between a collection of inclusions whose size ε is comparable with the spacing parameter d and the classical situation with $\varepsilon \sim \text{Const}\, d^3$ appearing in some classical solutions in the homogenization theory (see, for example, Cioranescu and Murat [5], Marchenko and Khruslov [21]). Here and in the next chapter, such intermediate cases will be referred to as *meso-scale*.

We derive the asymptotic formula for Green's function $G_N(\mathbf{x}, \mathbf{y})$, uniform with respect to $\mathbf{x}$ and $\mathbf{y}$. The following is a specially simple form in the case of $\Omega = \mathbb{R}^3$:

$$G_N(\mathbf{x}, \mathbf{y}) = \frac{1 - N}{4\pi|\mathbf{x} - \mathbf{y}|} + \sum_{j=1}^{N} g^{(j)}(\mathbf{x}, \mathbf{y})$$

$$+ \sum_{1 \le i,j \le N,\ i \ne j} C_{ij} P^{(i)}(\mathbf{x}) P^{(j)}(\mathbf{y}) + O(\varepsilon d^{-2}),$$

where $g^{(j)}$ are Green's functions in $\mathbb{R}^3 \setminus F^{(j)}$, and the matrix $\boldsymbol{C} = (C_{ij})_{i,j=1}^N$ is defined by $\boldsymbol{C} = (\mathbf{I} + \mathbf{SD})^{-1}\mathbf{S}$.

9.1 Main Notations and Formulation of the Problem in the Perforated Region

Let Ω be an arbitrary domain in $\mathbb{R}^3$, and let $\{\mathbf{O}^{(j)}\}_{j=1}^N$ and $\{F^{(j)}\}_{j=1}^N$ be collections of points and disjoint contractible compact subsets of Ω such that $\mathbf{O}^{(j)} \in F^{(j)}$, and $F^{(j)}$ have positive harmonic capacity. Assume that the diameter ε_j of $F^{(j)}$ is small compared to the diameter of Ω. We shall also use the notations

V. Maz'ya et al., *Green's Kernels and Meso-Scale Approximations in Perforated Domains*, Lecture Notes in Mathematics 2077, DOI 10.1007/978-3-319-00357-3_9,

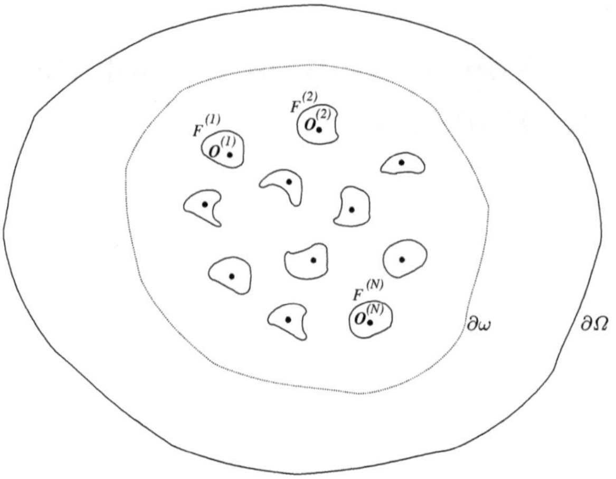

Fig. 9.1 Perforated domain containing many holes

$$d = 2^{-1} \min_{i \neq j, 1 \leq i,j \leq N} |\mathbf{O}^{(j)} - \mathbf{O}^{(i)}|, \ \varepsilon = \max_{1 \leq j \leq N} \varepsilon_j. \tag{9.1}$$

It is assumed that $\varepsilon < c\, d$, with c being a sufficiently small constant.

We require that there exists an open set ω such that

$$\begin{aligned} &\bigcup_{j=1}^{N} F^{(j)} \subset \omega, \ \text{diam}(\omega) = 1, \ \text{dist}\,(\partial\omega, \partial\Omega) \geq 2d, \\ &\text{and} \quad \text{dist}\Big\{\bigcup_{j=1}^{N} F^{(j)}, \partial\omega\Big\} \geq 2d. \end{aligned} \tag{9.2}$$

Let us introduce the complimentary domain

$$\Omega_N = \Omega \setminus \cup_{j=1}^{N} F^{(j)}, \tag{9.3}$$

as shown in Fig. 9.1.

Let u denote the variational solution of the Dirichlet problem

$$-\Delta u(\mathbf{x}) = f(\mathbf{x}), \ \mathbf{x} \in \Omega_N, \tag{9.4}$$

$$u(\mathbf{x}) = 0, \ \mathbf{x} \in \partial\Omega_N, \tag{9.5}$$

where f is assumed to be a smooth function with a compact support in Ω, such that diam(supp f) $\leq C$ with C being an absolute constant.

We seek an asymptotic approximation of u as $N \to \infty$. In the estimates contained below, when we use "Const" this means a constant independent of N and ε.

9.2 Auxiliary Problems

We collect here solutions of some boundary value problems to be used in the asymptotic approximation of u.

9.2.1 Solution of the Unperturbed Problem

By v_f we mean the variational solution of the Dirichlet problem

$$-\Delta v_f(\mathbf{x}) = f(\mathbf{x}), \ \mathbf{x} \in \Omega, \tag{9.6}$$

$$v_f(\mathbf{x}) = 0, \ \mathbf{x} \in \partial\Omega, \tag{9.7}$$

where f is the same smooth function as in (9.4).

9.2.2 Capacitary Potentials of $F^{(j)}$

The harmonic capacitary potential of $F^{(j)}$ will be denoted by $P^{(j)}$, and it is defined as a unique variational solution of the Dirichlet problem

$$\Delta P^{(j)}(\mathbf{x}) = 0 \ \text{on} \ \mathbb{R}^3 \setminus F^{(j)}, \tag{9.8}$$

$$P^{(j)}(\mathbf{x}) = 1 \ \text{for} \ \mathbf{x} \in \partial(\mathbb{R}^3 \setminus F^{(j)}), \tag{9.9}$$

$$P^{(j)}(\mathbf{x}) = O(\varepsilon|\mathbf{x} - \mathbf{O}^{(j)}|^{-1}) \ \text{as} \ \varepsilon^{-1}|\mathbf{x} - \mathbf{O}^{(j)}| \to \infty. \tag{9.10}$$

It is well known (see, for example, Pólya and Szegö [38]), that these functions have the following asymptotic representations:

$$P^{(j)}(\mathbf{x}) = \frac{\text{cap}(F^{(j)})}{|\mathbf{x} - \mathbf{O}^{(j)}|} + O(\varepsilon\, \text{cap}(F^{(j)})|\mathbf{x} - \mathbf{O}^{(j)}|^{-2}) \ \text{for} \ |\mathbf{x} - \mathbf{O}^{(j)}| > 2\varepsilon. \tag{9.11}$$

The normalised harmonic capacity of the set $F^{(j)}$ in $\mathbb{R}^3$ can be defined by

$$\mathrm{cap}(F^{(j)}) = \frac{1}{4\pi} \int_{\mathbb{R}^3 \setminus F^{(j)}} |\nabla P^{(j)}(\boldsymbol{\xi})|^2 d\boldsymbol{\xi}. \tag{9.12}$$

9.2.3 Green's Function for the Unperturbed Domain

Green's function for the unperturbed domain is denoted by $G(\mathbf{x}, \mathbf{y})$, and it satisfies the boundary value problem

$$\Delta_x G(\mathbf{x}, \mathbf{y}) + \delta(\mathbf{x} - \mathbf{y}) = 0, \quad \mathbf{x}, \mathbf{y} \in \Omega, \tag{9.13}$$

$$G(\mathbf{x}, \mathbf{y}) = 0 \text{ as } \mathbf{x} \in \partial\Omega \text{ and } \mathbf{y} \in \Omega. \tag{9.14}$$

The regular part of Green's function is defined by

$$H(\mathbf{x}, \mathbf{y}) = (4\pi|\mathbf{x} - \mathbf{y}|)^{-1} - G(\mathbf{x}, \mathbf{y}). \tag{9.15}$$

9.3 Formal Asymptotic Algorithm

Let the solution u of (9.4), (9.5) be written as

$$u(\mathbf{x}) = v_f(\mathbf{x}) + R^{(1)}(\mathbf{x}), \tag{9.16}$$

where v_f solves the auxiliary Dirichlet problem (9.6), (9.7) in the unperturbed domain, whereas the function $R^{(1)}$ is harmonic in Ω_N and satisfies the boundary conditions

$$R^{(1)}(\mathbf{x}) = 0 \text{ when } \mathbf{x} \in \partial\Omega, \tag{9.17}$$

and

$$R^{(1)}(\mathbf{x}) = -v_f(\mathbf{x}) = -v_f(\mathbf{O}^{(k)}) + O(\varepsilon) \text{ when } \mathbf{x} \in \partial(\mathbb{R}^3 \setminus F^{(k)}). \tag{9.18}$$

Let us approximate the function $R^{(1)}$ in the form

$$R^{(1)}(\mathbf{x}) \sim \sum_{j=1}^{N} C_j \Big(P^{(j)}(\mathbf{x}) - 4\pi \,\mathrm{cap}(F^{(j)})\, H(\mathbf{x}, \mathbf{O}^{(j)}) \Big), \tag{9.19}$$

where C_j are unknown constant coefficients, and $P^{(j)}$ and H are the same as in (9.8)–(9.11) and (9.15), respectively.

By (9.11), (9.15) and (9.14), we deduce

$$P^{(j)}(\mathbf{x}) - 4\pi \,\mathrm{cap}(F^{(j)})\, H(\mathbf{x}, \mathbf{O}^{(j)}) = O(\varepsilon \,\mathrm{cap}(F^{(j)})|\mathbf{x} - \mathbf{O}^{(j)}|^{-2}), \tag{9.20}$$

for all $\mathbf{x} \in \partial\Omega,\ j = 1, \dots, N$.

On the boundary of a small inclusion $F^{(k)}$ $(k = 1, \dots, N)$ we have

$$v_f(\mathbf{O}^{(k)}) + O(\varepsilon) + C_k(1 + O(\varepsilon)) \tag{9.21}$$

$$+ \sum_{1 \le j \le N,\ j \ne k} C_j \Big(4\pi \operatorname{cap}(F^{(j)})\, G(\mathbf{O}^{(k)}, \mathbf{O}^{(j)}) + O(\varepsilon \operatorname{cap}(F^{(j)}) |\mathbf{x} - \mathbf{O}^{(j)}|^{-2}) \Big) = 0,$$

for all $\mathbf{x} \in \partial(\mathbb{R}^3 \setminus F^{(k)})$.

Equation (9.21) suggests that the constant coefficients $C_j,\ j = 1, \dots, N$, should be chosen to satisfy the system of linear algebraic equations

$$v_f(\mathbf{O}^{(k)}) + C_k + 4\pi \sum_{1 \le j \le N,\ j \ne k} C_j \operatorname{cap}(F^{(j)})\, G(\mathbf{O}^{(k)}, \mathbf{O}^{(j)}) = 0, \tag{9.22}$$

where $k = 1, \dots, N$.

Then within certain constraints on the small parameters ε and d (see (9.1)), it will be shown in the sequel that the above system of algebraic equations is solvable and that the harmonic function

$$R^{(2)}(\mathbf{x}) = R^{(1)}(\mathbf{x}) - \sum_{j=1}^{N} C_j \Big(P^{(j)}(\mathbf{x}) - 4\pi \operatorname{cap}(F^{(j)})\, H(\mathbf{x}, \mathbf{O}^{(j)}) \Big)$$

is small on $\partial\Omega_N$. Further application of the maximum principle for harmonic functions leads to an estimate of the remainder $R^{(2)}$ in Ω_N.

Hence, the solution (9.16) takes the form

$$u(\mathbf{x}) = v_f(\mathbf{x}) + \sum_{j=1}^{N} C_j \Big(P^{(j)}(\mathbf{x}) - 4\pi \operatorname{cap}(F^{(j)})\, H(\mathbf{x}, \mathbf{O}^{(j)}) \Big) + R^{(2)}(\mathbf{x}), \tag{9.23}$$

where C_j are obtained from the algebraic system (9.22).

9.4 Algebraic System

In this section we analyse the solvability of the system (9.22), and subject to certain constraints on ε and d, derive estimates for the coefficients $C_j,\ j = 1, \dots, N$.

The following matrices $\mathbf{S}$ and $\mathbf{D}$ will be used here:

$$\mathbf{S} = \Big\{ (1 - \delta_{ik}) G(\mathbf{O}^{(k)}, \mathbf{O}^{(i)}) \Big\}_{i,k=1}^{N}, \tag{9.24}$$

and

$$\mathbf{D} = 4\pi \operatorname{diag}\{\operatorname{cap}(F^{(1)}), \dots, \operatorname{cap}(F^{(N)})\}. \tag{9.25}$$

If the matrix $\mathbf{I}+\mathbf{SD}$ is non-degenerate, then the components of the column vector $\mathbf{C}=(C_1,\dots,C_N)^T$ are defined by

$$\mathbf{C}=-(\mathbf{I}+\mathbf{SD})^{-1}\mathbf{V}_f, \tag{9.26}$$

where

$$\mathbf{V}_f=(v_f(\mathbf{O}^{(1)}),\dots,v_f(\mathbf{O}^{(N)}))^T. \tag{9.27}$$

Prior to the formulation of the result on the uniform asymptotic approximation of the solution to problem (9.4)–(9.5), we formulate and prove auxiliary statements incorporating the invertibility of the matrix $\mathbf{I}+\mathbf{SD}$ and estimates for components of the vector (9.26).

Lemma 9.1. *If* $\max_{1\le j\le N}\text{cap}(F^{(j)}) < 5d/(24\pi)$, *then the matrix* $\mathbf{I}+\mathbf{SD}$ *is invertible and the column vector* $\mathbf{C}$ *in (9.26) satisfies the estimate*

$$\sum_{j=1}^{N}\text{cap}(F^{(j)})\,C_j^2\le\Big(1-\frac{24\pi}{5d}\max_{1\le j\le N}\text{cap}(F^{(j)})\Big)^{-2}\sum_{j=1}^{N}\text{cap}(F^{(j)})\,(v_f(\mathbf{O}^{(j)}))^2. \tag{9.28}$$

Proof. According to (9.26), we have $(\mathbf{I}+\mathbf{SD})\mathbf{C}=-\mathbf{V}_f$. Hence

$$\langle\mathbf{C},\mathbf{DC}\rangle+\langle\mathbf{SDC},\mathbf{DC}\rangle=-\langle\mathbf{V}_f,\mathbf{DC}\rangle. \tag{9.29}$$

Obviously, the right-hand side in (9.29) does not exceed

$$\langle\mathbf{C},\mathbf{DC}\rangle^{1/2}\langle\mathbf{V}_f,\mathbf{DV}_f\rangle^{1/2}. \tag{9.30}$$

Consider the second term in the left-hand side of (9.29). Using the mean value theorem for harmonic functions we deduce

$$\langle\mathbf{SDC},\mathbf{DC}\rangle=(4\pi)^2\sum_{i\ne j,1\le i,j\le N}G(\mathbf{O}^{(i)},\mathbf{O}^{(j)})\text{cap}(F^{(i)})\text{cap}(F^{(j)})\,C_iC_j$$

$$=(4\pi)^2\sum_{i\ne j,1\le i,j\le N}\frac{\text{cap}(F^{(i)})\text{cap}(F^{(j)})\,C_iC_j}{|B^{(i)}|\,|B^{(j)}|}\int_{B^{(i)}}\int_{B^{(j)}}G(\mathbf{X},\mathbf{Y})d\mathbf{X}d\mathbf{Y},$$

where $B^{(j)}=\{\mathbf{x}:|\mathbf{x}-\mathbf{O}^{(j)}|<d\}$, $j=1,\dots,N$, are non-overlapping balls of radius d with the centers at $\mathbf{O}^{(j)}$, and $|B^{(j)}|=4\pi d^3/3$ are the volumes of the balls. Also, the notation B_d is used here for the ball of radius d with the center at the origin.

Let $\Xi(\mathbf{x})$ be a piecewise function defined on Ω as

$$\Xi(\mathbf{x}) = \begin{cases} C_j \operatorname{cap}(F^{(j)}) \text{ in } & B^{(j)}, j = 1, \dots, N, \\ 0 & \text{otherwise.} \end{cases}$$

Then

$$\langle \mathbf{SDC}, \mathbf{DC} \rangle = \frac{9}{d^6} \Big(\int_\Omega \int_\Omega G(\mathbf{X}, \mathbf{Y}) \Xi(\mathbf{X}) \Xi(\mathbf{Y}) d\mathbf{X} d\mathbf{Y}$$

$$- \sum_{j=1}^{N} (\operatorname{cap}(F^{(j)}))^2 C_j^2 \int_{B^{(j)}} \int_{B^{(j)}} G(\mathbf{X}, \mathbf{Y}) d\mathbf{X} d\mathbf{Y} \Big). \tag{9.31}$$

The first term in the right-hand side of (9.31) is non-negative, which follows from the relation

$$\int_\Omega \int_\Omega G(\mathbf{X}, \mathbf{Y}) \Xi(\mathbf{X}) \Xi(\mathbf{Y}) d\mathbf{X} d\mathbf{Y} = \int_\Omega \Big| \nabla_X \int_\Omega G(\mathbf{X}, \mathbf{Y}) \Xi(\mathbf{Y}) d\mathbf{Y} \Big|^2 d\mathbf{X} \geq 0. \tag{9.32}$$

The integral

$$\int_{B^{(j)}} \int_{B^{(j)}} G(\mathbf{X}, \mathbf{Y}) d\mathbf{X} d\mathbf{Y}$$

in the right-hand side of (9.31) allows for the estimate

$$\int_{B^{(j)}} \int_{B^{(j)}} G(\mathbf{X}, \mathbf{Y}) d\mathbf{X} d\mathbf{Y} \leq \frac{1}{4\pi} \int_{B_d} \int_{B_d} \frac{d\mathbf{X} d\mathbf{Y}}{|\mathbf{X} - \mathbf{Y}|}$$
$$= \frac{1}{4\pi} \int_{B_d} d\mathbf{X} \Big\{ \int_{|\mathbf{Y}| < |\mathbf{X}|} \frac{d\mathbf{Y}}{|\mathbf{X} - \mathbf{Y}|} + \int_{d > |\mathbf{Y}| > |\mathbf{X}|} \frac{d\mathbf{Y}}{|\mathbf{X} - \mathbf{Y}|} \Big\}$$
$$= \frac{1}{4\pi} \int_{B_d} d\mathbf{X} \Big\{ \int_0^{|\mathbf{X}|} d\rho \int_{\{\mathbf{Y}: |\mathbf{Y}| = \rho\}} \frac{dS_Y}{|\mathbf{X} - \mathbf{Y}|} + \int_{|\mathbf{X}|}^{d} d\rho \int_{\{\mathbf{Y}: |\mathbf{Y}| - \rho\}} \frac{dS_Y}{|\mathbf{X} - \mathbf{Y}|} \Big\}. \tag{9.33}$$

Using the mean value theorem for harmonic functions we deduce

$$\int_{\{\mathbf{Y}: |\mathbf{Y}| = \rho\}} \frac{dS_Y}{|\mathbf{X} - \mathbf{Y}|} = 4\pi\rho^2 |\mathbf{X}|^{-1} \text{ when } |\mathbf{X}| > \rho. \tag{9.34}$$

On the other hand,

$$\int_{\{\mathbf{Y}: |\mathbf{Y}| = \rho\}} \frac{dS_Y}{|\mathbf{X} - \mathbf{Y}|} = 4\pi\rho \text{ when } |\mathbf{X}| < \rho, \tag{9.35}$$

which follows from the relation

$$\int_{\{\mathbf{Y}:|\mathbf{Y}|=\rho\}} \frac{dS_Y}{\rho|\mathbf{X}-\mathbf{Y}|} = -\int_{\{\mathbf{Y}:|\mathbf{Y}|=\rho\}} \frac{\partial}{\partial|\mathbf{Y}|}\frac{1}{|\mathbf{X}-\mathbf{Y}|}dS_Y$$
$$= -\int_{\{\mathbf{Y}:|\mathbf{Y}|<\rho\}} \Delta_Y \frac{1}{|\mathbf{X}-\mathbf{Y}|}d\mathbf{Y} = 4\pi \quad \text{when } |\mathbf{X}|<\rho.$$

It follows from (9.33), (9.34) and (9.35) that

$$\int_{B^{(j)}}\int_{B^{(j)}} G(\mathbf{X},\mathbf{Y})d\mathbf{X}d\mathbf{Y} \le \frac{1}{2}\int_{B_d}\Big(d^2-\frac{|\mathbf{X}|^2}{3}\Big)d\mathbf{X} = \frac{8\pi d^5}{15}. \tag{9.36}$$

Next, (9.29), (9.30), (9.31) and (9.36) lead to

$$\langle \mathbf{SDC},\mathbf{DC}\rangle \ge \frac{9}{d^6}\int_\omega\int_\omega G(\mathbf{X},\mathbf{Y})\Xi(\mathbf{X})\Xi(\mathbf{Y})d\mathbf{X}d\mathbf{Y} - \frac{9\alpha}{d}\sum_{j=1}^N (\mathrm{cap}(F^{(j)}))^2 C_j^2, \tag{9.37}$$

where $\alpha = \frac{8\pi}{15}$. Then (9.29) and (9.37) imply

$$\Big(1-\frac{9\alpha}{d}\max_{1\le j\le N}\mathrm{cap}(F^{(j)})\Big)\sum_{j=1}^N C_j^2\mathrm{cap}(F^{(j)})$$
$$\le \Big(\sum_{j=1}^N C_j^2\mathrm{cap}(F^{(j)})\Big)^{1/2}\Big(\sum_{j=1}^N (v_f(\mathbf{O}^{(j)}))^2\mathrm{cap}(F^{(j)})\Big)^{1/2},$$

which yields

$$(1-\frac{24\pi}{5d}\max_{1\le j\le N}\mathrm{cap}(F^{(j)}))\Big(\sum_{j=1}^N C_j^2\mathrm{cap}(F^{(j)})\Big)^{1/2}$$
$$\le \Big(\sum_{j=1}^N (v_f(\mathbf{O}^{(j)}))^2\mathrm{cap}(F^{(j)})\Big)^{1/2}. \tag{9.38}$$

Thus, if $\max_{1\le j\le N}\mathrm{cap}(F^{(j)}) < \frac{5}{24\pi}d$, then the matrix $\mathbf{I}+\mathbf{SD}$ is invertible and the estimate (9.28) holds. The proof is complete. □

Replacement of the inequality $\varepsilon < cd$ by the stronger constraint $\varepsilon < cd^2$ leads to the statement

Lemma 9.2. *Let the small parameters ε and d, defined in (9.1), satisfy*

$$\varepsilon < cd^2, \tag{9.39}$$

where c is a sufficiently small absolute constant. Then the components C_j of vector $\mathbf{C}$ in (9.26) allow for the estimate

$$|C_k| \le c \max_{1\le j\le N} |v_f(\mathbf{O}^{(j)})|. \tag{9.40}$$

Proof. Let us write the system (9.22) as

$$C_k + 4\pi \sum_{1\le j\le N,\ j\ne k} C_j \frac{\mathrm{cap}(F^{(j)})}{|B^{(j)}_{d/4}|} \int_{B^{(j)}_{d/4}} G(\mathbf{O}^{(k)}, \mathbf{y}) d\mathbf{y} = -v_f(\mathbf{O}^{(k)}), \tag{9.41}$$

for $k = 1, \dots, N$, where $B^{(j)}_{d/4}$ is the ball of radius $d/4$ with the centre at $\mathbf{O}^{(j)}$. Also let σ be a piece-wise constant function such that

$$\sigma(\mathbf{x}) = \begin{cases} C_j \ \mathrm{cap}(F^{(j)}), & \mathbf{x} \in B^{(j)}_{d/4}, \\ 0, & \mathbf{x} \in \mathbb{R}^3 \setminus \cup_{m=1}^N B^{(m)}_{d/4}. \end{cases} \tag{9.42}$$

Multiplying (9.41) by $\mathrm{cap}(F^{(k)})$ and writing the equations obtained in terms of σ we get

$$\sigma(\mathbf{O}^{(k)}) + \frac{192}{d^3} \mathrm{cap}(F^{(k)}) \int_{\cup_{1\le j\le N,\ j\ne k} B^{(j)}_{d/4}} \sigma(\mathbf{y}) G(\mathbf{O}^{(k)}, \mathbf{y}) d\mathbf{y}$$
$$= -v_f(\mathbf{O}^{(k)}) \ \mathrm{cap}(F^{(k)}),$$

which is equivalent to

$$\sigma(\mathbf{O}^{(k)}) + \frac{192}{d^3} \mathrm{cap}(F^{(k)}) \int_{\cup_{1\le j\le N} B^{(j)}_{d/4}} G(\mathbf{y}, \mathbf{z}) \sigma(\mathbf{y}) d\mathbf{y} = \mathrm{cap}(F^{(k)}) \ \Phi^{(k)}(\mathbf{z}), \tag{9.43}$$

for $k = 1, \dots, N$, where

$$\Phi^{(k)}(\mathbf{z}) = -v_f(\mathbf{O}^{(k)}) + \frac{192}{d^3} \int_{B^{(k)}_{d/4}} G(\mathbf{y}, \mathbf{z}) \sigma(\mathbf{y}) d\mathbf{y} \tag{9.44}$$
$$+ \frac{192}{d^3} \int_{\cup_{1\le j\le N,\ j\ne k} B^{(j)}_{d/4}} \sigma(\mathbf{y}) \Big(H(\mathbf{O}^{(k)}, \mathbf{y}) - H(\mathbf{z}, \mathbf{y}) \Big) d\mathbf{y}$$
$$+ \frac{48}{\pi d^3} \int_{\cup_{1\le j\le N,\ j\ne k} B^{(j)}_{d/4}} \sigma(\mathbf{y}) \Big\{ \frac{1}{|\mathbf{y} - \mathbf{z}|} - \frac{1}{|\mathbf{O}^{(k)} - \mathbf{y}|} \Big\} d\mathbf{y}, \quad \text{for all } \mathbf{z} \in B^{(k)}_{d/4}.$$

Next, we multiply (9.43) by

$$\left(\int_{\bigcup_{1\le j\le N} B^{(j)}_{d/4}} G(\mathbf{y},\mathbf{z})\sigma(\mathbf{y})d\mathbf{y}\right)^{2M-1},$$

where M is a positive integer. Also, taking into account that $\sigma(\mathbf{O}^{(k)}) = \sigma(\mathbf{z})$ for all $\mathbf{z}\in B^{(k)}_{d/4}$ we write

$$\begin{aligned}&\left(\int_{\bigcup_{1\le j\le N} B^{(j)}_{d/4}} G(\mathbf{y},\mathbf{z})\sigma(\mathbf{y})d\mathbf{y}\right)^{2M-1}\sigma(\mathbf{z})\\ &+\frac{192}{d^3}\,\mathrm{cap}(F^{(k)})\left(\int_{\bigcup_{1\le j\le N} B^{(j)}_{d/4}} G(\mathbf{y},\mathbf{z})\sigma(\mathbf{y})d\mathbf{y}\right)^{2M}\\ &=\mathrm{cap}(F^{(k)})\,\Phi^{(k)}(\mathbf{z})\left(\int_{\bigcup_{1\le j\le N} B^{(j)}_{d/4}} G(\mathbf{y},\mathbf{z})\sigma(\mathbf{y})d\mathbf{y}\right)^{2M-1},\quad \mathbf{z}\in B^{(k)}_{d/4}.\end{aligned}$$

Since $\sigma = 0$ outside the balls $B^{(k)}_{d/4}$, it follows that the integration of the above equation over $B^{(k)}_{d/4}$ and summation with respect to $k = 1,\dots,N$ lead to

$$\begin{aligned}&\int_\Omega\left(\int_\Omega G(\mathbf{y},\mathbf{z})\sigma(\mathbf{y})d\mathbf{y}\right)^{2M-1}\sigma(\mathbf{z})d\mathbf{z}\\ &+\frac{192}{d^3}\sum_{k=1}^{N}\mathrm{cap}(F^{(k)})\int_{B^{(k)}_{d/4}}\left(\int_\Omega G(\mathbf{y},\mathbf{z})\sigma(\mathbf{y})d\mathbf{y}\right)^{2M}d\mathbf{z}\\ &=\sum_{k=1}^{N}\mathrm{cap}(F^{(k)})\int_{B^{(k)}_{d/4}}\Phi^{(k)}(\mathbf{z})\left(\int_\Omega G(\mathbf{y},\mathbf{z})\sigma(\mathbf{y})d\mathbf{y}\right)^{2M-1}d\mathbf{z}.\end{aligned}\tag{9.45}$$

The identity

$$\begin{aligned}&\int_\Omega\left(\int_\Omega G(\mathbf{y},\mathbf{z})\sigma(\mathbf{y})d\mathbf{y}\right)^{2M-1}\sigma(\mathbf{z})d\mathbf{z}\\ &=\frac{2M-1}{M^2}\int_\Omega\left|\nabla_z\left(\int_\Omega G(\mathbf{y},\mathbf{z})\sigma(\mathbf{y})d\mathbf{y}\right)^{M}\right|^2d\mathbf{z}\end{aligned}$$

shows that the first term in the left-hand side of (9.45) is non-negative. By Hölder's inequality, the right-hand side of (9.45) does not exceed

$$\left(\sum_{k=1}^{N}\operatorname{cap}(F^{(k)})\int_{B_{d/4}^{(k)}}(\Phi^{(k)}(\mathbf{z}))^{2M}d\mathbf{z}\right)^{1/(2M)}$$
$$\times\left(\sum_{k=1}^{N}\operatorname{cap}(F^{(k)})\int_{B_{d/4}^{(k)}}\left(\int_{\Omega}G(\mathbf{y},\mathbf{z})\sigma(\mathbf{y})d\mathbf{y}\right)^{2M}d\mathbf{z}\right)^{(2M-1)/(2M)},\tag{9.46}$$

and hence (9.45) yields

$$\frac{192}{d^3}\left(\sum_{k=1}^{N}\operatorname{cap}(F^{(k)})\int_{B_{d/4}^{(k)}}\left(\int_{\Omega}G(\mathbf{y},\mathbf{z})\sigma(\mathbf{y})d\mathbf{y}\right)^{2M}d\mathbf{z}\right)^{1/(2M)}$$
$$\le\left(\sum_{k=1}^{N}\operatorname{cap}(F^{(k)})\int_{B_{d/4}^{(k)}}(\Phi^{(k)}(\mathbf{z}))^{2M}d\mathbf{z}\right)^{1/(2M)}.\tag{9.47}$$

After the limit passage as $M\to\infty$ we arrive at

$$d^{-3}\sup_{\mathbf{z}\in\bigcup_{1\le k\le N}B_{d/4}^{(k)}}\left|\int_{\Omega}G(\mathbf{y},\mathbf{z})\sigma(\mathbf{y})d\mathbf{y}\right|\le c\max_{1\le k\le N}\sup_{\mathbf{z}\in B_{d/4}^{(k)}}|\Phi^{(k)}(\mathbf{z})|,$$

and by (9.43) we deduce

$$|\sigma(\mathbf{O}^{(k)})|\le c\operatorname{cap}(F^{(k)})\max_{1\le j\le N}\sup_{\mathbf{z}\in B_{d/4}^{(j)}}|\Phi^{(j)}(\mathbf{z})|.\tag{9.48}$$

In turn, it follows from the definition (9.44) of the functions $\Phi^{(k)}$ that

$$\sup_{\mathbf{z}\in B_{d/4}^{(k)}}|\Phi^{(k)}(\mathbf{z})|\le|v_f(\mathbf{O}^{(k)})|+\frac{192}{d^3}\max_{1\le q\le N}|\sigma(\mathbf{O}^{(q)})|\sup_{\mathbf{z}\in B_{d/4}^{(k)}}\int_{B_{d/4}^{(k)}}G(\mathbf{y},\mathbf{z})d\mathbf{y}$$
$$+\frac{192}{d^3}\max_{1\le q\le N}|\sigma(\mathbf{O}^{(q)})|\sup_{\mathbf{z}\in B_{d/4}^{(k)}}\sum_{1\le j\le N,\ j\ne k}\int_{B_{d/4}^{(j)}}|H(\mathbf{O}^{(k)},\mathbf{y})-H(\mathbf{z},\mathbf{y})|d\mathbf{y}$$
$$+\frac{48}{\pi d^3}\max_{1\le q\le N}|\sigma(\mathbf{O}^{(q)})|\sup_{\mathbf{z}\in B_{d/4}^{(k)}}\sum_{1\le j\le N,\ j\ne k}\int_{B_{d/4}^{(j)}}\frac{|\mathbf{z}-\mathbf{O}^{(k)}|}{|\mathbf{y}-\mathbf{z}||\mathbf{O}^{(k)}-\mathbf{y}|}d\mathbf{y},$$

which, together with (9.48), yields

$$|\sigma(\mathbf{O}^{(k)})|\le c\operatorname{cap}(F^{(k)})\Big\{\max_{1\le j\le N}|v_f(\mathbf{O}^{(j)})|+d^{-2}\max_{1\le j\le N}|\sigma(\mathbf{O}^{(j)})|\Big\}.$$

If $\max_{1\le k\le N} \text{cap}(F^{(k)}) < cd^2$, with c being a sufficiently small constant, then referring to the definition (9.42) of the function σ we deduce (9.40), which completes the proof. □

9.5 Meso-scale Uniform Approximation of u

We obtain the next theorem, which is one of the principal results of this chapter, under an additional assumption on the smallness of the capacities of $F^{(j)}$.

Theorem 9.1. *Let the parameters ε and d, introduced in (9.1), satisfy the inequality*

$$\varepsilon < c\, d^{7/4}, \tag{9.49}$$

where c is a sufficiently small absolute constant.

Then the matrix $\mathbf{I}+\mathbf{SD}$, defined according to (9.24), (9.25), is invertible, and the solution $u(\mathbf{x})$ to the boundary value problem (9.4)–(9.5) is defined by the asymptotic formula

$$u(\mathbf{x}) = v_f(\mathbf{x}) + \sum_{j=1}^{N} C_j\Big(P^{(j)}(\mathbf{x}) - 4\pi\ \text{cap}(F^{(j)})\ H(\mathbf{x},\mathbf{O}^{(j)})\Big) + R(\mathbf{x}), \tag{9.50}$$

where the column vector $\mathbf{C} = (C_1,\dots,C_N)^T$ is given by (9.26) and the remainder $R(\mathbf{x})$ is a function harmonic in Ω_N, which satisfies the estimate

$$|R(\mathbf{x})| \le C\Big\{\varepsilon\|\nabla v_f\|_{L_\infty(\omega)} + \varepsilon^2 d^{-7/2}\|v_f\|_{L_\infty(\omega)}\Big\}. \tag{9.51}$$

Here and elsewhere the notation C is used for different positive constants independent of ε and d.

Proof. The harmonicity of R follows directly from (9.50).

If $\mathbf{x} \in \partial\Omega$, then

$$R(\mathbf{x}) = -4\pi \sum_{j=1}^{N} C_j \text{cap}(F^{(j)})\Big(\frac{1}{4\pi|\mathbf{x}-\mathbf{O}^{(j)}|} - H(\mathbf{x},\mathbf{O}^{(j)})\Big)$$
$$+\sum_{j=1}^{N} |C_j|\, O(\varepsilon\ \text{cap}(F^{(j)})|\mathbf{x}-\mathbf{O}^{(j)}|^{-2}).$$

Since $G(\mathbf{x},\mathbf{O}^{(j)}) = 0$ on $\partial\Omega$, and $P^{(j)}$ satisfies (9.11) we deduce

$$R(\mathbf{x}) = \sum_{j=1}^{N} O(\varepsilon\ \text{cap}(F^{(j)})|C_j||\mathbf{x}-\mathbf{O}^{(j)}|^{-2}), \tag{9.52}$$

where $|\mathbf{x} - \mathbf{O}^{(j)}| \geq C\ d$, and C is a sufficiently large constant.

If $\mathbf{x} \in \partial(\mathbb{R}^3 \setminus F^{(k)})$ then

$$\begin{aligned} R(\mathbf{x}) = & -v_f(\mathbf{O}^{(k)}) + O(\varepsilon \|\nabla v_f\|_{L_\infty(\omega)}) \\ & +4\pi \sum_{j=1}^{N} C_j \mathrm{cap}(F^{(j)})\Big(H(\mathbf{O}^{(k)}, \mathbf{O}^{(j)}) + O(\varepsilon)\Big) \\ & -C_k - \sum_{1 \leq j \leq N,\ j \neq k} C_j \Big\{ \frac{\mathrm{cap}(F^{(j)})}{|\mathbf{O}^{(k)} - \mathbf{O}^{(j)}|} + O(\frac{\varepsilon\ \mathrm{cap}(F^{(j)})}{|\mathbf{O}^{(k)} - \mathbf{O}^{(j)}|^2}) \Big\}. \end{aligned} \tag{9.53}$$

Noting that (9.26) can be written as the algebraic system

$$\begin{aligned} & C_k + 4\pi \sum_{j=1}^{N} C_j (1 - \delta_{jk})\ \mathrm{cap}(F^{(j)})\Big(\frac{1}{4\pi |\mathbf{O}^{(k)} - \mathbf{O}^{(j)}|} \\ & -H(\mathbf{O}^{(k)}, \mathbf{O}^{(j)})\Big) + v_f(\mathbf{O}^{(k)}) = 0, \end{aligned} \tag{9.54}$$

which, along with (9.53) and the obvious inequality $\mathrm{cap}(F^{(j)}) \leq \varepsilon$, implies

$$\begin{aligned} R(\mathbf{x}) = & \ O(\varepsilon \|\nabla v_f\|_{L_\infty(\omega)}) + 4\pi C_k \mathrm{cap}(F^{(k)}) H(\mathbf{O}^{(k)}, \mathbf{O}^{(k)}) \\ & + \sum_{j=1}^{N} O(\varepsilon\ \mathrm{cap}(F^{(j)})|C_j|) + \sum_{1 \leq j \leq N,\ j \neq k} O(\frac{\varepsilon\ \mathrm{cap}(F^{(j)})}{|\mathbf{O}^{(k)} - \mathbf{O}^{(j)}|^2}|C_j|). \end{aligned} \tag{9.55}$$

It suffices to estimate the sums

$$\sum_{1 \leq j \leq N,\ j \neq k} \frac{\varepsilon\ \mathrm{cap}(F^{(j)})|C_j|}{|\mathbf{O}^{(k)} - \mathbf{O}^{(j)}|^2}$$

and

$$\sum_{1 \leq j \leq N} \frac{\varepsilon\ \mathrm{cap}(F^{(j)})|C_j|}{|\mathbf{x} - \mathbf{O}^{(j)}|^2}, \quad \mathbf{x} \in \partial\Omega.$$

When $\varepsilon < c\ d^{7/4}$ we refer to Lemma 9.1, and using the inequality (9.28) we derive

$$
\begin{aligned}
&\sum_{j\neq k, 1\le j\le N} \frac{\varepsilon\, \mathrm{cap}(F^{(j)})|C_j|}{|\mathbf{O}^{(k)}-\mathbf{O}^{(j)}|^2} \\
&\le \Big(\sum_{j\neq k, 1\le j\le N} \frac{\varepsilon^2\, \mathrm{cap}(F^{(j)})}{|\mathbf{O}^{(k)}-\mathbf{O}^{(j)}|^4}\Big)^{1/2} \Big(\sum_{1\le j\le N} \mathrm{cap}(F^{(j)}) C_j^2\Big)^{1/2} \\
&\le \mathrm{Const}\, d^{-1/2} \Big(\sum_{1\le j\le N} \mathrm{cap}(F^{(j)})(v_f(\mathbf{O}^{(j)}))^2\Big)^{1/2} \Big(\max_{1\le j\le N} \varepsilon^2 d^{-3} \mathrm{cap}(F^{(j)})\Big)^{1/2} \\
&\le \mathrm{Const}\, \frac{\varepsilon^2}{d^{7/2}} \|v_f\|_{L_\infty(\omega)}. \qquad (9.56)
\end{aligned}
$$

Similarly, when $\mathbf{x} \in \partial\Omega$ we deduce

$$
\sum_{1\le j\le N} \frac{\varepsilon\, \mathrm{cap}(F^{(j)})|C_j|}{|\mathbf{x}-\mathbf{O}^{(j)}|^2} \le \mathrm{Const}\, \frac{\varepsilon^2}{d^{7/2}} \|v_f\|_{L_\infty(\omega)}. \qquad (9.57)
$$

Combining (9.55), (9.56), (9.57) and (9.52) we complete the proof by referring to the classical maximum principle for harmonic functions. □

Under the stronger constraint (9.39) on ε and d, Lemma 9.2 and representations (9.52), (9.53) lead to the following

Theorem 9.2. *If the inequality (9.49) is replaced by (9.39), then the remainder term from (9.50) satisfies the estimate*

$$
|R(\mathbf{x})| \le C\Big\{\varepsilon \|\nabla v_f\|_{L_\infty(\omega)} + \varepsilon^2 d^{-3} \|v_f\|_{L_\infty(\omega)}\Big\}. \qquad (9.58)
$$

9.6 The Energy Estimate

Under the constraint (9.39) on ε and d, which is stronger than (9.49), we derive the energy estimate for the remainder R. This result is important, since it allows for the generalization to general elliptic systems, and in particular to elasticity where the classical maximum principle cannot be applied.

Theorem 9.3. *Let the parameters ε and d, introduced in (9.1), satisfy the inequality*

$$
\varepsilon < c\, d^2, \qquad (9.59)
$$

where c is a sufficiently small absolute constant. Then the remainder R in (9.50) satisfies the estimate

$$\|\nabla R\|_{L_2(\Omega_N)} \leq \text{Const}\, \frac{\varepsilon^2}{d^4} \|f\|_{L_\infty(\Omega_N)}, \tag{9.60}$$

where f is the same as in (9.4).

Proof. For every $k = 1, \dots, N$, we introduce the function

$$\begin{aligned}\Psi_k(\mathbf{x}) = \; & v_f(\mathbf{x}) - v_f(\mathbf{O}^{(k)}) + \sum_{1 \leq j \leq N, j \neq k} C_j \Big(P^{(j)}(\mathbf{x}) - \frac{\text{cap}(F^{(j)})}{|\mathbf{O}^{(k)} - \mathbf{O}^{(j)}|} \Big) \\ & -4\pi \sum_{j=1}^{N} C_j \, \text{cap}(F^{(j)}) \Big(H(\mathbf{x}, \mathbf{O}^{(j)}) - H(\mathbf{O}^{(k)}, \mathbf{O}^{(j)}) \Big) \\ & -4\pi C_k \, \text{cap}(F^{(k)}) H(\mathbf{O}^{(k)}, \mathbf{O}^{(k)}),\end{aligned} \tag{9.61}$$

where the coefficients C_j satisfy the system (9.22).

By (9.50) and (9.61), for quasi-every $\mathbf{x} \in \partial(\mathbb{R}^3 \setminus F^{(k)})$

$$\begin{aligned}R(\mathbf{x}) + \Psi_k(\mathbf{x}) = & -v_f(\mathbf{O}^{(k)}) - C_k \\ & - \sum_{1 \leq j \leq N, \; j \neq k} C_j \Big(\frac{\text{cap}(F^{(j)})}{|\mathbf{O}^{(k)} - \mathbf{O}^{(j)}|} - 4\pi \, \text{cap}(F^{(j)}) H(\mathbf{O}^{(k)}, \mathbf{O}^{(j)}) \Big),\end{aligned}$$

which together with (9.22) implies

$$R(\mathbf{x}) + \Psi_k(\mathbf{x}) = 0$$

quasi-everywhere on $\partial(\mathbb{R}^3 \setminus F^{(k)})$ (i.e. outside of a set with zero capacity).

The function Ψ_0, defined by

$$\Psi_0(\mathbf{x}) = \sum_{j=1}^{N} C_j \Big(P^{(j)}(\mathbf{x}) - \frac{\text{cap}\, F^{(j)}}{|\mathbf{x} - \mathbf{O}^{(j)}|} \Big), \tag{9.62}$$

satisfies

$$R(\mathbf{x}) + \Psi_0(\mathbf{x}) = 0$$

quasi-everywhere on $\partial\Omega$, which follows from (9.50) and (9.5), (9.7).

We set $B_\rho^{(k)} = \{\mathbf{x} : \; |\mathbf{x} - \mathbf{O}^{(k)}| < \rho\}$, and define the capacitary potential of $F^{(k)}$ relative to $B_{d/4}^{(k)}$, that is a unique variational solution of the Dirichlet problem

$$\Delta \tilde{P}_k(\mathbf{x}) = 0, \;\; \mathbf{x} \in B_{d/4}^{(k)} \setminus F^{(k)}, \tag{9.63}$$

$$\tilde{P}_k(\mathbf{x}) = 1, \;\; \mathbf{x} \in \partial(\mathbb{R}^3 \setminus F^{(k)}), \tag{9.64}$$

$$\tilde{P}_k(\mathbf{x}) = 0, \;\; |\mathbf{x} - \mathbf{O}^{(k)}| = d/4. \tag{9.65}$$

Also, let a surface S_d be a smooth perturbation of $\partial\Omega$ such that

$$S_d \subset \Omega \ \text{ and } \ d/4 \leq \text{dist}(S_d, \mathbf{x}) \leq d/2 \ \text{ for all } \ \mathbf{x} \in \partial\Omega.$$

In turn, the set of all points placed between the surfaces $\partial\Omega$ and S_d is denoted by Π_d, and the function $\tilde{P}_0$ is defined as a unique variational solution of the Dirichlet problem

$$\Delta \tilde{P}_0(\mathbf{x}) = 0, \ \mathbf{x} \in \Pi_d, \tag{9.66}$$

$$\tilde{P}_0(\mathbf{x}) = 1, \ \mathbf{x} \in \partial\Omega, \tag{9.67}$$

$$\tilde{P}_0(\mathbf{x}) = 0, \ \mathbf{x} \in S_d. \tag{9.68}$$

We note that

$$R(\mathbf{x}) + \sum_{k=0}^{N} \tilde{P}_k(\mathbf{x})\Psi_k(\mathbf{x}) \tag{9.69}$$

vanishes quasi-everywhere on $\partial\Omega_N$ and that the Dirichlet integral of (9.69) over Ω_N is finite. Therefore, by harmonicity of R

$$\int_{\Omega_N} \nabla R(\mathbf{x}) \cdot \nabla\Big(R(\mathbf{x}) + \sum_{0\leq k\leq N} \tilde{P}_k(\mathbf{x})\Psi_k(\mathbf{x})\Big) d\mathbf{x} = 0.$$

Hence

$$\|\nabla R\|^2_{L_2(\Omega_N)} \leq \|\nabla R\|_{L_2(\Omega_N)} \, \|\nabla \sum_{0\leq k\leq N} \tilde{P}_k \Psi_k\|_{L_2(\Omega_N)},$$

which is equivalent to the estimate

$$\| \nabla R \|_{L_2(\Omega_N)} \leq \left(\sum_{k=1}^{N} \| \nabla(\tilde{P}_k \Psi_k) \|^2_{L_2(B^{(k)}_{d/4})} + \| \nabla(\tilde{P}_0 \Psi_0) \|^2_{L_2(\Pi_d)} \right)^{1/2}. \tag{9.70}$$

In the remaining part of the proof, we obtain an upper estimate for the right-hand side in (9.70).

The inequality (9.70) and the definition of Ψ_k lead to

$$\| \nabla R \|^2_{L_2(\Omega_N)} \leq 2\Big(\mathcal{K}^{(1)} + \mathcal{K}^{(2)} + \mathcal{L}^{(1)} + \mathcal{L}^{(2)} + \mathcal{M}^{(1)} + \mathcal{M}^{(2)} + \mathcal{N} + \mathcal{Q}\Big),$$

where

$$\mathcal{K}^{(1)} = \sum_{k=1}^{N} \|\nabla\Big(\tilde{P}_k(v_f(\cdot) - v_f(\mathbf{O}^{(k)}))\Big)\|^2_{L_2(B^{(k)}_{3\varepsilon})}, \tag{9.71}$$

$$\mathcal{L}^{(1)} = \sum_{k=1}^{N} \| \sum_{1 \le j \le N,\ j \ne k} C_j \nabla\Big(\tilde{P}_k(P^{(j)}(\cdot) - \frac{\mathrm{cap}(F^{(j)})}{|\mathbf{O}^{(k)} - \mathbf{O}^{(j)}|})\Big) \|^2_{L_2(B^{(k)}_{3\varepsilon})}, \tag{9.72}$$

$$\mathcal{M}^{(1)} = (4\pi)^2 \sum_{k=1}^{N} \Big\| \sum_{j=1}^{N} C_j \ \mathrm{cap}(F^{(j)}) \ \nabla\Big(\tilde{P}_k(H(\cdot, \mathbf{O}^{(j)}) - H(\mathbf{O}^{(k)}, \mathbf{O}^{(j)}))\Big) \Big\|^2_{L_2(B^{(k)}_{3\varepsilon})}, \tag{9.73}$$

$$\mathcal{N} = (4\pi)^2 \sum_{k=1}^{N} |C_k|^2 \left(\mathrm{cap}(F^{(k)})\right)^2 \left(H(\mathbf{O}^{(k)}, \mathbf{O}^{(k)})\right)^2 \| \nabla \tilde{P}_k \|^2_{L_2(B^{(k)}_{d/4})}, \tag{9.74}$$

$$\mathcal{Q} = \| \sum_{1 \le j \le N} C_j \nabla\Big(\tilde{P}_0(P^{(j)}(\cdot) - \frac{\mathrm{cap}(F^{(j)})}{|\mathbf{x} - \mathbf{O}^{(j)}|})\Big) \|^2_{L_2(\Pi_d)}, \tag{9.75}$$

and $\mathcal{K}^{(2)}$, $\mathcal{L}^{(2)}$, $\mathcal{M}^{(2)}$ are defined by replacing $B^{(k)}_{3\varepsilon}$ in the definitions of $\mathcal{K}^{(1)}$, $\mathcal{L}^{(1)}$, $\mathcal{M}^{(1)}$ by $B^{(k)}_{d/4} \setminus B^{(k)}_{3\varepsilon}$.

We start with the sum $\mathcal{K}^{(1)}$. Clearly,

$$\begin{aligned}\mathcal{K}^{(1)} &\le C \|\nabla v_f\|^2_{L_\infty(\omega)} \sum_{k=1}^{N} \int_{B^{(k)}_{3\varepsilon}} \Big\{ |\nabla \tilde{P}_k(\mathbf{x})|^2 \, |\mathbf{x} - \mathbf{O}^{(k)}|^2 + \left(\tilde{P}_k(\mathbf{x})\right)^2 \Big\} d\mathbf{x} \\ &\le C \|\nabla v_f\|^2_{L_\infty(\omega)} \sum_{k=1}^{N} \varepsilon^2 \, \mathrm{cap}(F^{(k)})\end{aligned} \tag{9.76}$$

and hence

$$\mathcal{K}^{(1)} \le C \varepsilon^3 d^{-3} \|\nabla v_f\|^2_{L_\infty(\omega)}. \tag{9.77}$$

Furthermore, by Green's formula and by (9.6) we deduce

$$\begin{aligned}\mathcal{K}^{(2)} = &- \sum_{k=1}^{N} \int_{B^{(k)}_{d/4} \setminus B^{(k)}_{3\varepsilon}} \tilde{P}_k(\mathbf{x}) \Big(v_f(\mathbf{x}) - v_f(\mathbf{O}^{(k)}) \Big) \Big\{ - \tilde{P}_k(\mathbf{x}) f(\mathbf{x}) \\ &+ 2 \nabla \tilde{P}_k(\mathbf{x}) \cdot \nabla v_f(\mathbf{x}) \Big\} d\mathbf{x} \\ &- \sum_{k=1}^{N} \int_{\partial B^{(k)}_{3\varepsilon}} \tilde{P}_k(\mathbf{x}) \Big(v_f(\mathbf{x}) - v_f(\mathbf{O}^{(k)}) \Big) \Big\{ \tilde{P}_k(\mathbf{x}) \frac{\partial v_f}{\partial |\mathbf{x}|}(\mathbf{x}) \\ &+ (v_f(\mathbf{x}) - v_f(\mathbf{O}^{(k)})) \frac{\partial \tilde{P}_k}{\partial |\mathbf{x}|}(\mathbf{x}) \Big\} dS\end{aligned} \tag{9.78}$$

By the mean value theorem for harmonic functions and the inequality $\tilde{P}_k(\mathbf{x}) \le P^{(k)}(\mathbf{x})$, we have

$$|\nabla \tilde{P}_k(\mathbf{x})| \le \frac{C}{|\mathbf{x}-\mathbf{O}^{(k)}|} \max_{\mathbf{y}\in B} P^{(k)}(\mathbf{y}),$$

where $B = \{\mathbf{y} : |\mathbf{y}-\mathbf{x}| < |\mathbf{x}-\mathbf{O}^{(k)}|/4\}$. Making use of the asymptotics (9.11) far from $\mathbf{O}^{(k)}$ we deduce

$$|\nabla \tilde{P}_k(\mathbf{x})| \le C \frac{\text{cap}(F^{(k)})}{|\mathbf{x}-\mathbf{O}^{(k)}|^2}, \quad \mathbf{x} \in B^{(k)}_{d/4} \setminus B^{(k)}_{3\varepsilon}. \tag{9.79}$$

Now we turn to the estimate of (9.78). The volume integral in the right-hand side of (9.78) does not exceed

$$\begin{aligned} C \sum_{k=1}^{N} \|\nabla v_f\|_{L_\infty(\omega)} \text{cap}(F^{(k)}) \int_{B^{(k)}_{d/4}\setminus B^{(k)}_{3\varepsilon}} \Big\{ \frac{\text{cap}(F^{(k)})}{|\mathbf{x}-\mathbf{O}^{(k)}|} |f(\mathbf{x})| \\ + \frac{\text{cap}(F^{(k)})}{|\mathbf{x}-\mathbf{O}^{(k)}|^2} \|\nabla v_f\|_{L_\infty(\omega)} \Big\} d\mathbf{x} \\ \le C\varepsilon d^{-3} \|\nabla v_f\|_{L_\infty(\omega)} \Big\{ \varepsilon d^2 \|f\|_{L_\infty(\Omega_N)} + \varepsilon d \|\nabla v_f\|_{L_\infty(\omega)} \Big\} \\ \le C\varepsilon^2 d^{-2} \|f\|^2_{L_\infty(\Omega_N)}. \end{aligned} \tag{9.80}$$

By $0 \le \tilde{P}_k(\mathbf{x}) \le 1$ and (9.79), the surface integral in (9.78) is dominated by

$$C\varepsilon^3 d^{-3} \|\nabla v_f\|^2_{L_\infty(\omega)} \le C\varepsilon^3 d^{-3} \|f\|^2_{L_\infty(\Omega_N)}. \tag{9.81}$$

Combining (9.80) and (9.81) we arrive at the estimate

$$\mathcal{K}^{(2)} \le C\varepsilon^2 d^{-2} \|f\|^2_{L_\infty(\Omega_N)}. \tag{9.82}$$

Let us estimate $\mathcal{L}^{(1)}$ (see (9.72)). Obviously,

$$\mathcal{L}^{(1)} \le \sum_{k=1}^{N} \Bigg(\sum_{1\le j\le N,\ j\ne k} |C_j| \, \Big\| \nabla \Big(\tilde{P}_k (P^{(j)}(\cdot) - \frac{\text{cap}(F^{(j)})}{|\mathbf{O}^{(k)}-\mathbf{O}^{(j)}|}) \Big) \Big\|_{L_2(B^{(k)}_{3\varepsilon})} \Bigg)^2.$$

Furthermore, when $j \ne k$ we have

$$\begin{aligned} \Big\| \nabla \Big(\tilde{P}_k (P^{(j)}(\cdot) - \frac{\text{cap}(F^{(j)})}{|\mathbf{O}^{(k)}-\mathbf{O}^{(j)}|}) \Big) \Big\|_{L_2(B^{(k)}_{3\varepsilon})} \\ \le \|(\nabla \tilde{P}_k) \Big(P^{(j)}(\cdot) - \frac{\text{cap}(F^{(j)})}{|\mathbf{O}^{(k)}-\mathbf{O}^{(j)}|} \Big) \|_{L_2(B^{(k)}_{3\varepsilon})} + \|\tilde{P}_k \nabla P^{(j)}\|_{L_2(B^{(k)}_{3\varepsilon})}, \end{aligned}$$

which does not exceed

$$C\left\{\frac{\varepsilon\,\mathrm{cap}(F^{(j)})(\mathrm{cap}(F^{(k)}))^{1/2}}{|\mathbf{O}^{(k)}-\mathbf{O}^{(j)}|^2}+\frac{\varepsilon^{3/2}\mathrm{cap}(F^{(j)})}{|\mathbf{O}^{(k)}-\mathbf{O}^{(j)}|^2}\right\}.$$

Hence, using Lemma 9.1 we deduce

$$\mathcal{L}^{(1)}\le C\sum_{k=1}^{N}\left(\varepsilon^2\sum_{1\le j\le N,\ j\ne k}|C_j|\frac{(\mathrm{cap}(F^{(j)}))^{1/2}}{|\mathbf{O}^{(k)}-\mathbf{O}^{(j)}|^2}\right)^2$$

$$\le C\,\varepsilon^4\sum_{k=1}^{N}\left\{\sum_{1\le j\le N,\ j\ne k}C_j^2\,\mathrm{cap}(F^{(j)})\sum_{1\le j\le N,\ j\ne k}\frac{1}{|\mathbf{O}^{(k)}-\mathbf{O}^{(j)}|^4}\right\},$$

and therefore

$$\mathcal{L}^{(1)}\le C\varepsilon^5 d^{-10}\|v_f\|^2_{L_\infty(\omega)}. \tag{9.83}$$

Similar steps can be followed to estimate $\mathcal{Q}$ in (9.75). We have

$$\left\|\nabla\Big(\tilde{P}_0(P^{(j)}(\cdot)-\frac{\mathrm{cap}(F^{(j)})}{|\mathbf{x}-\mathbf{O}^{(j)}|})\Big)\right\|_{L_2(\Pi_d)}$$

$$\le\left\|(\nabla\tilde{P}_0)\Big(P^{(j)}(\cdot)-\frac{\mathrm{cap}(F^{(j)})}{|\mathbf{x}-\mathbf{O}^{(j)}|}\Big)\right\|_{L_2(\Pi_d)}$$

$$+\left\|\tilde{P}_0\Big(\nabla P^{(j)}+\mathrm{cap}(F^{(j)})\frac{\mathbf{x}-\mathbf{O}^{(j)}}{|\mathbf{x}-\mathbf{O}^{(j)}|^3}\Big)\right\|_{L_2(\Pi_d)},$$

which does not exceed

$$C\varepsilon\,\mathrm{cap}(F^{(j)})|\mathbf{x}-\mathbf{O}^{(j)}|^{-2},\ \ \mathbf{x}\in\Pi_d.$$

As above, we use Lemma 9.1 to deduce

$$\mathcal{Q}\le C\,\varepsilon^3\left(\sum_{1\le j\le N}|C_j|\frac{(\mathrm{cap}(F^{(j)}))^{1/2}}{|\mathbf{x}-\mathbf{O}^{(j)}|^2}\right)^2$$

$$\le C\,\varepsilon^3\sum_{1\le j\le N}C_j^2\,\mathrm{cap}(F^{(j)})\sum_{1\le j\le N}\frac{1}{|\mathbf{x}-\mathbf{O}^{(j)}|^4}\le C\varepsilon^4 d^{-7}\|v_f\|^2_{L_\infty(\omega)}$$

$$\text{for }\ \mathbf{x}\in\Pi_d. \tag{9.84}$$

Next, we estimate $\mathcal{L}^{(2)}$. Integration by parts gives

$$\int_{B_{d/4}^{(k)}\setminus B_{3\varepsilon}^{(k)}} \nabla\Big(\tilde{P}_k(\mathbf{x})(P^{(j)}(\mathbf{x}) - \frac{\text{cap}(F^{(j)})}{|\mathbf{O}^{(k)}-\mathbf{O}^{(j)}|})\Big)$$

$$\cdot\nabla\Big(\tilde{P}_k(\mathbf{x})(P^{(m)}(\mathbf{x}) - \frac{\text{cap}(F^{(m)})}{|\mathbf{O}^{(k)}-\mathbf{O}^{(m)}|})\Big)d\mathbf{x}$$

$$= -2\int_{B_{d/4}^{(k)}\setminus B_{3\varepsilon}^{(k)}} \tilde{P}_k(\mathbf{x})\Big(P^{(j)}(\mathbf{x}) - \frac{\text{cap}(F^{(j)})}{|\mathbf{O}^{(k)}-\mathbf{O}^{(j)}|}\Big)\Big(\nabla\tilde{P}_k(\mathbf{x})\cdot\nabla P^{(m)}(\mathbf{x})\Big)d\mathbf{x}$$

$$-\int_{\partial B_{3\varepsilon}^{(k)}} \tilde{P}_k(\mathbf{x})\,\Big(P^{(j)}(\mathbf{x}) - \frac{\text{cap}(F^{(j)})}{|\mathbf{O}^{(k)}-\mathbf{O}^{(j)}|}\Big)\Big\{\tilde{P}_k(\mathbf{x})\,\frac{\partial}{\partial|\mathbf{x}|}P^{(m)}(\mathbf{x})$$

$$+(P^{(m)}(\mathbf{x}) - \frac{\text{cap}(F^{(m)})}{|\mathbf{O}^{(k)}-\mathbf{O}^{(m)}|})\frac{\partial\tilde{P}_k}{\partial|\mathbf{x}|}(\mathbf{x})\Big\}dS, \tag{9.85}$$

When $j \neq k$ and $m \neq k$, the volume integral in the right-hand side of (9.85) is estimated as follows

$$\Big|\int_{B_{d/4}^{(k)}\setminus B_{3\varepsilon}^{(k)}} \tilde{P}_k(\mathbf{x})\Big(P^{(j)}(\mathbf{x}) - \frac{\text{cap}(F^{(j)})}{|\mathbf{O}^{(k)}-\mathbf{O}^{(j)}|}\Big)\Big(\nabla\tilde{P}_k(\mathbf{x})\cdot\nabla P^{(m)}(\mathbf{x})\Big)d\mathbf{x}\Big|$$

$$\leq C\,\frac{\text{cap}(F^{(k)})\,\text{cap}(F^{(j)})}{|\mathbf{O}^{(k)}-\mathbf{O}^{(j)}|^2}\int_{B_{d/4}^{(k)}\setminus B_{3\varepsilon}^{(k)}} \frac{\text{cap}(F^{(k)})}{|\mathbf{x}-\mathbf{O}^{(k)}|^2}\,\frac{\text{cap}(F^{(m)})}{|\mathbf{x}-\mathbf{O}^{(m)}|^2}d\mathbf{x} \tag{9.86}$$

$$\leq C\frac{(\text{cap}(F^{(k)}))^2\text{cap}(F^{(j)})\text{cap}(F^{(m)})d}{|\mathbf{O}^{(k)}-\mathbf{O}^{(j)}|^2\,|\mathbf{O}^{(k)}-\mathbf{O}^{(m)}|^2} \leq C\frac{\varepsilon^2 d\;\text{cap}(F^{(j)})\;\text{cap}(F^{(m)})}{|\mathbf{O}^{(k)}-\mathbf{O}^{(j)}|^2\,|\mathbf{O}^{(k)}-\mathbf{O}^{(m)}|^2}.$$

In turn, when $j \neq k$ and $m \neq k$ the modulus of the surface integral in the right-hand side of (9.85) does not exceed

$$C\frac{\varepsilon\;\text{cap}(F^{(j)})}{|\mathbf{O}^{(k)}-\mathbf{O}^{(j)}|^2}\int_{\partial B_{3\varepsilon}^{(k)}}\Big\{\frac{\text{cap}(F^{(m)})}{|\mathbf{x}-\mathbf{O}^{(m)}|^2} + \frac{\varepsilon\;\text{cap}(F^{(m)})}{|\mathbf{O}^{(m)}-\mathbf{O}^{(k)}|^2}\Big|\frac{\partial\tilde{P}_k}{\partial|\mathbf{x}|}(\mathbf{x})\Big|\Big\}\,dS$$

$$\leq C\frac{\varepsilon\;\text{cap}(F^{(j)})\;\text{cap}(F^{(m)})}{|\mathbf{O}^{(k)}-\mathbf{O}^{(j)}|^2|\mathbf{O}^{(k)}-\mathbf{O}^{(m)}|^2}\Big\{\varepsilon^2 + \varepsilon\int_{\partial B_{3\varepsilon}^{(k)}}\frac{\text{cap}(F^{(k)})}{|\mathbf{x}-\mathbf{O}^{(k)}|^2}dS\Big\}$$

$$\leq C\frac{\varepsilon^3\;\text{cap}(F^{(j)})\;\text{cap}(F^{(m)})}{|\mathbf{O}^{(k)}-\mathbf{O}^{(j)}|^2|\mathbf{O}^{(k)}-\mathbf{O}^{(m)}|^2}. \tag{9.87}$$

We have

$$\mathcal{L}^{(2)} = \sum_{1\le m,j\le N} C_m C_j \times \sum_{1\le k\le N,\, k\ne m, k\ne j} \int_{B^{(k)}_{d/4}\setminus B^{(k)}_{3\varepsilon}} \nabla\Big(\tilde{P}_k(\mathbf{x})(P^{(j)}(\mathbf{x}) - \frac{\text{cap}(F^{(j)})}{|\mathbf{O}^{(k)}-\mathbf{O}^{(j)}|})\Big) \cdot\nabla\Big(\tilde{P}_k(\mathbf{x})(P^{(m)}(\mathbf{x}) - \frac{\text{cap}(F^{(m)})}{|\mathbf{O}^{(k)}-\mathbf{O}^{(m)}|})\Big)d\mathbf{x},$$

and by (9.85), (9.86) and (9.87)

$$\mathcal{L}^{(2)} \le C\varepsilon^2 d \sum_{1\le m,j\le N} |C_m||C_j| \sum_{1\le k\le N,\, k\ne m,k\ne j} \frac{\text{cap}(F^{(j)})\text{cap}(F^{(m)})}{|\mathbf{O}^{(k)}-\mathbf{O}^{(j)}|^2\,|\mathbf{O}^{(k)}-\mathbf{O}^{(m)}|^2}$$

$$= C\,\frac{\varepsilon^2}{d^2}\sum_{1\le m,j\le N} |C_m||C_j|\text{cap}(F^{(j)})\,\text{cap}(F^{(m)}) \times \sum_{1\le k\le N,\, k\ne m,k\ne j} \frac{d^3}{|\mathbf{O}^{(k)}-\mathbf{O}^{(j)}|^2\,|\mathbf{O}^{(k)}-\mathbf{O}^{(m)}|^2},$$

and therefore

$$\mathcal{L}^{(2)} \le C\,\frac{\varepsilon^2}{d^2}\sum_{1\le m,j\le N} \frac{|C_m||C_j|\text{cap}(F^{(j)})\,\text{cap}(F^{(m)})}{d+|\mathbf{O}^{(j)}-\mathbf{O}^{(m)}|}. \tag{9.88}$$

Let us introduce a piece-wise constant function

$$\xi(\mathbf{x}) = \begin{cases} |C_m|(\text{cap}(F^{(m)}))^{1/2}, & \text{when } \mathbf{x}\in B^{(m)}_{d/4}, \\ 0, & \text{otherwise.}\end{cases}$$

Then the inequality (9.88) leads to

$$\mathcal{L}^{(2)} \le C\,\frac{\varepsilon^3}{d^8}\sum_{1\le m,j\le N}\frac{\Big(|C_m|(\text{cap}(F^{(m)}))^{1/2}\Big)\Big(|C_j|(\text{cap}(F^{(j)}))^{1/2}\Big)d^6}{d+|\mathbf{O}^{(j)}-\mathbf{O}^{(m)}|}$$

$$\le C\,\frac{\varepsilon^3}{d^8}\int_\omega\int_\omega \frac{\xi(\mathbf{X})\xi(\mathbf{Y})}{d+|\mathbf{X}-\mathbf{Y}|}d\mathbf{X}d\mathbf{Y} \le C\,\frac{\varepsilon^3}{d^8}\,\|\xi\|^2_{L_2(\omega)},$$

where the constant C depends on ω, and using Lemma 9.1 we deduce

$$\mathcal{L}^{(2)} \le C\,\frac{\varepsilon^3}{d^8}\sum_{1\le j\le N} C_j^2\,\text{cap}(F^{(j)})d^3 \le C\,\frac{\varepsilon^4}{d^8}\|v_f\|^2_{L_2(\omega)}. \tag{9.89}$$

To evaluate $\mathcal{M}^{(1)} + \mathcal{M}^{(2)}$ we apply the result of Lemma 9.2 and use the same algorithm as for $\mathcal{K}^{(1)}$ and $\mathcal{K}^{(2)}$ to deduce

$$\mathcal{M}^{(1)} + \mathcal{M}^{(2)} \leq C\|v_f\|^2_{L_\infty(\omega)}\varepsilon d^{-3}\Big(\varepsilon^3 d^{-3} + \varepsilon^2 d^{-2}\Big) \leq C\varepsilon^3 d^{-5}\|v_f\|^2_{L_\infty(\omega)}. \tag{9.90}$$

Similarly, applying Lemma 9.2, we derive the estimate for the term $\mathcal{N}$

$$\mathcal{N} \leq C\varepsilon^3 d^{-3}\|v_f\|^2_{L_\infty(\omega)}. \tag{9.91}$$

The proof is completed by the reference to (9.76), (9.82), (9.83), (9.84), (9.89), (9.91). □

9.7 Meso-scale Approximation of Green's Function in Ω_N

Let $G_N(\mathbf{x},\mathbf{y})$ be Green's function of the Dirichlet problem for the operator $-\Delta$ in Ω_N. In this section, we derive the asymptotic approximation of $G_N(\mathbf{x},\mathbf{y})$ and estimate the remainder term. In the asymptotic algorithm, we will refer to the algebraic system similar to that of Sect. 9.4. We need here Green's functions $g^{(j)}(\mathbf{x},\mathbf{y})$ of the Dirichlet problem for the operator $-\Delta$ in $\mathbb{R}^3 \setminus F^{(j)}$, $j = 1,\dots,N$. The notation $h^{(j)}$ will be used for the regular part of $g^{(j)}$, that is

$$h^{(j)}(\mathbf{x},\mathbf{y}) = (4\pi|\mathbf{x}-\mathbf{y}|)^{-1} - g^{(j)}(\mathbf{x},\mathbf{y}),\ \ \mathbf{x},\mathbf{y} \in \mathbb{R}^3 \setminus F^{(j)}. \tag{9.92}$$

According to Lemma 1.2, the functions $h^{(j)}$ allow for the following estimate:

$$\left|h^{(j)}(\mathbf{x},\mathbf{y}) - \frac{P^{(j)}(\mathbf{y})}{4\pi|\mathbf{x}-\mathbf{O}^{(j)}|}\right| \leq \text{Const}\,\frac{\varepsilon P^{(j)}(\mathbf{y})}{|\mathbf{x}-\mathbf{O}^{(j)}|^2}, \tag{9.93}$$

for all $\mathbf{y} \in \mathbb{R}^3 \setminus F^{(j)}$ and $|\mathbf{x}-\mathbf{O}^{(j)}| > 2\varepsilon$.

The principal result of this section is

Theorem 9.4. *Let the small parameters ε and d, introduced in (9.1), satisfy the inequality $\varepsilon < c\,d^2$, where c is a sufficiently small absolute constant. Then*

$$G_N(\mathbf{x},\mathbf{y}) = G(\mathbf{x},\mathbf{y}) - \sum_{j=1}^{N}\Big\{h^{(j)}(\mathbf{x},\mathbf{y}) - P^{(j)}(\mathbf{y})H(\mathbf{x},\mathbf{O}^{(j)}) \tag{9.94}$$

$$-P^{(j)}(\mathbf{x})H(\mathbf{O}^{(j)},\mathbf{y}) + 4\pi\ \text{cap}(F^{(j)})H(\mathbf{x},\mathbf{O}^{(j)})H(\mathbf{O}^{(j)},\mathbf{y})$$

$$+H(\mathbf{O}^{(j)},\mathbf{O}^{(j)})\,T^{(j)}(\mathbf{x})T^{(j)}(\mathbf{y}) - \sum_{i=1}^{N} C_{ij}T^{(i)}(\mathbf{x})T^{(j)}(\mathbf{y})\Big\} + \mathcal{R}(\mathbf{x},\mathbf{y}),$$

where

$$T^{(j)}(\mathbf{y}) = P^{(j)}(\mathbf{y}) - 4\pi \operatorname{cap}(F^{(j)}) H(\mathbf{O}^{(j)}, \mathbf{y}), \tag{9.95}$$

with the capacitary potentials $P^{(j)}$ *and the regular part* H *of Green's function* G *of* Ω *being the same as in Sect. 9.2. The matrix* $\mathcal{C} = (C_{ij})_{i,j=1}^N$ *is defined by*

$$\mathcal{C} = (\mathbf{I} + \mathbf{SD})^{-1}\mathbf{S}, \tag{9.96}$$

where $\mathbf{S}$ *and* $\mathbf{D}$ *are the same as in (9.24), (9.25). The remainder* $\mathcal{R}(\mathbf{x}, \mathbf{y})$ *is a harmonic function, both in* $\mathbf{x}$ *and* $\mathbf{y}$, *and satisfies the estimate*

$$|\mathcal{R}(\mathbf{x}, \mathbf{y})| \le \text{Const}\, \varepsilon d^{-2} \tag{9.97}$$

uniformly with respect to $\mathbf{x}$ *and* $\mathbf{y}$ *in* Ω_N.

Prior to the proof of the theorem, we formulate an auxiliary result.

Lemma 9.3. *Let the small parameters* ε *and* d, *defined in (9.1), obey the inequality (9.39). Then the matrix* $\mathcal{C}$ *in (9.96) satisfies the estimate*

$$\|\mathcal{C}\|_{\mathbb{R}^N \to \mathbb{R}^N} \le c d^{-3}, \tag{9.98}$$

where c *is an absolute constant.*

Proof. First, we note that

$$\|\mathcal{C}\|_{\mathbb{R}^N \to \mathbb{R}^N} \le \text{Const}\, \|\mathbf{S}\|_{\mathbb{R}^N \to \mathbb{R}^N}, \tag{9.99}$$

which follows from Lemma 9.2, where $\mathbf{V}_f$ should be replaced by the columns of the matrix $\mathbf{S}$.

Additionally,

$$\|\mathbf{S}\|_{\mathbb{R}^N \to \mathbb{R}^N} \le \text{Const}\, d^{-3}. \tag{9.100}$$

To verify this estimate we introduce a vector $\boldsymbol{\xi} = (\xi_j)_{j=1}^N$, $\|\boldsymbol{\xi}\| = 1$, and a function $\xi(\mathbf{x})$ defined in Ω by

$$\xi(\mathbf{x}) = \begin{cases} \xi_j & \text{in } B^{(j)} = \{\mathbf{x} : |\mathbf{x} - \mathbf{O}^{(j)}| < d\},\ j = 1, \dots, N, \\ 0 & \text{otherwise.} \end{cases} \tag{9.101}$$

Then

$$\langle \mathbf{S}\boldsymbol{\xi}, \boldsymbol{\xi} \rangle \le \text{Const}\, d^{-6} \int_\omega \int_\omega \frac{\xi(\mathbf{X})\xi(\mathbf{Y}) d\mathbf{X} d\mathbf{Y}}{4\pi |\mathbf{X} - \mathbf{Y}|}$$

$$\le \text{Const}\, d^{-6} \int_\omega |\xi(\mathbf{X})|^2 dX \le \text{Const}\, d^{-3},$$

which yields (9.100). Then (9.99) together with (9.100) lead to (9.98). □

Proof of Theorem 9.4. The harmonicity of $\mathcal{R}$ follows directly from (9.94).

Let us estimate the boundary values of $\mathcal{R}$ on $\partial\Omega_N$.

If $\mathbf{x} \in \partial\Omega$ and $\mathbf{y} \in \Omega_N$, then according to the definitions of Sect. 9.2.3 for Green's function of Ω and its regular part the remainder term $\mathcal{R}$ in (9.94) takes the form

$$\mathcal{R}(\mathbf{x},\mathbf{y}) = \sum_{j=1}^{N} \Big\{ h^{(j)}(\mathbf{x},\mathbf{y}) - \frac{P^{(j)}(\mathbf{y})}{4\pi|\mathbf{x}-\mathbf{O}^{(j)}|} - H(\mathbf{O}^{(j)},\mathbf{y})\Big(P^{(j)}(\mathbf{x}) - \frac{\mathrm{cap}(F^{(j)})}{|\mathbf{x}-\mathbf{O}^{(j)}|}\Big)$$

$$+H(\mathbf{O}^{(j)},\mathbf{O}^{(j)})\, T^{(j)}(\mathbf{y})\Big(P^{(j)}(\mathbf{x}) - \frac{\mathrm{cap}(F^{(j)})}{|\mathbf{x}-\mathbf{O}^{(j)}|}\Big)$$

$$-\sum_{i=1}^{N} C_{ij} T^{(j)}(\mathbf{y})\Big(P^{(i)}(\mathbf{x}) - \frac{\mathrm{cap}(F^{(i)})}{|\mathbf{x}-\mathbf{O}^{(i)}|}\Big)\Big\}.$$

Taking into account the estimate (9.93) for $h^{(j)}$ together with the asymptotic representation (9.11) of $P^{(j)}$ we obtain

$$\mathcal{R}(\mathbf{x},\mathbf{y}) = \sum_{j=1}^{N} O\Big(\frac{\varepsilon P^{(j)}(\mathbf{y})}{|\mathbf{x}-\mathbf{O}^{(j)}|^2}\Big) + \sum_{j=1}^{N}\sum_{i=1}^{N} C_{ij} T^{(j)}(\mathbf{y}) O\Big(\frac{\varepsilon^2}{|\mathbf{x}-\mathbf{O}^{(i)}|^2}\Big) \tag{9.102}$$

for all $\mathbf{x} \in \partial\Omega$.

Here $\mathcal{R}(\mathbf{x},\mathbf{y})$ is harmonic as a function of $\mathbf{y}$. Next, we estimate (9.102) for $\mathbf{y} \in \partial\Omega_N$.

If $\mathbf{y}, \mathbf{x} \in \partial\Omega$ then (9.102), (9.95) and (9.11) lead to

$$\mathcal{R}(\mathbf{x},\mathbf{y}) = \sum_{j=1}^{N} O\Big(\frac{\varepsilon^2}{|\mathbf{y}-\mathbf{O}^{(j)}||\mathbf{x}-\mathbf{O}^{(j)}|^2}\Big) + \sum_{j=1}^{N}\sum_{i=1}^{N} C_{ij} O\Big(\frac{\varepsilon^4}{|\mathbf{x}-\mathbf{O}^{(i)}|^2|\mathbf{y}-\mathbf{O}^{(j)}|^2}\Big). \tag{9.103}$$

Using (9.98) we can estimate the double sum from (9.103). For a fixed $\mathbf{x} \in \partial\Omega$, let us introduce a vector

$$\mathbf{V} = \Big(\frac{\varepsilon^2}{|\mathbf{x}-\mathbf{O}^{(i)}|^2}\Big)_{i=1}^{N},$$

and a function $V(\mathbf{X})$ defined in Ω by

$$V(\mathbf{X}) = \begin{cases} V_j & \text{when } \mathbf{X} \in B^{(j)}, \\ 0 & \text{otherwise,} \end{cases}$$

where the balls $B^{(j)}$ are the same as in (9.101). It follows from Lemma 9.3 that the double sum in (9.103) does not exceed

$$c\, d^{-3}\|\mathbf{V}\|^2 \le \frac{\text{Const}}{d^6}\int_\omega (V(\mathbf{X}))^2 d\mathbf{X} \le \frac{\text{Const}\ \varepsilon^4}{d^7}.$$

The above estimate together with (9.103) imply

$$\mathcal{R}(\mathbf{x},\mathbf{y}) = O(\varepsilon^2 d^{-3}|\log d| + \varepsilon^4 d^{-7}) \text{ when } \mathbf{x},\mathbf{y}\in\partial\Omega. \tag{9.104}$$

Now, we estimate (9.102) for $\mathbf{y}\in\partial(\mathbb{R}^3\setminus F^{(m)})$. In this case we have

$$\begin{aligned}\mathcal{R}(\mathbf{x},\mathbf{y}) = {}& O\Big(\frac{\varepsilon}{|\mathbf{x}-\mathbf{O}^{(j)}|^2}\Big) + \sum_{1\le j\le N, j\ne m} O\Big(\frac{\varepsilon^2}{|\mathbf{y}-\mathbf{O}^{(j)}||\mathbf{x}-\mathbf{O}^{(j)}|^2}\Big)\\ &+\sum_{i=1}^N O\Big(\frac{\varepsilon^2}{|\mathbf{x}-\mathbf{O}^{(i)}|^2}\Big)\Big\{C_{im}\Big(1-4\pi\ \text{cap}(F^{(m)})H(\mathbf{O}^{(m)},\mathbf{y})\Big)\\ &+4\pi\sum_{1\le j\le N, j\ne m} C_{ij}\Big(\text{cap}(F^{(j)})G(\mathbf{O}^{(j)},\mathbf{y}) + O(\frac{\varepsilon^2}{|\mathbf{y}-\mathbf{O}^{(j)}|^2})\Big)\Big\}.\end{aligned} \tag{9.105}$$

We also note that according to (9.96) the coefficients C_{ij} satisfy the system of algebraic equations

$$\begin{aligned}&(1-\delta_{im})G(\mathbf{O}^{(m)},\mathbf{O}^{(i)}) - C_{im}\\ &-4\pi\sum_{1\le j\le N,\ j\ne m} C_{ij}\ \text{cap}(F^{(i)})\ G(\mathbf{O}^{(m)},\mathbf{O}^{(i)}) = 0,\end{aligned} \tag{9.106}$$

for $m, i = 1,\dots,N$. Hence, in the above formula (9.105) the expression in curly brackets can be written as

$$\begin{aligned}&C_{im} + O(|C_{im}|\varepsilon) + 4\pi\sum_{1\le j\le N, j\ne m} C_{ij}\Big(\text{cap}(F^{(j)})G(\mathbf{O}^{(j)},\mathbf{y}) + O(\frac{\varepsilon^2}{|\mathbf{y}-\mathbf{O}^{(j)}|^2})\Big)\\ &= (1-\delta_{im})G(\mathbf{O}^{(m)},\mathbf{O}^{(i)}) + O(\varepsilon d^{-1})\\ &+\sum_{1\le j\le N, j\ne m} C_{ij} O(\frac{\varepsilon^2}{|\mathbf{y}-\mathbf{O}^{(j)}|^2}),\ \ \mathbf{y}\in\partial(\mathbb{R}^3\setminus F^{(m)}),\end{aligned} \tag{9.107}$$

and then formulae (9.105) and (9.107) imply

$$\mathcal{R}(\mathbf{x},\mathbf{y}) = O\Big(\varepsilon d^{-2} + \varepsilon^2 |\log d| d^{-3} + \varepsilon^3 d^{-4}\Big)$$

$$+ \sum_{1\le i\le N} \sum_{1\le j\le N, j\ne m} C_{ij} O\Big(\frac{\varepsilon^4}{|\mathbf{y}-\mathbf{O}^{(j)}|^2 |\mathbf{x}-\mathbf{O}^{(i)}|^2}\Big),$$

where the estimate of the double sum is similar to (9.103). Thus, we obtain

$$\mathcal{R}(\mathbf{x},\mathbf{y}) = O\Big(\varepsilon d^{-2} + \varepsilon^2 |\log d| d^{-3} + \varepsilon^3 d^{-4} + \varepsilon^4 d^{-7}\Big) = O(\varepsilon d^{-2}), \quad (9.108)$$

for all $\mathbf{x} \in \partial\Omega$ and $\mathbf{y} \in \partial(\mathbb{R}^3 \setminus F^{(m)}),\ m = 1, \dots, N$.

Using the estimates (9.104) and (9.108) and applying the maximum principle for harmonic functions we deduce that

$$\mathcal{R}(\mathbf{x},\mathbf{y}) = O(\varepsilon d^{-2}), \quad (9.109)$$

for all $\mathbf{x} \in \partial\Omega$ and $\mathbf{y} \in \Omega_N$.

In turn, when $\mathbf{x} \in \partial(\mathbb{R}^3 \setminus F^{(k)})$, the formula (9.94) and the definition (9.92) of $h^{(j)}$ lead to the expression for the remainder term on the boundary of the inclusion

$$\begin{aligned}\mathcal{R}(\mathbf{x},\mathbf{y}) = {} & H(\mathbf{x},\mathbf{y}) - H(\mathbf{O}^{(k)},\mathbf{y}) \\ & + \sum_{1\le j\le N,\ j\ne k} \Big(h^{(j)}(\mathbf{x},\mathbf{y}) - P^{(j)}(\mathbf{x}) H(\mathbf{O}^{(j)},\mathbf{y})\Big) \\ & + \sum_{j=1}^{N} T^{(j)}(\mathbf{y}) \Bigg(H(\mathbf{O}^{(j)},\mathbf{O}^{(j)})\, T^{(j)}(\mathbf{x}) - H(\mathbf{x},\mathbf{O}^{(j)}) - \sum_{j=1}^{N} C_{ij} T^{(i)}(\mathbf{x}) \Bigg).\end{aligned} \quad (9.110)$$

Using the formulae (9.11) and (9.93) for $P^{(j)}$ and $h^{(j)}$ together with the definition (9.95) of $T^{(j)}$ and the definition of Sect. 9.2.3 of the regular part of Green's function of Ω we deduce that

$$h^{(j)}(\mathbf{x},\mathbf{y}) - P^{(j)}(\mathbf{x}) H(\mathbf{O}^{(j)},\mathbf{y}) = \frac{T^{(j)}(\mathbf{y})}{4\pi|\mathbf{x}-\mathbf{O}^{(j)}|} + O\Big(\frac{\varepsilon^2 + \varepsilon P^{(j)}(\mathbf{y})}{|\mathbf{O}^{(k)}-\mathbf{O}^{(j)}|^2}\Big), \quad j \ne k, \quad (9.111)$$

and

$$H(\mathbf{x},\mathbf{y}) = H(\mathbf{O}^{(k)},\mathbf{y}) + O(\varepsilon), \quad (9.112)$$

for $\mathbf{x} \in \partial(\mathbb{R}^3 \setminus F^{(k)})$ and $\mathbf{y} \in \Omega_N$. The representations (9.95) together with (9.110)–(9.112) imply

$$\mathcal{R}(\mathbf{x},\mathbf{y}) = \sum_{1\le j\le N,\ j\ne k} \Big\{ \frac{T^{(j)}(\mathbf{y})}{4\pi|\mathbf{x}-\mathbf{O}^{(j)}|} + O\Big(\frac{\varepsilon^2+\varepsilon P^{(j)}(\mathbf{y})}{|\mathbf{O}^{(k)}-\mathbf{O}^{(j)}|^2}\Big)\Big\}$$
$$-\sum_{j=1}^{N} T^{(j)}(\mathbf{y})\Big(H(\mathbf{O}^{(k)},\mathbf{O}^{(j)}) - H(\mathbf{O}^{(j)},\mathbf{O}^{(j)})T^{(j)}(\mathbf{x})$$
$$+\sum_{i=1}^{N} C_{ij}T^{(i)}(\mathbf{x})\Big) + \sum_{j=1}^{N} O(\varepsilon|T^{(j)}(\mathbf{y})|). \tag{9.113}$$

Bearing in mind the asymptotic formula (9.11) for the capacitary potentials and the definition (9.95) we deduce that for $\mathbf{x}\in\partial(\mathbb{R}^3\setminus F^{(k)})$

$$T^{(j)}(\mathbf{x}) = \frac{\text{cap}(F^{(j)})}{|\mathbf{x}-\mathbf{O}^{(j)}|} - 4\pi\ \text{cap}(F^{(j)})H(\mathbf{x},\mathbf{O}^{(j)}) + O\Big(\frac{\varepsilon\ \text{cap}(F^{(j)})}{|\mathbf{x}-\mathbf{O}^{(j)}|^2}\Big) \tag{9.114}$$
$$= 4\pi\ \text{cap}(F^{(j)})G(\mathbf{x},\mathbf{O}^{(j)}) + O\Big(\frac{\varepsilon^2}{|\mathbf{x}-\mathbf{O}^{(j)}|^2}\Big),\ j\ne k.$$

Thus, (9.113) can be rearranged in the form

$$\mathcal{R}(\mathbf{x},\mathbf{y}) = \sum_{1\le j\le N,\ j\ne k} T^{(j)}(\mathbf{y})\Big\{ G(\mathbf{O}^{(k)},\mathbf{O}^{(j)}) - C_{kj}$$
$$-4\pi \sum_{1\le i\le N,\ i\ne k} C_{ij}\ \text{cap}(F^{(i)})\ G(\mathbf{O}^{(k)},\mathbf{O}^{(i)})\Big\} + \mathcal{R}^{(1)}(\mathbf{x},\mathbf{y}), \tag{9.115}$$

where

$$\mathcal{R}^{(1)}(\mathbf{x},\mathbf{y}) = O(\varepsilon) + \sum_{1\le j\le N,\ j\ne k}\Big\{ O\Big(\frac{\varepsilon\,|T^{(j)}(\mathbf{y})|}{|\mathbf{O}^{(k)}-\mathbf{O}^{(j)}|}\Big) + O(\varepsilon d^{-1}|T^{(j)}(\mathbf{y})|)\Big\}$$
$$+ \sum_{1\le j\le N,\ j\ne k} O\Big(\varepsilon|T^{(j)}(\mathbf{y})| + \frac{\varepsilon^2+\varepsilon P^{(j)}(\mathbf{y})}{|\mathbf{O}^{(k)}-\mathbf{O}^{(j)}|^2}\Big)$$
$$+ \sum_{1\le i\le N,\ i\ne k} O\Big(\frac{\varepsilon\ \text{cap}(F^{(i)})}{d\ |\mathbf{O}^{(k)}-\mathbf{O}^{(i)}|^2}\Big) \tag{9.116}$$
$$+\sum_{j=1}^{N}\sum_{1\le i\le N,\ i\ne k} C_{ij}T^{(j)}(\mathbf{y})O\Big(\frac{\varepsilon\ \text{cap}(F^{(i)})}{|\mathbf{O}^{(k)}-\mathbf{O}^{(i)}|^2}\Big).$$

It follows from (9.96) that the coefficients C_{ij} satisfy the system of algebraic equations

$$(1-\delta_{kj})G(\mathbf{O}^{(k)},\mathbf{O}^{(j)}) - C_{kj} - 4\pi \sum_{1\le i\le N,\, i\ne k} C_{ij}\,\mathrm{cap}(F^{(i)})\; G(\mathbf{O}^{(k)},\mathbf{O}^{(i)}) = 0, \tag{9.117}$$

for $k, j = 1,\dots,N$, and hence using (9.115)–(9.117), we arrive at

$$\mathcal{R}(\mathbf{x},\mathbf{y}) = \mathcal{R}^{(1)}(\mathbf{x},\mathbf{y}) \tag{9.118}$$

for all $\mathbf{x} \in \partial(\mathbb{R}^3 \setminus F^{(k)})$ and $\mathbf{y} \in \Omega_N$.

Let us consider the case when $\mathbf{y} \in \partial(\mathbb{R}^3 \setminus F^{(m)})$. Then

$$T^{(j)}(\mathbf{y}) = 4\pi\; \mathrm{cap}(F^{(j)})\; G(\mathbf{O}^{(j)},\mathbf{y}) + O\Big(\frac{\varepsilon\; \mathrm{cap}(F^{(j)})}{|\mathbf{y}-\mathbf{O}^{(j)}|^2}\Big),\; j\ne m,$$

and

$$T^{(m)}(\mathbf{y}) = 1 - 4\pi\; \mathrm{cap}(F^{(m)})\; H(\mathbf{O}^{(m)},\mathbf{y}).$$

The double sum in (9.116) can be rearranged according to (9.106)

$$\begin{aligned}
&\sum_{1\le i\le N,\, i\ne k} O\Big(\frac{\varepsilon\; \mathrm{cap}(F^{(i)})}{|\mathbf{O}^{(k)}-\mathbf{O}^{(i)}|^2}\Big)\sum_{j=1}^{N} C_{ij}T^{(j)}(\mathbf{y})\\
&= \sum_{1\le i\le N,\, i\ne k} O\Big(\frac{\varepsilon\; \mathrm{cap}(F^{(i)})}{|\mathbf{O}^{(k)}-\mathbf{O}^{(i)}|^2}\Big)\Big\{C_{im} + O(|C_{im}|\varepsilon)\\
&+4\pi \sum_{1\le j\le N, j\ne m} C_{ij}\Big(\mathrm{cap}(F^{(j)})G(\mathbf{O}^{(j)},\mathbf{y}) + O(\frac{\varepsilon^2}{|\mathbf{y}-\mathbf{O}^{(j)}|^2})\Big)\Big\}\\
&= \sum_{1\le i\le N,\, i\ne k, i\ne m} O\Big(\frac{\varepsilon^2}{|\mathbf{O}^{(k)}-\mathbf{O}^{(i)}|^2}\Big)\Big\{G(\mathbf{O}^{(m)},\mathbf{O}^{(i)}) + O(\varepsilon d^{-1})\Big\}\\
&+ \sum_{1\le i\le N,\, i\ne k}\; \sum_{1\le j\le N, j\ne m} C_{ij}O(\frac{\varepsilon^4}{|\mathbf{O}^{(m)}-\mathbf{O}^{(j)}|^2|\mathbf{O}^{(k)}-\mathbf{O}^{(i)}|^2}) \qquad (9.119)\\
&= O(\varepsilon^2|\log d|d^{-3} + \varepsilon^3 d^{-4} + \varepsilon^4 d^{-7}),
\end{aligned}$$

for $\mathbf{x} \in \partial(\mathbb{R}^3 \setminus F^{(k)})$, $\mathbf{y} \in \partial(\mathbb{R}^3 \setminus F^{(m)})$, where the estimate of the last double sum in (9.119) is similar to (9.103). Combining (9.116), (9.118) and (9.119), we deduce that $\mathcal{R}(\mathbf{x},\mathbf{y}) = O(\varepsilon d^{-2})$ for $\mathbf{x} \in \partial(\mathbb{R}^3 \setminus F^{(k)})$, $\mathbf{y} \in \partial(\mathbb{R}^3 \setminus F^{(m)})$, $m, k = 1,\dots,N$. Using the symmetry of $\mathcal{R}(\mathbf{x},\mathbf{y})$ together with (9.108) we also obtain that $\mathcal{R}(\mathbf{x},\mathbf{y}) = O(\varepsilon d^{-2})$ for $\mathbf{x} \in \partial(\mathbb{R}^3 \setminus F^{(k)})$, $k = 1,\dots,N$, $\mathbf{y} \in \partial\Omega$. Applying the maximum principle for harmonic functions we get

$$\mathcal{R}(\mathbf{x}, \mathbf{y}) = O(\varepsilon d^{-2}) \text{ for } \mathbf{x} \in \partial(\mathbb{R}^3 \setminus F^{(k)}), k = 1, \dots, N, \ \mathbf{y} \in \Omega_N. \tag{9.120}$$

Finally, formulae (9.109) and (9.120) imply that $\mathcal{R}(\mathbf{x}, \mathbf{y}) = O(\varepsilon d^{-2})$ for $\mathbf{x} \in \partial\Omega_N$ and $\mathbf{y} \in \Omega_N$, and then applying the maximum principle for harmonic functions we complete the proof. □

Chapter 10
Mixed Boundary Value Problems in Multiply-Perforated Domains

In this chapter we discuss meso-scale approximations of solutions to mixed problems for the Poisson equation for domains containing a large number of small perforations of arbitrary shape. The Dirichlet condition is set on the exterior boundary of the perforated body, and the Neumann conditions are specified on the boundaries of small holes.

The asymptotic methods, presented here and in Maz'ya and Movchan [27], can be applied to modelling of dilute composites in problems of mechanics, electromagnetism, heat conduction and phase transition.

Asymptotic approximations applied to solutions of boundary value problems of mixed type in domains containing many small spherical inclusions were considered in Figari and Teta [9]. The point interaction approximations to solutions of diffusion problems in domains with many small spherical holes were analysed in Figari, Papanicolaou and Rubinstein [10]. Modelling of multi-particle interaction in problems of phase transition was considered in Hönig, Niethammer and Otto [14] where the evolution of a large number of small spherical particles embedded into an ambient medium takes place during the last stage of phase transformation; such a phenomenon where particles in a melt are subjected to growth is referred to as Ostwald ripening.

The asymptotic approximations introduced in this chapter are uniform with respect to the independent variable. The boundary layers near individual inclusions incorporate the dipole fields characterising the shape of the inclusions and their orientation. A model algebraic problem is solved to evaluate the coefficients in the meso-scale asymptotic approximations.

V. Maz'ya et al., *Green's Kernels and Meso-Scale Approximations in Perforated Domains*, Lecture Notes in Mathematics 2077, DOI 10.1007/978-3-319-00357-3_10,

10.1 An Outline

As in Chapter 9, the notation Ω_N is used here for a domain containing small voids $F^{(j)}$, $j = 1, \dots, N$, while the unperturbed domain is denoted by Ω. The small parameters ε and d have the same meaning as in Section 9.1 and characterise the small size of inclusions and the distance between inclusions, respectively.

Our goal is to obtain an asymptotic approximation to a unique solution $u_N \in L^{1,2}(\Omega_N)$ of the problem

$$-\Delta u_N(\mathbf{x}) = f(\mathbf{x}) , \quad \mathbf{x} \in \Omega_N , \tag{10.1}$$

$$u_N(\mathbf{x}) = \phi(\mathbf{x}) , \quad \mathbf{x} \in \partial\Omega , \tag{10.2}$$

$$\frac{\partial u_N}{\partial n}(\mathbf{x}) = 0 , \quad \mathbf{x} \in \partial F^{(j)} , j = 1, \dots, N , \tag{10.3}$$

where $\phi \in L^{1/2,2}(\partial\Omega)$ and $f(\mathbf{x})$ is a function in $L_\infty(\Omega)$ with compact support at a positive distance from the cloud of small perforations.

We need solutions to certain model problems in order to construct the approximation to u_N; these include

1. v as the solution of the unperturbed problem in Ω (without voids).
2. $\boldsymbol{\mathcal{D}}^{(k)}$ as the vector function whose components are the dipole fields for the void $F^{(k)}$.
3. H as the regular part of Green's function G in Ω.

The approximation relies upon a certain algebraic system, incorporating the field v_f and integral characteristics associated with the small voids. We define

$$\boldsymbol{\Theta} = \left((\nabla v(\mathbf{O}^{(1)}))^T, \dots, (\nabla v(\mathbf{O}^{(N)}))^T\right)^T$$

and $\mathfrak{S} = [\mathfrak{S}_{ij}]_{i,j=1}^N$ which is a $3N \times 3N$ matrix with 3×3 block entries

$$\mathfrak{S}_{ij} = \begin{cases} (\nabla_{\mathbf{z}} \otimes \nabla_{\mathbf{w}}) \left(G(\mathbf{z}, \mathbf{w})\right) \Big|_{\substack{\mathbf{z}=\mathbf{O}^{(i)} \\ \mathbf{w}=\mathbf{O}^{(j)}}} & \text{if } i \neq j \\ 0 I_3 & \text{otherwise} \end{cases} ,$$

where G is Green's function in Ω, and I_3 is the 3×3 identity matrix. We also use the block-diagonal matrix

$$\mathbf{Q} = \text{diag}\{\boldsymbol{\mathcal{Q}}^{(1)}, \dots, \boldsymbol{\mathcal{Q}}^{(N)}\}, \tag{10.4}$$

where $\boldsymbol{\mathcal{Q}}^{(k)}$ is the so-called 3×3 polarization matrix for the small void $F^{(k)}$ (see Maz'ya and Movchan [26] and Appendix G of Pólya and Szegö [38]). The shapes of the voids $F^{(j)}$, $j = 1, \dots, N$, are constrained in such a way that the maximal and minimal eigenvalues $\lambda_{max}^{(j)}$, $\lambda_{min}^{(j)}$ of the matrices $-\boldsymbol{\mathcal{Q}}^{(j)}$ satisfy the inequalities

$$A_1\varepsilon^3 > \max_{1\le j\le N} \lambda_{max}^{(j)}, \quad \min_{1\le j\le N} \lambda_{min}^{(j)} > A_2\varepsilon^3, \tag{10.5}$$

where A_1 and A_2 are positive and independent of ε.

One of the results, for the case when $\Omega = \mathbb{R}^3$, $H \equiv 0$, and when (10.2) is replaced by the condition of decay of u_N at infinity, can be formulated as follows

Theorem 10.1. *Let*

$$\varepsilon < c\,d\ ,$$

where c is a sufficiently small absolute constant. Then the solution $u_N(\mathbf{x})$ admits the asymptotic representation

$$u_N(\mathbf{x}) = v(\mathbf{x}) + \sum_{k=1}^{N} \boldsymbol{C}^{(k)} \cdot \boldsymbol{\mathcal{D}}^{(k)}(\mathbf{x}) + \mathcal{R}_N(\mathbf{x})\ , \tag{10.6}$$

where $\boldsymbol{C}^{(k)} = (C_1^{(k)}, C_2^{(k)}, C_3^{(k)})^T$ and the column vector $\mathbf{C} = (C_1^{(1)}, C_2^{(1)}, C_3^{(1)}, \dots, C_1^{(N)}, C_2^{(N)}, C_3^{(N)})^T$ satisfies the invertible linear algebraic system

$$(\mathbf{I} + \mathfrak{S}\mathbf{Q})\mathbf{C} = -\boldsymbol{\Theta}\ . \tag{10.7}$$

The remainder $\mathcal{R}_N$ satisfies the energy estimate

$$\|\nabla \mathcal{R}_N\|^2_{L_2(\Omega_N)} \le \text{Const}\left\{\varepsilon^{11} d^{-11} + \varepsilon^5 d^{-3}\right\} \|\nabla v\|^2_{L_2(\Omega)} \tag{10.8}$$

We remark that since ε and d are non-dimensional parameters, there is no dimensional mismatch in the right-hand side of (10.8).

We now describe the plan of the chapter. In Sect. 10.2, we introduce the multiply-perforated geometry and consider the above model problems. The formal asymptotic algorithm for a cloud of small perforations in the infinite space and the analysis of the algebraic system (10.7) are given in Sects. 10.3 and 10.4. Section 10.5 presents the proof of Theorem 10.1. The problem for a cloud of small perforations in a general domain is considered in Sect. 10.6. Finally, in Sect. 10.7 we give an illustrative example accompanied by the numerical simulation.

10.2 Main Notations and Model Boundary Value Problems

Let Ω be a bounded domain in $\mathbb{R}^3$ with a smooth boundary $\partial\Omega$. We shall also consider the case when $\Omega = \mathbb{R}^3$.

The perforated domain Ω_N, is given by

$$\Omega_N = \Omega \setminus \overline{\cup_{j=1}^N F^{(j)}}\ ,$$

where $F^{(j)}$ are small voids introduced in the previous section. Also in the previous section we introduced the notations ε and d for two small parameters, characterizing the maximum of the diameters of $F^{(j)}, j = 1, \dots, N$, and the minimal distance between the small voids, respectively.

In sections where we are concerned with the energy estimates of the remainders produced by asymptotic approximations we frequently use the obvious estimate

$$N \leq \text{Const}\, d^{-3} . \tag{10.9}$$

We consider the approximation of the function u_N which is a variational solution of the mixed problem (10.1)–(10.3).

Before constructing the approximation to u_N, we introduce model auxiliary functions which the asymptotic scheme relies upon.

1. *Solution v in the unperturbed domain Ω.* Let $v \in L^{1,2}(\Omega)$ denote a unique variational solution of the problem

$$-\Delta v(\mathbf{x}) = f(\mathbf{x}) , \quad \mathbf{x} \in \Omega , \tag{10.10}$$

$$v(\mathbf{x}) = \phi(\mathbf{x}) , \quad \mathbf{x} \in \partial\Omega . \tag{10.11}$$

2. *Regular part of Green's function in Ω.* By H we mean the regular part of Green's function G in Ω defined by the formula

$$H(\mathbf{x}, \mathbf{y}) = (4\pi|\mathbf{x} - \mathbf{y}|)^{-1} - G(\mathbf{x}, \mathbf{y}) . \tag{10.12}$$

Then H is a variational solution of

$$\Delta_{\mathbf{x}} H(\mathbf{x}, \mathbf{y}) = 0 , \quad \mathbf{x}, \mathbf{y} \in \Omega ,$$

$$H(\mathbf{x}, \mathbf{y}) = (4\pi|\mathbf{x} - \mathbf{y}|)^{-1} , \quad \mathbf{x} \in \partial\Omega, \mathbf{y} \in \Omega .$$

3. *The dipole fields $\mathcal{D}_i^{(j)}$, $i = 1, 2, 3$, associated with the void $F^{(j)}$.* The vector functions $\boldsymbol{\mathcal{D}}^{(j)} = \{\mathcal{D}_i^{(j)}\}_{i=1}^3$, which are called the dipole fields, are variational solutions of the exterior Neumann problems

$$\left.\begin{aligned} \Delta\boldsymbol{\mathcal{D}}^{(j)}(\mathbf{x}) &= \mathbf{O} , \quad \mathbf{x} \in \mathbb{R}^3 \setminus \bar{F}^{(j)} , \\ \frac{\partial\boldsymbol{\mathcal{D}}^{(j)}}{\partial n}(\mathbf{x}) &= \mathbf{n}^{(j)} , \quad \mathbf{x} \in \partial F^{(j)} , \\ \boldsymbol{\mathcal{D}}^{(j)}(\mathbf{x}) &= O(\varepsilon^3|\mathbf{x} - \mathbf{O}^{(j)}|^{-2}) \quad \text{as} \quad |\mathbf{x}| \to \infty , \end{aligned}\right\} \tag{10.13}$$

where $\mathbf{n}^{(j)}$ is the unit outward normal with respect to $F^{(j)}$. In the text below we also use the negative definite polarization matrix $\mathcal{Q}^{(j)} = \{\mathcal{Q}_{ik}^{(j)}\}_{i,k=1}^3$, as well as the following asymptotic result (see Maz'ya and Movchan [26] and Appendix G of Pólya and Szegö [38]), for every void $F^{(j)}$:

Lemma 10.1. *For* $|\mathbf{x} - \mathbf{O}^{(j)}| > 2\varepsilon$, *the dipole fields admit the asymptotic representation*

$$\mathcal{D}_i^{(j)}(\mathbf{x}) = \frac{1}{4\pi} \sum_{m=1}^{3} \mathcal{Q}_{im}^{(j)} \frac{x_m - O_m^{(j)}}{|\mathbf{x} - \mathbf{O}^{(j)}|^3} + O\left(\varepsilon^4 |\mathbf{x} - \mathbf{O}^{(j)}|^{-3}\right), \quad i = 1, 2, 3\,. \tag{10.14}$$

The shapes of the voids $F^{(j)}$, $j = 1, \ldots, N$, are constrained in such a way that the maximal and minimal eigenvalues $\lambda_{max}^{(j)}$, $\lambda_{min}^{(j)}$ of the matrices $-\mathcal{Q}^{(j)}$ satisfy the inequalities (10.5).

10.3 The Formal Approximation of u_N for the Infinite Space Containing Many Voids

In this section we deduce formally the uniform asymptotic approximation of u_N:

$$u_N(\mathbf{x}) \sim v(\mathbf{x}) + \sum_{k=1}^{N} \boldsymbol{C}^{(k)} \cdot \boldsymbol{\mathcal{D}}^{(k)}(\mathbf{x})\,,$$

for the case $\Omega = \mathbb{R}^3$ and derive an algebraic system for the coefficients $\boldsymbol{C}^{(k)} = \{C_i^{(k)}\}_{i=1}^{3}$, $k = 1, \ldots, N$.

The function u_N satisfies

$$-\Delta u_N(\mathbf{x}) = f(\mathbf{x})\,, \quad \mathbf{x} \in \Omega_N\,, \tag{10.15}$$

$$\frac{\partial u_N}{\partial n}(\mathbf{x}) = 0\,, \quad \mathbf{x} \in \partial F^{(j)}, j = 1, \ldots, N\,, \tag{10.16}$$

$$u_N(\mathbf{x}) \to 0\,, \quad \text{as } |\mathbf{x}| \to \infty\,. \tag{10.17}$$

We begin by constructing the asymptotic representation for u_N in this way

$$u_N(\mathbf{x}) = v(\mathbf{x}) + \sum_{k=1}^{N} \boldsymbol{C}^{(k)} \cdot \boldsymbol{\mathcal{D}}^{(k)}(\mathbf{x}) + \mathcal{R}_N(\mathbf{x}) \tag{10.18}$$

where $\mathcal{R}_N$ is the remainder, and $v(\mathbf{x})$ satisfies

$$-\Delta v(\mathbf{x}) = f(\mathbf{x})\,, \quad \mathbf{x} \in \mathbb{R}^3\,,$$

$$v(\mathbf{x}) \to 0 \quad \text{as} \quad |\mathbf{x}| \to \infty\,,$$

and $\mathcal{D}^{(k)}$ are the dipole fields defined as solutions of problems (10.13). The function $\mathcal{R}_N$ is harmonic in Ω_N and

$$\mathcal{R}_N(\mathbf{x}) = O(|\mathbf{x}|^{-1}) \quad \text{as } |\mathbf{x}| \to \infty \, . \tag{10.19}$$

Placement of (10.18) into (10.16) together with (10.13) gives the boundary condition on $\partial F^{(j)}$:

$$\frac{\partial \mathcal{R}_N}{\partial n}(\mathbf{x}) = -\mathbf{n}^{(j)} \cdot \Big\{ \nabla v(\mathbf{O}^{(j)}) + \boldsymbol{C}^{(j)} + O(\varepsilon) + \sum_{\substack{k \neq j \\ 1 \le k \le N}} \nabla(\boldsymbol{C}^{(k)} \cdot \mathcal{D}^{(k)}(\mathbf{x})) \Big\}.$$

Now we use (10.14), for $\mathcal{D}^{(k)}$, $k \neq j$, so that this boundary condition becomes

$$\frac{\partial \mathcal{R}_N}{\partial n}(\mathbf{x}) \sim -\mathbf{n}^{(j)} \cdot \Big\{ \nabla v(\mathbf{O}^{(j)}) + \boldsymbol{C}^{(j)} + \sum_{\substack{k \neq j \\ 1 \le k \le N}} T(\mathbf{x}, \mathbf{O}^{(k)}) \boldsymbol{\mathcal{Q}}^{(k)} \boldsymbol{C}^{(k)} \Big\} ,$$

$$\mathbf{x} \in \partial F^{(j)}, \, j = 1, \dots, N \, ,$$

where

$$T(\mathbf{x}, \mathbf{y}) = (\nabla_{\mathbf{z}} \otimes \nabla_{\mathbf{w}}) \left(\frac{1}{4\pi |\mathbf{z} - \mathbf{w}|} \right) \Big|_{\substack{\mathbf{z}=\mathbf{x} \\ \mathbf{w}=\mathbf{y}}} \, . \tag{10.20}$$

Finally, Taylor's expansion of $T(\mathbf{x}, \mathbf{O}^{(k)})$ about $\mathbf{x} = \mathbf{O}^{(j)}$, $j \neq k$, leads to

$$\frac{\partial \mathcal{R}_N}{\partial n}(\mathbf{x}) \sim -\mathbf{n}^{(j)} \cdot \Big\{ \nabla v(\mathbf{O}^{(j)}) + \boldsymbol{C}^{(j)} + \sum_{\substack{k \neq j \\ 1 \le k \le N}} T(\mathbf{O}^{(j)}, \mathbf{O}^{(k)}) \boldsymbol{\mathcal{Q}}^{(k)} \boldsymbol{C}^{(k)} \Big\} ,$$

$$\mathbf{x} \in \partial F^{(j)}, \, j = 1, \dots, N \, .$$

To remove the leading order discrepancy in the above boundary condition, we require that the vector coefficients $\mathbf{C}^{(j)}$ satisfy the algebraic system

$$\nabla v(\mathbf{O}^{(j)}) + \boldsymbol{C}^{(j)} + \sum_{\substack{k \neq j \\ 1 \le k \le N}} T(\mathbf{O}^{(j)}, \mathbf{O}^{(k)}) \boldsymbol{\mathcal{Q}}^{(k)} \boldsymbol{C}^{(k)} = \mathbf{O} \, , \quad \text{for } j = 1, \dots, N \, , \tag{10.21}$$

where the polarization matrices $\boldsymbol{\mathcal{Q}}^{(j)}$ characterize the geometry of $F^{(j)}$, $j = 1, \dots, N$. Upon solving the above algebraic system, the formal asymptotic approximation of u_N is complete. The next section addresses the solvability of the system (10.21), together with estimates for the vector coefficients $\boldsymbol{C}^{(j)}$.

10.4 Algebraic System for the Coefficients in the Meso-scale Approximation

The algebraic system for the coefficients $\boldsymbol{C}^{(j)}$ can be written in the form

$$\mathbf{C} + \mathcal{S}\mathbf{Q}\mathbf{C} = -\boldsymbol{\Theta}, \tag{10.22}$$

where

$$\mathbf{C} = ((\boldsymbol{C}^{(1)})^T, \dots, (\boldsymbol{C}^{(N)})^T)^T, \quad \boldsymbol{\Theta} = ((\nabla v(\mathbf{O}^{(1)}))^T, \dots, (\nabla v(\mathbf{O}^{(N)}))^T)^T,$$

are vectors of the dimension $3N$, and

$$\mathcal{S} = [\mathcal{S}_{ij}]_{i,j=1}^N, \; \mathcal{S}_{ij} = \begin{cases} (\nabla_{\mathbf{z}} \otimes \nabla_{\mathbf{w}}) \left(\dfrac{1}{4\pi|\mathbf{z}-\mathbf{w}|} \right) \Big|_{\substack{\mathbf{z}=\mathbf{O}^{(i)} \\ \mathbf{w}=\mathbf{O}^{(j)}}} & \text{if } i \neq j \\ 0 I_3 & \text{otherwise,} \end{cases} \tag{10.23}$$

$$\mathbf{Q} = \text{diag}\{\mathcal{Q}^{(1)}, \dots, \mathcal{Q}^{(N)}\} \text{ is negative definite.} \tag{10.24}$$

These are $3N \times 3N$ matrices whose entries are 3×3 blocks. The notation in (10.23) is interpreted as

$$\mathcal{S}_{ij} = \left\{ \frac{1}{4\pi} \frac{\partial}{\partial z_q} \left(\frac{z_r - O_r^{(j)}}{|\mathbf{z} - \mathbf{O}^{(j)}|^3} \right) \Big|_{\mathbf{z}=\mathbf{O}^{(i)}} \right\}_{q,r=1}^3 \quad \text{when } i \neq j.$$

We use the piecewise constant vector function

$$\boldsymbol{\Xi}(\mathbf{x}) = \begin{cases} \mathcal{Q}^{(j)} \boldsymbol{C}^{(j)}, & \text{when } \mathbf{x} \in \overline{B}_{d/4}^{(j)}, \; j = 1, \dots, N, \\ 0, & \text{otherwise,} \end{cases} \tag{10.25}$$

where $B_r^{(j)} = \{\mathbf{x} : |\mathbf{x} - \mathbf{O}^{(j)}| < r\}$.

Theorem 10.2. *Assume that $\lambda_{max} < \text{Const}\; d^3$, where λ_{max} is the largest eigenvalue of the positive definite matrix $-\mathbf{Q}$ and the constant is independent of d. Then the algebraic system (10.22) is solvable and the vector coefficients $\boldsymbol{C}^{(j)}$ satisfy the estimate*

$$\sum_{j=1}^N |(\boldsymbol{C}^{(j)})^T \mathcal{Q}^{(j)} \boldsymbol{C}^{(j)}| \leq (1 - \text{Const}\, \frac{\lambda_{max}}{d^3})^{-2} \sum_{j=1}^N |(\nabla v(\mathbf{O}^{(j)}))^T \mathcal{Q}^{(j)} \nabla v(\mathbf{O}^{(j)})|. \tag{10.26}$$

We consider the scalar product of (10.22) and the vector **QC**:

$$\langle \mathbf{C}, \mathbf{QC}\rangle + \langle \mathcal{S}\mathbf{QC}, \mathbf{QC}\rangle = -\langle \boldsymbol{\Theta}, \mathbf{QC}\rangle. \tag{10.27}$$

Prior to the proof of Theorem we formulate and prove the following identity.

Lemma 10.2. *(a) The scalar product* $\langle \mathcal{S}\mathbf{QC}, \mathbf{QC}\rangle$ *admits the representation*

$$\begin{aligned}\langle \mathcal{S}\mathbf{QC}, \mathbf{QC}\rangle &= \frac{576}{\pi^3 d^6}\int_{\mathbb{R}^3}\int_{\mathbb{R}^3}\frac{1}{|\mathbf{X}-\mathbf{Y}|}(\nabla\cdot\boldsymbol{\Xi}(\mathbf{X}))(\nabla\cdot\boldsymbol{\Xi}(\mathbf{Y}))d\mathbf{Y}d\mathbf{X}\\ &-\frac{16}{\pi d^3}\sum_{j=1}^{N}|\mathcal{Q}^{(j)}C^{(j)}|^2.\end{aligned} \tag{10.28}$$

(b) The following estimate holds

$$|\langle \mathcal{S}\mathbf{QC}, \mathbf{QC}\rangle| \le \text{Const}\, d^{-3}\sum_{1\le j\le N}|\mathbf{Q}^{(j)}C^{(j)}|^2\,,$$

where the constant in the right-hand side does not depend on d.

Remark. Using the notation $\mathcal{N}(\nabla\cdot\boldsymbol{\Xi})$ for the Newton's potential acting on $\nabla\cdot\boldsymbol{\Xi}$ we can interpret the integral in (10.28) as

$$\Big(\mathcal{N}(\nabla\cdot\boldsymbol{\Xi}), \nabla\cdot\boldsymbol{\Xi}\Big)_{L_2(\mathbb{R}^3)},$$

since obviously $\nabla\cdot\boldsymbol{\Xi}\in W^{-1,2}(\mathbb{R}^3)$ and $\mathcal{N}(\nabla\cdot\boldsymbol{\Xi})\in W^{1,2}(\mathbb{R}^3)$. Here and in the sequel we use the notation (φ,ψ) for the extension of the integral $\int_{\mathbb{R}^3}\varphi(\mathbf{X})\psi(\mathbf{X})d\mathbf{X}$ onto the Cartesian product $W^{1,2}(\mathbb{R}^3)\times W^{-1,2}(\mathbb{R}^3)$.

Proof of Lemma 10.2.

(a) By (10.23), (10.24), the following representation holds

$$\begin{aligned}\langle \mathcal{S}\mathbf{QC}, \mathbf{QC}\rangle &= \frac{1}{4\pi}\sum_{j=1}^{N}\Big(\mathcal{Q}^{(j)}C^{(j)}\Big)^T\\ &\sum_{1\le k\le N, k\ne j}(\nabla_{\mathbf{z}}\otimes\nabla_{\mathbf{w}})\Big(\frac{1}{|\mathbf{z}-\mathbf{w}|}\Big)\Big|_{\substack{\mathbf{z}=\mathbf{O}^{(j)}\\ \mathbf{w}=\mathbf{O}^{(k)}}}\Big(\mathcal{Q}^{(k)}C^{(k)}\Big).\end{aligned} \tag{10.29}$$

Using the mean value theorem for harmonic functions we note that when $j\ne k$

$$(\nabla_{\mathbf{z}}\otimes\nabla_{\mathbf{w}})\Big(\frac{1}{|\mathbf{z}-\mathbf{w}|}\Big)\Big|_{\substack{\mathbf{z}=\mathbf{O}^{(j)}\\ \mathbf{w}=\mathbf{O}^{(k)}}} = \frac{3}{4\pi(d/4)^3}\int_{B^{(k)}_{d/4}}(\nabla_{\mathbf{z}}\otimes\nabla_{\mathbf{w}})\Big(\frac{1}{|\mathbf{z}-\mathbf{w}|}\Big)\Big|_{\mathbf{z}=\mathbf{O}^{(j)}}d\mathbf{w}.$$

Substituting this identity into (10.29) and using definition (10.25) we see that the inner sum on the right-hand side of (10.29) can be presented in the form

$$\frac{48}{\pi d^3} \lim_{\tau \to 0+} \int_{\mathbb{R}^3 \setminus B^{(j)}_{(d/4)-\tau}} \Big\{ \frac{\partial}{\partial Y_q} \Big(\frac{Y_r - O^{(j)}_r}{|\mathbf{Y} - \mathbf{O}^{(j)}|^3} \Big) \Big\}^3_{q,r=1} \boldsymbol{\Xi}(\mathbf{Y}) d\mathbf{Y},$$

and further integration by parts gives

$$\begin{aligned} \langle \mathcal{S}\mathbf{QC}, \mathbf{QC} \rangle &= -\frac{12}{\pi^2 d^3} \sum_{j=1}^{N} \Big(\mathcal{Q}^{(j)} \mathbf{C}^{(j)} \Big)^T \\ &\cdot \lim_{\tau \to 0+} \Bigg\{ \int_{\mathbb{R}^3 \setminus B^{(j)}_{(d/4)-\tau}} \Big\{ \frac{Y_r - O^{(j)}_r}{|\mathbf{Y} - \mathbf{O}^{(j)}|^3} \nabla \cdot \boldsymbol{\Xi}(\mathbf{Y}) \Big\}^3_{r=1} d\mathbf{Y} \\ &+ \int_{|\mathbf{Y} - \mathbf{O}^{(j)}| = (d/4) - \tau} \Bigg\{ \frac{(Y_r - O^{(j)}_r)(Y_q - O^{(j)}_q)}{|\mathbf{Y} - \mathbf{O}^{(j)}|^4} \Bigg\}^3_{r,q=1} dS_{\mathbf{Y}}\, \mathcal{Q}^{(j)} \mathbf{C}^{(j)} \Bigg\}, \end{aligned} \tag{10.30}$$

where the integral over $\mathbb{R}^3 \setminus B^{(j)}_{(d/4)-\tau}$ in (10.30) is understood in the sense of distributions. The surface integral in (10.30) can be evaluated explicitly, i.e.

$$\int_{|\mathbf{Y} - \mathbf{O}^{(j)}| = (d/4) - \tau} \Bigg\{ \frac{(Y_r - O^{(j)}_r)(Y_q - O^{(j)}_q)}{|\mathbf{Y} - \mathbf{O}^{(j)}|^4} \Bigg\}^3_{r,q=1} dS_{\mathbf{Y}}\, \mathcal{Q}^{(j)} \mathbf{C}^{(j)} = \frac{4\pi}{3} \mathcal{Q}^{(j)} \mathbf{C}^{(j)}. \tag{10.31}$$

Once again, applying the mean value theorem for harmonic functions in the outer sum of (10.30) and using (10.31) together with the definition (10.25) we arrive at

$$\begin{aligned} \langle \mathcal{S}\mathbf{QC}, \mathbf{QC} \rangle &= -\frac{16}{\pi d^3} \sum_{j=1}^{N} |\mathcal{Q}^{(j)} \mathbf{C}^{(j)}|^2 \\ -\frac{576}{\pi^3 d^6} \lim_{\tau \to 0+} \sum_{j=1}^{N} &\int_{B^{(j)}_{(d/4)+\tau}} \int_{\mathbb{R}^3 \setminus B^{(j)}_{(d/4)-\tau}} \sum_{r=1}^{3} \Xi_r(\mathbf{X}) \frac{\partial}{\partial X_r} \Big(\frac{1}{|\mathbf{Y} - \mathbf{X}|} \Big) \nabla \cdot \boldsymbol{\Xi}(\mathbf{Y}) d\mathbf{Y} d\mathbf{X}, \end{aligned} \tag{10.32}$$

where Ξ_r are the components of the vector function $\boldsymbol{\Xi}$ defined in (10.25).

The last integral is understood in the sense of distributions. Referring to the definition (10.25), integrating by parts, and taking the limit as $\tau \to 0+$ we deduce that the integral term in (10.32) can be written as

$$\frac{576}{\pi^3 d^6} \int_{\mathbb{R}^3} \int_{\mathbb{R}^3} \frac{1}{|\mathbf{Y} - \mathbf{X}|} \Big(\nabla \cdot \boldsymbol{\Xi}(\mathbf{X}) \Big) \Big(\nabla \cdot \boldsymbol{\Xi}(\mathbf{Y}) \Big) d\mathbf{Y} d\mathbf{X} \tag{10.33}$$

Using (10.32) and (10.33) we arrive at (10.28).

(b) Let us introduce a piece-wise constant function

$$\mathcal{C}(\mathbf{x}) = \begin{cases} \mathcal{C}^{(j)}, & \text{when } \mathbf{x} \in \overline{B^{(j)}_{d/4}}, \quad j = 1, \dots, N, \\ 0, & \text{otherwise}. \end{cases}$$

According to the system (10.21), $\nabla \times \mathcal{C}(\mathbf{x}) = \mathbf{O}$, and one can use the representation

$$\mathcal{C}(\mathbf{x}) = \nabla W(\mathbf{x}) \tag{10.34}$$

where W is a scalar function with compact support, and (10.34) is understood in the sense of distributions. We give a proof for the case when all voids are spherical, of diameter ε, and hence $\mathcal{Q}^{(j)} = -\frac{\pi}{4}\varepsilon^3 I_3$, where I_3 is the identity matrix. Then according to (10.32) we have

$$|\langle \mathcal{S}\mathbf{Q}\mathbf{C}, \mathbf{Q}\mathbf{C}\rangle| \leq \frac{16}{\pi d^3} \sum_{1\leq j\leq N} |\mathbf{Q}^{(j)}\mathcal{C}^{(j)}|^2$$

$$+\frac{36\varepsilon^6}{\pi d^3}\Big| \int_{\mathbb{R}^3}\int_{\mathbb{R}^3} \Big(\nabla_{\mathbf{X}} W(\mathbf{X}) \cdot \nabla_{\mathbf{X}}\Big(\frac{1}{|\mathbf{Y}-\mathbf{X}|}\Big)\Big)\Delta_{\mathbf{Y}} W(\mathbf{Y})\, d\mathbf{Y} d\mathbf{X}\Big|$$

$$\leq \frac{16}{\pi d^3} \sum_{1\leq j\leq N} |\mathbf{Q}^{(j)}\mathcal{C}^{(j)}|^2 + \frac{144\varepsilon^6}{d^3} \sum_{1\leq j\leq N} \int_{B^{(j)}_{d/4}} |\nabla W(\mathbf{Y})|^2 d\mathbf{Y}$$

$$\leq \frac{\text{Const}}{d^3} \sum_{1\leq j\leq N} |\mathcal{Q}^{(j)}\mathcal{C}^{(j)}|^2 .$$

□

Proof of Theorem 10.2. Consider the equation (10.27). The absolute value of its right-hand side does not exceed

$$\langle \mathbf{C}, -\mathbf{Q}\mathbf{C}\rangle^{1/2} \langle \boldsymbol{\Theta}, -\mathbf{Q}\boldsymbol{\Theta}\rangle^{1/2}.$$

Using Lemma 10.1 and part b) of Lemma 10.2 we derive

$$\langle \mathbf{C}, -\mathbf{Q}\mathbf{C}\rangle - \text{Const}\, d^{-3}\langle -\mathbf{Q}\mathbf{C}, -\mathbf{Q}\mathbf{C}\rangle \leq \langle \mathbf{C}, -\mathbf{Q}\mathbf{C}\rangle^{1/2} \langle \boldsymbol{\Theta}, -\mathbf{Q}\boldsymbol{\Theta}\rangle^{1/2},$$

leading to

$$\Big(1 - \frac{\text{Const}}{d^3}\frac{\langle -\mathbf{Q}\mathbf{C}, -\mathbf{Q}\mathbf{C}\rangle}{\langle \mathbf{C}, -\mathbf{Q}\mathbf{C}\rangle}\Big)\langle \mathbf{C}, -\mathbf{Q}\mathbf{C}\rangle^{1/2} \leq \langle \boldsymbol{\Theta}, -\mathbf{Q}\boldsymbol{\Theta}\rangle^{1/2},$$

which implies

$$\Big(1 - \text{Const}\frac{\lambda_{max}}{d^3}\Big)^2 \langle \mathbf{C}, -\mathbf{Q}\mathbf{C}\rangle \leq \langle \boldsymbol{\Theta}, -\mathbf{Q}\boldsymbol{\Theta}\rangle. \tag{10.35}$$

The proof is complete. □

Assuming that the eigenvalues of the matrices $-\boldsymbol{Q}^{(j)}$ are strictly positive and satisfy the inequality (10.5), we also find that Theorem 10.2 yields

Corollary 10.1. *Assume that the inequalities (10.5) hold for λ_{max} and λ_{min}. Then the vector coefficients $\boldsymbol{C}^{(j)}$ in the system (10.22) satisfy the estimate*

$$\sum_{1\le j\le N} |\boldsymbol{C}^{(j)}|^2 \le \text{Const}\, d^{-3} \|\nabla v\|^2_{L_2(\omega)}, \tag{10.36}$$

where the constant depends only on the coefficients A_1 and A_2 in (10.5).

Proof. According to the inequality (10.26) of Theorem 10.2 we deduce

$$\lambda_{min} \sum_{1\le j\le N} |\boldsymbol{C}^{(j)}|^2 \le (1 - \frac{\text{Const}}{d^3}\lambda_{max})^{-2} \lambda_{max} \sum_{1\le j\le N} |\nabla v(\mathbf{O}^{(j)})|^2. \tag{10.37}$$

We note that v is harmonic in a neighbourhood of $\overline{\omega}$. Applying the mean value theorem for harmonic functions together with the Cauchy inequality we write

$$|\nabla v(\mathbf{O}^{(j)})|^2 \le \frac{48}{\pi d^3} \|\nabla v\|^2_{L_2(B^{(j)}_{d/4})}.$$

Hence, it follows from (10.37) that

$$\begin{aligned}\sum_{1\le j\le N} |\boldsymbol{C}^{(j)}|^2 &\le d^{-3}(1 - \frac{\text{Const}}{d^3}\lambda_{max})^{-2} \frac{48}{\pi} \frac{\lambda_{max}}{\lambda_{min}} \sum_{1\le j\le N} \|\nabla v\|^2_{L_2(B^{(j)}_{d/4})} \\ &\le d^{-3}\left((1 - \frac{\text{Const}}{d^3}\lambda_{max})^{-2} \frac{48}{\pi} \frac{\lambda_{max}}{\lambda_{min}}\right) \|\nabla v\|^2_{L_2(\omega)}, \end{aligned} \tag{10.38}$$

which is the required estimate (10.36). □

10.5 Energy Estimate

In this section we prove the result concerning the asymptotic approximation of u_N for the perforated domain $\Omega_N = \mathbb{R}^3 \backslash \overline{\cup_{j=1}^N F^{(j)}}$. The changes in the argument, necessary for the treatment of a general domain, will be described in Sect. 10.6.

Proof of Theorem 10.1.

(a) *Neumann problem for the remainder.* The remainder term $\mathcal{R}_N$ in (10.6) is a harmonic function in Ω_N, which vanishes at infinity and satisfies the boundary conditions

$$\frac{\partial \mathcal{R}_N}{\partial n}(\mathbf{x}) = -\left(\nabla v(\mathbf{x}) + \boldsymbol{C}^{(j)}\right) \cdot \mathbf{n}^{(j)} - \sum_{\substack{k \neq j \\ 1 \leq k \leq N}} \boldsymbol{C}^{(k)} \cdot \frac{\partial}{\partial n} \boldsymbol{\mathcal{D}}^{(k)}(\mathbf{x}),$$

$$\text{when } \mathbf{x} \in \partial F^{(j)}, j = 1, \ldots, N. \tag{10.39}$$

Since supp$f$ is separated from $F^{(j)}, j = 1, \ldots, N$, and since $\boldsymbol{\mathcal{D}}^{(j)}$, $j = 1, \ldots, N$, satisfy (10.13) we have

$$\int_{\partial F^{(j)}} \frac{\partial \mathcal{R}_N}{\partial n}(\mathbf{x}) dS_{\mathbf{x}} = 0, \; j = 1, \ldots, N. \tag{10.40}$$

(b) *Auxiliary functions.* Throughout the proof we use the notation $B_\rho^{(k)} = \{\mathbf{x} : |\mathbf{x} - \mathbf{O}^{(k)}| < \rho\}$. We introduce auxiliary functions which will help us to obtain (10.8). Let

$$\Psi_k(\mathbf{x}) = v(\mathbf{x}) - v(\mathbf{O}^{(k)}) - (\mathbf{x} - \mathbf{O}^{(k)}) \cdot \nabla v(\mathbf{O}^{(k)}) + \sum_{\substack{1 \leq j \leq N \\ j \neq k}} \boldsymbol{C}^{(j)} \cdot \boldsymbol{\mathcal{D}}^{(j)}(\mathbf{x})$$

$$- \sum_{\substack{1 \leq j \leq N \\ j \neq k}} (\mathbf{x} - \mathbf{O}^{(j)}) \cdot T(\mathbf{O}^{(k)}, \mathbf{O}^{(j)}) \boldsymbol{\mathcal{Q}}^{(j)} \boldsymbol{C}^{(j)}, \tag{10.41}$$

for all $\mathbf{x} \in \Omega_N$ and $k = 1, \ldots, N$. Every function Ψ_k satisfies

$$-\Delta \Psi_k(\mathbf{x}) = f(\mathbf{x}), \quad \mathbf{x} \in \Omega_N, \tag{10.42}$$

and since $\omega \cap \text{supp } f = \varnothing$, we see that Ψ_k, $k = 1, \ldots, N$, are harmonic in ω. Since the coefficients $\boldsymbol{C}^{(j)}$ satisfy system (10.22), we obtain

$$\frac{\partial \Psi_k}{\partial n}(\mathbf{x}) + \frac{\partial \mathcal{R}_N}{\partial n}(\mathbf{x}) = 0, \quad \mathbf{x} \in \partial F^{(k)}, \tag{10.43}$$

and according to (10.40) the functions Ψ_k have zero flux through the boundaries of small voids $F^{(k)}$, i.e.

$$\int_{\partial F^{(k)}} \frac{\partial \Psi_k}{\partial n}(\mathbf{x})\, d\mathbf{x} = 0, k = 1, \ldots, N. \tag{10.44}$$

Next, we introduce smooth cutoff functions

$$\chi_\varepsilon^{(k)} : \mathbf{x} \to \chi((\mathbf{x} - \mathbf{O}^{(k)})/\varepsilon), \; k = 1, \ldots, N,$$

equal to 1 on $B_{2\varepsilon}^{(k)}$ and vanishing outside $B_{3\varepsilon}^{(k)}$. Then by (10.43) we have

$$\frac{\partial}{\partial n}\Big(\mathcal{R}_N(\mathbf{x}) + \sum_{1\le k\le N} \chi_\varepsilon^{(k)}(\mathbf{x})\Psi_k(\mathbf{x})\Big) = 0 \text{ on } \partial F^{(j)},\ j = 1, \ldots, N. \tag{10.45}$$

(c) *Estimate of the energy integral of* $\mathcal{R}_N$ *in terms of* Ψ_k. Integrating by parts in Ω_N and using the definition of $\chi_\varepsilon^{(k)}$, we write the identity

$$\int_{\Omega_N} \nabla\mathcal{R}_N \cdot \nabla\Big(\mathcal{R}_N + \sum_{1\le k\le N} \chi_\varepsilon^{(k)}\Psi_k\Big)\, d\mathbf{x}$$
$$= -\int_{\Omega_N} \mathcal{R}_N \Delta\Big(\mathcal{R}_N + \sum_{1\le k\le N} \chi_\varepsilon^{(k)}\Psi_k\Big)\, d\mathbf{x}, \tag{10.46}$$

which is equivalent to

$$\int_{\Omega_N} |\nabla\mathcal{R}_N|^2 d\mathbf{x} + \sum_{1\le k\le N} \int_{B_{3\varepsilon}^{(k)}\setminus\overline{F}^{(k)}} \nabla\mathcal{R}_N \cdot \nabla(\chi_\varepsilon^{(k)}\Psi_k)\, d\mathbf{x}$$
$$= -\sum_{1\le k\le N} \int_{B_{3\varepsilon}^{(k)}\setminus\overline{F}^{(k)}} \mathcal{R}_N \Delta(\chi_\varepsilon^{(k)}\Psi_k)\, d\mathbf{x}, \tag{10.47}$$

since $\mathcal{R}_N$ is harmonic in Ω_N.

We preserve the notation $\mathcal{R}_N$ for an extension of $\mathcal{R}_N$ onto the union of voids $F^{(k)}$ with preservation of the class $W^{1,2}$. Such an extension can be constructed by using only values of $\mathcal{R}_N$ on the sets $B_{2\varepsilon}^{(k)} \setminus \overline{F}^{(k)}$ in such a way that

$$\|\nabla\mathcal{R}_N\|_{L_2(B_{2\varepsilon}^{(k)})} \le \text{Const}\|\nabla\mathcal{R}_N\|_{L_2(B_{2\varepsilon}^{(k)}\setminus\overline{F}^{(k)})}. \tag{10.48}$$

The above fact follows by dilation $\mathbf{x} \to \mathbf{x}/\varepsilon$ from the well-known extension theorem for domains with Lipschitz boundaries (see Sect. 3 of Chap. 6 in Stein [42]). We shall use the notation $\overline{\mathcal{R}}^{(k)}$ for the mean value of $\mathcal{R}_N$ on $B_{3\varepsilon}^{(k)}$.

The integral on the right-hand side of (10.47) can be written as

$$-\sum_{1\le k\le N} \int_{B_{3\varepsilon}^{(k)}\setminus\overline{F}^{(k)}} \mathcal{R}_N \Delta(\chi_\varepsilon^{(k)}\Psi_k)\, d\mathbf{x}$$
$$= -\sum_{1\le k\le N} \int_{B_{3\varepsilon}^{(k)}\setminus\overline{F}^{(k)}} (\mathcal{R}_N - \overline{\mathcal{R}}^{(k)})\Delta(\chi_\varepsilon^{(k)}\Psi_k)\, d\mathbf{x}, \tag{10.49}$$

In the derivation of (10.49) we have used that

$$\int_{B_{3\varepsilon}^{(k)}\setminus\overline{F}^{(k)}} \Delta\Big(\chi_\varepsilon^{(k)}\Psi_k\Big)\, d\mathbf{x} = \int_{\partial F^{(k)}} \frac{\partial\Psi_k}{\partial n} dS_{\mathbf{x}} = 0 \tag{10.50}$$

according to (10.44) and the definition of $\chi_\varepsilon^{(k)}$.

Owing to (10.46) and (10.49), we can write

$$\|\nabla \mathcal{R}_N\|^2_{L_2(\Omega_N)} \le \Sigma_1 + \Sigma_2, \tag{10.51}$$

where

$$\Sigma_1 = \sum_{1\le k\le N} \Big| \int_{B^{(k)}_{3\varepsilon}\setminus F^{(k)}} \nabla \mathcal{R}_N \cdot \nabla\big(\chi^{(k)}_\varepsilon \Psi_k\big) d\mathbf{x} \Big|, \tag{10.52}$$

and

$$\Sigma_2 = \sum_{1\le k\le N} \Big| \int_{B^{(k)}_{3\varepsilon}\setminus \overline{F}^{(k)}} (\mathcal{R}_N - \overline{\mathcal{R}}^{(k)}) \Delta\big(\chi^{(k)}_\varepsilon (\Psi_k - \overline{\Psi}_k)\big)\, d\mathbf{x} \Big|, \tag{10.53}$$

where $\overline{\Psi}_k$ is the mean value of Ψ_k over the ball $B^{(k)}_{3\varepsilon}$. Here, we have taken into account that by harmonicity of $\mathcal{R}_N$, (10.40) and definition of $\chi^{(k)}_\varepsilon$

$$\int_{B^{(k)}_{3\varepsilon}\setminus \overline{F}^{(k)}} \Delta\Big(\mathcal{R}_N - \overline{\mathcal{R}}^{(k)}\Big)\chi^{(k)}_\varepsilon d\mathbf{x} = \int_{B^{(k)}_{3\varepsilon}} \Delta\Big(\mathcal{R}_N - \overline{\mathcal{R}}^{(k)}\Big)\chi^{(k)}_\varepsilon d\mathbf{x} = 0.$$

By the Cauchy inequality, the first sum in (10.51) allows for the estimate

$$\Sigma_1 \le \Big(\sum_{1\le k\le N} \|\nabla \mathcal{R}_N\|^2_{L_2(B^{(k)}_{3\varepsilon}\setminus \overline{F}^{(k)})}\Big)^{1/2}$$
$$\times \Big(\sum_{1\le k\le N} \Big\|\nabla\big(\chi^{(k)}_\varepsilon \Psi_k\big)\Big\|^2_{L_2(B^{(k)}_{3\varepsilon}\setminus \overline{F}^{(k)})}\Big)^{1/2}. \tag{10.54}$$

Furthermore, using the inequality

$$\sum_{1\le k\le N} \|\nabla \mathcal{R}_N\|^2_{L_2(B^{(k)}_{3\varepsilon}\setminus \overline{F}^{(k)})} \le \|\nabla \mathcal{R}_N\|^2_{L_2(\Omega_N)}, \tag{10.55}$$

together with (10.54), we deduce

$$\Sigma_1 \le \|\nabla \mathcal{R}_N\|_{L_2(\Omega_N)} \Big(\sum_{1\le k\le N} \|\nabla\big(\chi^{(k)}_\varepsilon \Psi_k\big)\|^2_{L_2(B^{(k)}_{3\varepsilon}\setminus \overline{F}^{(k)})}\Big)^{1/2}. \tag{10.56}$$

Similarly to (10.54), the second sum in (10.51) can be estimated as

$$\Sigma_2 \le \sum_{1\le k\le N} \Big(\int_{B^{(k)}_{3\varepsilon}} (\mathcal{R}_N - \overline{\mathcal{R}}^{(k)})^2 d\mathbf{x}\Big)^{1/2}$$
$$\times \Big(\int_{B^{(k)}_{3\varepsilon}\setminus \overline{F}^{(k)}} \big(\Delta(\chi^{(k)}_\varepsilon (\Psi_k - \overline{\Psi}_k))\big)^2\, d\mathbf{x}\Big)^{1/2}. \tag{10.57}$$

By the Poincaré inequality for the ball $B_{3\varepsilon}^{(k)}$

$$\|\mathcal{R}_N - \overline{\mathcal{R}}^{(k)}\|^2_{L_2(B_{3\varepsilon}^{(k)})} \le \text{Const } \varepsilon^2 \|\nabla \mathcal{R}_N\|^2_{L_2(B_{3\varepsilon}^{(k)})} \tag{10.58}$$

we obtain

$$\Sigma_2 \le \text{Const } \varepsilon \Big(\sum_{1\le k\le N} \|\nabla \mathcal{R}_N\|^2_{L_2(B_{3\varepsilon}^{(k)})} \Big)^{1/2}$$
$$\times \Big(\sum_{1\le k\le N} \int_{B_{3\varepsilon}^{(k)} \setminus \overline{F}^{(k)}} \big(\Delta(\chi_\varepsilon^{(k)}(\Psi_k - \overline{\Psi}_k))\big)^2 d\mathbf{x}\Big)^{1/2},$$

which does not exceed

$$\text{Const } \varepsilon \, \|\nabla \mathcal{R}_N\|_{L_2(\Omega_N)} \Big(\sum_{1\le k\le N} \int_{B_{3\varepsilon}^{(k)} \setminus \overline{F}^{(k)}} \big(\Delta\big(\chi_\varepsilon^{(k)}(\Psi_k - \overline{\Psi}_k)\big)\big)^2 d\mathbf{x}\Big)^{1/2}, \tag{10.59}$$

because of (10.48). Combining (10.51)–(10.59) and dividing both sides of (10.51) by $\|\nabla \mathcal{R}_N\|_{L_2(\Omega_N)}$ we arrive at

$$\|\nabla \mathcal{R}_N\|_{L_2(\Omega_N)} \le \Big(\sum_{1\le k\le N} \Big\|\nabla(\chi_\varepsilon^{(k)}(\Psi_k - \overline{\Psi}_k))\Big\|^2_{L_2(B_{3\varepsilon}^{(k)})} \Big)^{1/2}$$
$$+ \text{Const } \varepsilon \Big(\sum_{1\le k\le N} \int_{B_{3\varepsilon}^{(k)}} \{(\Psi_k - \overline{\Psi}_k)\Delta\chi_\varepsilon^{(k)} + 2\nabla\chi_\varepsilon^{(k)} \cdot \nabla\Psi_k\}^2 d\mathbf{x}\Big)^{1/2}, \tag{10.60}$$

which leads to

$$\|\nabla \mathcal{R}_N\|^2_{L_2(\Omega_N)} \le \text{Const} \sum_{1\le k\le N} \Big(\|\nabla \Psi_k\|^2_{L_2(B_{3\varepsilon}^{(k)})} + \varepsilon^{-2} \|\Psi_k - \overline{\Psi}_k\|^2_{L_2(B_{3\varepsilon}^{(k)})} \Big). \tag{10.61}$$

Applying the Poincaré inequality (see (10.58)) for Ψ_k in the ball $B_{3\varepsilon}^{(k)}$ and using (10.61), we deduce

$$\|\nabla \mathcal{R}_N\|^2_{L_2(\Omega_N)} \le \text{Const} \sum_{1\le k\le N} \|\nabla \Psi_k\|^2_{L_2(B_{3\varepsilon}^{(k)})}. \tag{10.62}$$

(d) *Final energy estimate.* Here we prove the inequality (10.8). Using definition (10.41) of Ψ_k, $k = 1, \dots, N$, we can replace the preceding inequality by

$$\|\nabla \mathcal{R}_N\|^2_{L_2(\Omega_N)} \le \text{Const} \left\{\mathcal{K} + \mathcal{L}\right\}, \tag{10.63}$$

where

$$\mathcal{K} = \sum_{1\le k\le N} \|\nabla v(\cdot) - \nabla v(\mathbf{O}^{(k)})\|^2_{L_2(B_{3\varepsilon}^{(k)})},$$

$$\mathcal{L} = \sum_{1\le k\le N} \Big\| \sum_{\substack{j\neq k\\ 1\le j\le N}} \Big[\nabla\Big(\boldsymbol{C}^{(j)}\cdot\boldsymbol{\mathcal{D}}^{(j)}(\cdot)\Big) - T(\mathbf{O}^{(k)},\mathbf{O}^{(j)})\boldsymbol{\mathcal{Q}}^{(j)}\boldsymbol{C}^{(j)}\Big]\Big\|^2_{L_2(B_{3\varepsilon}^{(k)})}. \tag{10.64}$$

The estimate for $\mathcal{K}$ is straightforward and it follows by Taylor's expansion of v in the vicinity of $\mathbf{O}^{(k)}$,

$$\mathcal{K} \le \text{Const}\ \varepsilon^5 d^{-3} \max_{\mathbf{x}\in\overline{\omega},1\le i,j\le 3} \Big|\frac{\partial^2 v}{\partial x_i \partial x_j}\Big|^2. \tag{10.65}$$

Since v is harmonic in a neighbourhood of $\overline{\omega}$, we obtain by the local regularity property of harmonic functions that

$$\mathcal{K} \le \text{Const}\ \varepsilon^5 d^{-3} \|\nabla v\|^2_{L_2(\mathbb{R}^3)}. \tag{10.66}$$

To estimate $\mathcal{L}$, we use Lemma 10.1 on the asymptotics of the dipole fields together with the definition (10.20) of the matrix function T, which lead to

$$|\nabla(\boldsymbol{C}^{(j)}\cdot\boldsymbol{\mathcal{D}}^{(j)}(\mathbf{x})) - T(\mathbf{O}^{(k)},\mathbf{O}^{(j)})\boldsymbol{\mathcal{Q}}^{(j)}\boldsymbol{C}^{(j)}| \le \text{Const}\ \varepsilon^4 |\boldsymbol{C}^{(j)}||\mathbf{x}-\mathbf{O}^{(j)}|^{-4}, \tag{10.67}$$

for $\mathbf{x}\in B_{3\varepsilon}^{(k)}$. Now, it follows from (10.64) and (10.67) that

$$\mathcal{L} \le \text{Const}\ \varepsilon^8 \sum_{k=1}^N \int_{B_{3\varepsilon}^{(k)}} \Big(\sum_{1\le j\le N, j\neq k} \frac{|\boldsymbol{C}^{(j)}|}{|\mathbf{x}-\mathbf{O}^{(j)}|^4}\Big)^2 d\mathbf{x}, \tag{10.68}$$

and by the Cauchy inequality the right-hand side does not exceed

$$\begin{aligned}
&\text{Const}\ \varepsilon^8 \sum_{p=1}^N |\boldsymbol{C}^{(p)}|^2 \sum_{k=1}^N \sum_{1\le j\le N, j\neq k} \int_{B_{3\varepsilon}^{(k)}} \frac{d\mathbf{x}}{|\mathbf{x}-\mathbf{O}^{(j)}|^8}\\
&\le \text{Const}\ \varepsilon^{11} \sum_{p=1}^N |\boldsymbol{C}^{(p)}|^2 \sum_{k=1}^N \sum_{1\le j\le N, j\neq k} \frac{1}{|\mathbf{O}^{(k)}-\mathbf{O}^{(j)}|^8}\\
&\le \text{Const}\ \frac{\varepsilon^{11}}{d^6} \sum_{p=1}^N |\boldsymbol{C}^{(p)}|^2 \int\int_{\{\omega\times\omega: |\mathbf{X}-\mathbf{Y}|>d\}} \frac{d\mathbf{X}d\mathbf{Y}}{|\mathbf{X}-\mathbf{Y}|^8}\\
&\le \text{Const}\ \frac{\varepsilon^{11}}{d^8} \sum_{p=1}^N |\boldsymbol{C}^{(p)}|^2.
\end{aligned} \tag{10.69}$$

Since the eigenvalues of the matrix $-\mathbf{Q}$ satisfy the constraint (10.5), we can apply Corollary 10.1 and use the estimate (10.36) for the right-hand side of (10.69) to obtain

$$\mathcal{L} \leq \text{Const}\, \varepsilon^{11} d^{-11} \|\nabla v\|^2_{L_2(\omega)}. \tag{10.70}$$

Combining (10.63), (10.66) and (10.70), we arrive at (10.8) and complete the proof. □

10.6 Approximation of u_N for a Perforated Domain

Now we seek an approximation of the solution u_N to the problem (10.1)–(10.3) assuming that Ω is an arbitrary domain in $\mathbb{R}^3$. We first describe the formal asymptotic algorithm and derive a system of algebraic equations, similar to (10.22), which is used for evaluation of the coefficients in the asymptotic representation of u_N.

10.6.1 *Formal Asymptotic Algorithm for the Perforated Domain Ω_N*

The solution $u_N \in L^{1,2}(\Omega_N)$ of (10.1)–(10.3) is sought in the form

$$u_N(\mathbf{x}) = v(\mathbf{x}) + \sum_{k=1}^{N} \boldsymbol{C}^{(k)} \cdot \left\{ \boldsymbol{\mathcal{D}}^{(k)}(\mathbf{x}) - \boldsymbol{\mathcal{Q}}^{(k)} \nabla_{\mathbf{y}} H(\mathbf{x}, \mathbf{y})\big|_{\mathbf{y}=\mathbf{O}^{(k)}} \right\} + R_N(\mathbf{x})\,, \tag{10.71}$$

where in this instance v solves problem (10.10), (10.11) in Sect. 10.2, and R_N is a harmonic function in Ω_N. Here $\boldsymbol{C}^{(k)}$, $k = 1, \ldots, N$ are the vector coefficients to be determined.

Owing to the definitions of $\boldsymbol{\mathcal{D}}^{(k)}$, $k = 1, \ldots, N$, and H as solutions of Problems 2 and 3 in Sect. 10.2, and taking into account Lemma 10.1 on the asymptotics of $\boldsymbol{\mathcal{D}}^{(k)}$ we deduce that $|R_N(\mathbf{x})|$ is small for $\mathbf{x} \in \partial\Omega$.

On the boundaries $\partial F^{(j)}$, the substitution of (10.71) into (10.3) yields

$$\begin{aligned} \frac{\partial R_N}{\partial n}(\mathbf{x}) = &-\mathbf{n}^{(j)} \cdot \Big\{ \nabla v(\mathbf{O}^{(j)}) + \boldsymbol{C}^{(j)} + O(\varepsilon) + O(\varepsilon^3 |\boldsymbol{C}^{(j)}|) \\ &+ \sum_{\substack{k \neq j \\ 1 \leq k \leq N}} \nabla \Big\{ \boldsymbol{C}^{(k)} \cdot \Big(\boldsymbol{\mathcal{D}}^{(k)}(\mathbf{x}) - \boldsymbol{\mathcal{Q}}^{(k)} \nabla_{\mathbf{y}} H(\mathbf{x}, \mathbf{y})\big|_{\mathbf{y}=\mathbf{O}^{(k)}} \Big) \Big\} \Big\}\,, \\ &\mathbf{x} \in \partial F^{(j)},\ j = 1, \ldots, N\,. \end{aligned} \tag{10.72}$$

Then, using the asymptotic representation (10.14) in Lemma 10.1 we deduce

$$\frac{\partial R_N}{\partial n}(\mathbf{x}) \sim -\mathbf{n}^{(j)} \cdot \Big\{\nabla v(\mathbf{O}^{(j)}) + \boldsymbol{C}^{(j)} + \sum_{\substack{k \neq j \\ 1 \leq k \leq N}} \mathfrak{T}(\mathbf{x}, \mathbf{O}^{(k)}) \boldsymbol{Q}^{(k)} \boldsymbol{C}^{(k)}\Big\} ,$$

$$\mathbf{x} \in \partial F^{(j)}, \; j = 1, \dots, N , \tag{10.73}$$

where $\mathfrak{T}(\mathbf{x}, \mathbf{y})$ is defined by

$$\mathfrak{T}(\mathbf{x}, \mathbf{y}) = (\nabla_{\mathbf{x}} \otimes \nabla_{\mathbf{y}}) G(\mathbf{x}, \mathbf{y}) , \tag{10.74}$$

with $G(\mathbf{x}, \mathbf{y})$ being Green's function for the domain Ω, as defined in Sect. 10.2. To compensate for the leading discrepancy in the boundary conditions (10.73), we choose the coefficients $\boldsymbol{C}^{(m)}$, $m = 1, \dots, N$, subject to the algebraic system

$$\nabla v(\mathbf{O}^{(j)}) + \boldsymbol{C}^{(j)} + \sum_{\substack{k \neq j \\ 1 \leq k \leq N}} \mathfrak{T}(\mathbf{O}^{(j)}, \mathbf{O}^{(k)}) \boldsymbol{Q}^{(k)} \boldsymbol{C}^{(k)} = 0, \;\; j \,, = 1, \dots, N, \tag{10.75}$$

where $\boldsymbol{Q}^{(k)}, k = 1, \dots, N$, are polarization matrices of small voids $F^{(k)}$, as in Lemma 10.1.

Provided system (10.75) has been solved for the vector coefficients $\boldsymbol{C}^{(k)}$, formula (10.71) leads to the formal asymptotic approximation of u_N:

$$u_N(\mathbf{x}) \sim v(\mathbf{x}) + \sum_{k=1}^{N} \boldsymbol{C}^{(k)} \cdot \Big\{\boldsymbol{\mathcal{D}}^{(k)}(\mathbf{x}) - \boldsymbol{Q}^{(k)} \nabla_{\mathbf{y}} H(\mathbf{x}, \mathbf{y})\big|_{\mathbf{y}=\mathbf{O}^{(k)}}\Big\}. \tag{10.76}$$

10.6.2 Algebraic System

The system (10.75) can be written in the matrix form

$$\mathbf{C} + \mathfrak{S}\mathbf{Q}\mathbf{C} = -\boldsymbol{\Theta}, \tag{10.77}$$

where

$$\mathfrak{S} = [\mathfrak{S}_{ij}]_{i,j=1}^{N}, \; \mathfrak{S}_{ij} = \begin{cases} (\nabla_{\mathbf{z}} \otimes \nabla_{\mathbf{w}}) G(\mathbf{z}, \mathbf{w})\Big|_{\substack{\mathbf{z}=\mathbf{O}^{(i)} \\ \mathbf{w}=\mathbf{O}^{(j)}}} & \text{if } i \neq j \\ 0 I_3 & \text{otherwise} \end{cases} \tag{10.78}$$

with $G(\mathbf{z}, \mathbf{w})$ standing for Green's function in the limit domain Ω, and the block-diagonal matrix $\mathbf{Q}$ being the same as in (10.4). The system (10.77) is similar to that in Sect. 10.4, with the only change of the matrix $\mathcal{S}$ for $\mathfrak{S}$. The elements of $\mathfrak{S}$ are given via the second-order derivatives of Green's function in Ω, as defined in (10.74). The next assertion is similar to Corollary 10.1.

Lemma 10.3. *Assume that inequalities (10.5) hold for λ_{max} and λ_{min}. Also let v be a unique solution of problem (10.10), (10.11) in the domain Ω. Then the vector coefficients $\mathbf{C}^{(j)}$ in the system (10.75) satisfy the estimate*

$$\sum_{1\le j\le N} |\mathbf{C}^{(j)}|^2 \le \text{Const}\, d^{-3} \|\nabla v\|^2_{L_2(\Omega)}, \tag{10.79}$$

where the constant depends on the shape of the voids $F^{(j)}$, $j = 1, \dots, N$.

Proof. The proof of the theorem is very similar to the one given in Sect. 10.4. We consider the scalar product of (10.77) and the vector $\mathbf{QC}$:

$$\langle \mathbf{C}, \mathbf{QC}\rangle + \langle \mathfrak{S}\mathbf{QC}, \mathbf{QC}\rangle = -\langle \boldsymbol{\Theta}, \mathbf{QC}\rangle, \tag{10.80}$$

and similarly to (10.28) derive

$$\begin{aligned} \langle \mathfrak{S}\mathbf{QC}, \mathbf{QC}\rangle &= 48^2\, \pi^{-2}\, d^{-6} \int_\Omega \int_\Omega G(\mathbf{X}, \mathbf{Y})(\nabla \cdot \boldsymbol{\Xi}\,(\mathbf{X}))(\nabla \cdot \boldsymbol{\Xi}\,(\mathbf{Y}))d\mathbf{Y}d\mathbf{X} \\ &\quad -16\pi^{-1} d^{-3} \sum_{1\le j\le N} |\mathcal{Q}^{(j)}\mathbf{C}^{(j)}|^2 \\ &\quad + \sum_{1\le j\le N} \left(\mathcal{Q}^{(j)}\mathbf{C}^{(j)}\right)^T (\nabla_{\mathbf{z}} \otimes \nabla_{\mathbf{w}})\,(H(\mathbf{z}, \mathbf{w}))\Big|_{\substack{\mathbf{z}=\mathbf{O}^{(j)}\\ \mathbf{w}=\mathbf{O}^{(j)}}} \left(\mathcal{Q}^{(j)}\mathbf{C}^{(j)}\right), \end{aligned} \tag{10.81}$$

where the integral in the right-hand side is positive, and it is understood in the sense of distributions, in the same way as in the proof of Lemma 10.2, while the magnitude of the last sum in (10.81) is small compared to the magnitude of the second sum.

Now, the right-hand side in (10.80) does not exceed

$$\langle \mathbf{C}, -\mathbf{QC}\rangle^{1/2} \langle \boldsymbol{\Theta}, -\mathbf{Q}\boldsymbol{\Theta}\rangle^{1/2}.$$

Following the same pattern as in the proof of Theorem 10.2, we deduce

$$\langle \mathbf{C}, -\mathbf{QC}\rangle - \text{Const}\, d^{-3} \langle -\mathbf{QC}, -\mathbf{QC}\rangle \le \langle \mathbf{C}, -\mathbf{QC}\rangle^{1/2} \langle \boldsymbol{\Theta}, -\mathbf{Q}\boldsymbol{\Theta}\rangle^{1/2},$$

where the constant is independent of d. Furthermore, this leads to

$$\left(1 - \text{Const}\, d^{-3} \frac{\langle -\mathbf{QC}, -\mathbf{QC}\rangle}{\langle \mathbf{C}, -\mathbf{QC}\rangle}\right) \langle \mathbf{C}, -\mathbf{QC}\rangle^{1/2} \le \langle \boldsymbol{\Theta}, -\mathbf{Q}\boldsymbol{\Theta}\rangle^{1/2},$$

which implies

$$\left(1 - \text{Const}\, d^{-3}\lambda_{max}\right)^2 \langle \mathbf{C}, -\mathbf{Q}\mathbf{C}\rangle \le \langle \boldsymbol{\Theta}, -\mathbf{Q}\boldsymbol{\Theta}\rangle, \tag{10.82}$$

where λ_{max} is the largest eigenvalue of the positive definite matrix $-\mathbf{Q}$. Then using the same estimates (10.37) and (10.38) as in the proof of Corollary 10.1 we arrive at (10.79). □

10.6.3 Energy Estimate for the Remainder

Theorem 10.3. *Let the parameters ε and d satisfy the inequality*

$$\varepsilon < c\, d\ ,$$

where c is a sufficiently small absolute constant. Then the solution $u_N(\mathbf{x})$ of (10.1)–(10.3) is represented by the asymptotic formula

$$u_N(\mathbf{x}) = v(\mathbf{x}) + \sum_{k=1}^{N} \boldsymbol{C}^{(k)} \cdot \{\boldsymbol{\mathcal{D}}^{(k)}(\mathbf{x}) - \boldsymbol{\mathcal{Q}}^{(k)}\nabla_{\mathbf{y}} H(\mathbf{x},\mathbf{y})\Big|_{\mathbf{y}=\mathbf{O}^{(k)}}\} + R_N(\mathbf{x})\ , \tag{10.83}$$

where $\boldsymbol{C}^{(k)} = (C_1^{(k)}, C_2^{(k)}, C_3^{(k)})^T$ solve the linear algebraic system (10.75). The remainder R_N in (10.83) satisfies the energy estimate

$$\|\nabla R_N\|^2_{L_2(\Omega_N)} \le \text{Const}\left\{\varepsilon^{11} d^{-11} + \varepsilon^5 d^{-3}\right\} \|\nabla v\|^2_{L_2(\Omega)}\ . \tag{10.84}$$

Proof. Essentially, the proof follows the same steps as in Theorem 10.1. Thus, we give an outline indicating the obvious modifications, which are brought by the boundary $\partial\Omega$.

(a) *Auxiliary functions.* Let us preserve the notations $\chi_\varepsilon^{(k)}$ for cutoff functions used in the proof of Theorem 10.1. We also need a new cutoff function χ_0 to isolate $\partial\Omega$ from the cloud of holes. Namely, let $(1-\chi_0) \in C_0^\infty(\Omega)$ and $\chi_0 = 0$ on a neighbourhood of $\overline{\omega}$. A neighbourhood of $\partial\Omega$ containing supp χ_0 will be denoted by $\mathcal{V}$. Instead of the functions Ψ_k defined in (10.41), we introduce

$$\Psi_k^{(\Omega)}(\mathbf{x}) = v(\mathbf{x}) - v(\mathbf{O}^{(k)}) - (\mathbf{x} - \mathbf{O}^{(k)}) \cdot \nabla v(\mathbf{O}^{(k)}) + \sum_{\substack{j \ne k \\ 1 \le j \le N}} \boldsymbol{C}^{(j)} \cdot \boldsymbol{\mathcal{D}}^{(j)}(\mathbf{x})$$
$$- \sum_{\substack{j \ne k \\ 1 \le j \le N}} (\mathbf{x} - \mathbf{O}^{(j)}) \cdot \mathfrak{T}(\mathbf{O}^{(k)}, \mathbf{O}^{(j)})\boldsymbol{\mathcal{Q}}^{(j)}\boldsymbol{C}^{(j)}$$
$$- \sum_{j=1}^{N} \boldsymbol{C}^{(j)} \cdot \boldsymbol{\mathcal{Q}}^{(j)}\nabla_{\mathbf{y}} H(\mathbf{x},\mathbf{y})\Big|_{\mathbf{y}=\mathbf{O}^{(j)}}\ , \tag{10.85}$$

where the matrix $\mathfrak{T}$ is defined in (10.74) via second-order derivatives of Green's function in Ω. Owing to (10.83) and the algebraic system (10.75) we have

$$\frac{\partial}{\partial n}\left(\Psi_k^{(\Omega)}(\mathbf{x}) + R_N(\mathbf{x})\right) = 0, \;\; \mathbf{x} \in \partial F^{(k)}. \tag{10.86}$$

We also use the function

$$\Psi_0(\mathbf{x}) = \sum_{j=1}^{N} \boldsymbol{C}^{(j)} \cdot \{\mathcal{D}^{(j)}(\mathbf{x}) - \mathcal{Q}^{(j)} \frac{(\mathbf{x} - \mathbf{O}^{(j)})}{4\pi|\mathbf{x} - \mathbf{O}^{(j)}|^3}\}\,, \tag{10.87}$$

which is harmonic in Ω_N. It follows from (10.83) that

$$R_N(\mathbf{x}) + \Psi_0(\mathbf{x}) \tag{10.88}$$

$$= - \sum_{1\le j\le N} \boldsymbol{C}^{(j)} \cdot \mathcal{Q}^{(j)} \left\{\frac{(\mathbf{x} - \mathbf{O}^{(j)})}{4\pi|\mathbf{x} - \mathbf{O}^{(j)}|^3} - \nabla_{\mathbf{y}} H(\mathbf{x}, \mathbf{Y})\Big|_{\mathbf{Y}=\mathbf{O}^{(j)}}\right\} = 0\,,$$

$$\mathbf{x} \in \partial\Omega\,. \tag{10.89}$$

(b) *The energy estimate for R_N*. We start with the identity

$$\int_{\Omega_N} \nabla\Big(R_N + \chi_0\Psi_0\Big) \cdot \nabla\Big(R_N + \sum_{1\le k\le N} \chi_\varepsilon^{(k)}\Psi_k^{(\Omega)}\Big)\, d\mathbf{x}$$

$$= -\int_{\Omega_N} \Big(R_N + \chi_0\Psi_0\Big)\Delta\Big(R_N + \sum_{1\le k\le N} \chi_\varepsilon^{(k)}\Psi_k^{(\Omega)}\Big)\, d\mathbf{x}, \tag{10.90}$$

which follows from (10.86), (10.89) by Green's formula. According to the definitions of χ_0 and $\chi_\varepsilon^{(k)}$, we have supp $\chi_0 \cap \text{supp}\chi_\varepsilon^{(k)} = \varnothing$ for all $k = 1, \dots, N$. Hence the integrals in (10.90) involving the products of χ_0 and $\chi_\varepsilon^{(k)}$ or their derivatives are equal to zero. Thus, using that $\Delta R_N = 0$ on Ω_N, we reduce (10.90) to the equality

$$\int_{\Omega_N} |\nabla R_N|^2 d\mathbf{x} + \sum_{1\le k\le N} \int_{B_{3\varepsilon}\setminus \overline{F}^{(k)}} \nabla R_N \cdot \nabla\Big(\chi_\varepsilon^{(k)}\Psi_k^{(\Omega)}\Big) d\mathbf{x} \tag{10.91}$$

$$+ \int_{\Omega_N \cap \mathcal{V}} \nabla R_N \cdot \nabla\Big(\chi_0\Psi_0\Big) d\mathbf{x} = -\sum_{1\le k\le N} \int_{\Omega_N} R_N \Delta\left(\chi_\varepsilon^{(k)}\Psi_k^{(\Omega)}\right) d\mathbf{x},$$

which differs in the left-hand side from (10.47) only by the integral over $\Omega_N \cap \mathcal{V}$.

Similarly to the part (b) of the proof of Theorem 10.1 we deduce

$$\|\nabla \mathcal{R}_N\|^2_{L_2(\Omega_N)} \le \text{Const}\Big\{\|\nabla \Psi_0\|^2_{L_2(\Omega\cap\mathcal{V})} + \|\Psi_0\|^2_{L_2(\Omega\cap\mathcal{V})} + \sum_{1\le k\le N} \|\nabla \Psi_k\|^2_{L_2(B^{(k)}_{3\varepsilon})}\Big\}. \tag{10.92}$$

Similar to the steps of part (d) of the proof in Theorem 10.1, the last sum is majorized by

$$\text{Const}\,(\varepsilon^{11} d^{-11} + \varepsilon^5 d^{-3})\|\nabla v\|^2_{L_2(\Omega)}. \tag{10.93}$$

It remains to estimate two terms in (10.92) containing Ψ_0. Using (10.67), together with (10.79) we deduce

$$\|\Psi_0\|^2_{L_2(\Omega\cap\mathcal{V})} \le \text{Const}\,\varepsilon^8 \sum_{1\le j\le N} \int_{\Omega\cap\mathcal{V}} \frac{|\mathbf{C}^{(j)}|^2 d\mathbf{x}}{|\mathbf{x}-\mathbf{O}^{(j)}|^6}$$
$$\le \text{Const}\,\varepsilon^8 \sum_{1\le j\le N} |\mathbf{C}^{(j)}|^2 \le \text{Const}\frac{\varepsilon^8}{d^3}\|\nabla v\|^2_{L_2(\Omega)}, \tag{10.94}$$

and

$$\|\nabla\Psi_0\|^2_{L_2(\Omega\cap\mathcal{V})} \le \text{Const}\,\varepsilon^8 \sum_{1\le j\le N} \int_{\Omega\cap\mathcal{V}} \frac{|\mathbf{C}^{(j)}|^2 d\mathbf{x}}{|\mathbf{x}-\mathbf{O}^{(j)}|^8}$$
$$\le \text{Const}\,\varepsilon^8 \sum_{1\le j\le N} |\mathbf{C}^{(j)}|^2 \le \text{Const}\frac{\varepsilon^8}{d^3}\|\nabla v\|^2_{L_2(\Omega)}. \tag{10.95}$$

Combining (10.92)–(10.95) we complete the proof. □

10.7 Illustrative Example

Now, the asymptotic approximation derived in the previous section is applied to the case of a relatively simple geometry, where all the terms in the formula (10.83) can be written explicitly.

10.7.1 *The Case of a Domain with a Cloud of Spherical Voids*

Let Ω_N be a ball of a finite radius R, with the centre at the origin, containing N spherical voids $F^{(j)}$ of radii ρ_j with the centres at $\mathbf{O}^{(j)}$, $j = 1,\dots,N$, as shown in

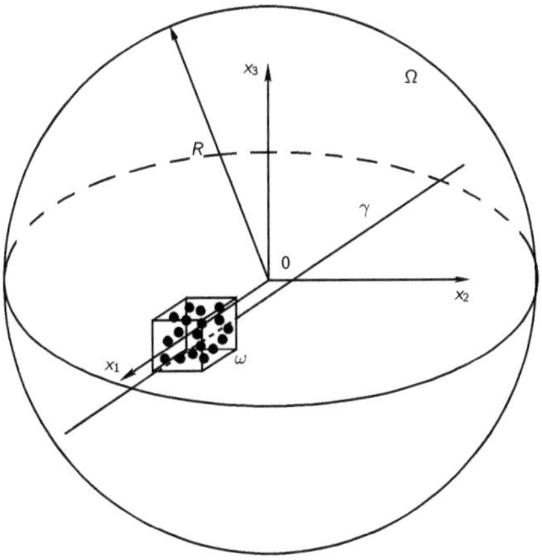

Fig. 10.1 Example configuration of a sphere containing a cloud of spherical voids in the cube ω

Fig. 10.1. The radii of the voids are assumed to be smaller than the distance between nearest neighbors. We put $\phi \equiv 0$ and

$$f(\mathbf{x}) = \begin{cases} 6 & \text{when } |\mathbf{x}| < \rho, \\ 0 & \text{when } \rho < |\mathbf{x}| < R. \end{cases} \tag{10.96}$$

Here, it is assumed that $\rho + b < |\mathbf{O}^{(j)}| < R - b,\ 1 \le j \le N$, where ρ and b are positive constants independent of ε and d.

The function u_N is the solution of the mixed boundary value problem for the Poisson equation:

$$\Delta u_N(\mathbf{x}) + f(\mathbf{x}) = 0, \ \text{ when } \mathbf{x} \in \Omega_N, \tag{10.97}$$

$$u_N(\mathbf{x}) = 0, \ \text{ when } |\mathbf{x}| = R, \tag{10.98}$$

$$\frac{\partial u_N}{\partial n}(\mathbf{x}) = 0, \ \text{ when } |\mathbf{x} - \mathbf{O}^{(j)}| = \rho_j,\ j = 1, \dots, N. \tag{10.99}$$

In this case, u_N is approximated by (10.83), where the solution of the Dirichlet problem in Ω is given by

$$v(\mathbf{x}) = \begin{cases} \rho^2(3 - 2\rho R^{-1}) - |\mathbf{x}|^2 & \text{when } |\mathbf{x}| < \rho, \\ 2\rho^3(|\mathbf{x}|^{-1} - R^{-1}) & \text{when } \rho < |\mathbf{x}| < R. \end{cases} \tag{10.100}$$

In turn, the dipole fields $\mathcal{D}^{(j)}$ and the dipole matrices $\mathcal{Q}^{(j)}$ have the form

$$\mathcal{D}^{(j)}(\mathbf{x}) = -\rho_j^3 \frac{\mathbf{x} - \mathbf{O}^{(j)}}{|\mathbf{x} - \mathbf{O}^{(j)}|^3}, \ \ \mathcal{Q}^{(j)} = -4\pi\rho_j^3 I_3, \tag{10.101}$$

where I_3 is the 3×3 identity matrix.

The regular part $H(\mathbf{x}, \mathbf{y})$ of Green's function in the domain Ω (see (10.12)) is

$$H(\mathbf{x}, \mathbf{y}) = \frac{R}{4\pi |\mathbf{y}||\mathbf{x} - \hat{\mathbf{y}}|}, \quad \hat{\mathbf{y}} = \frac{R^2}{|\mathbf{y}|^2}\mathbf{y}. \tag{10.102}$$

The coefficients $\boldsymbol{C}^{(j)}$, $j = 1, \dots, N$, in (10.83) are defined from the algebraic system (10.75), where Green's function $G(\mathbf{x}, \mathbf{y})$ is given by

$$G(\mathbf{x}, \mathbf{y}) = \frac{1}{4\pi |\mathbf{x} - \mathbf{y}|} - \frac{R}{4\pi |\mathbf{y}||\mathbf{x} - \hat{\mathbf{y}}|}. \tag{10.103}$$

10.7.2 Finite Elements Simulation Versus the Asymptotic Approximation

The explicit representations of the fields $v, \boldsymbol{\mathcal{D}}^{(j)}, H, G$, given above, are used in the asymptotic formula (10.83). Here, we present a comparison between the results of an independent Finite Element computation, produced in COMSOL, and the meso-scale asymptotic approximation (10.83).

For the computational example, we set $R = 120$, and consider a cloud of $N = 18$ spherical voids arranged into a cloud of a parallelepiped shape. The position of the centre and radius of each void is included in Table 10.1. The support of the function f (see (10.96)), is chosen to be inside the sphere with radius $\rho = 30$ and centre at the origin, as stated in (10.96).

Figure 10.2 shows the asymptotic solution u_N of the mixed boundary value problem (part (b) of the figure) and its numerical counterpart obtained in COMSOL 3.5 (part (a) of the figure). This computation has been produced for a spherical body containing 18 small voids defined in Table 10.1. The relative error for the chosen configuration does not exceed 2%, which confirms a very good agreement between the asymptotic and numerical results, which are visually indistinguishable in Fig. 10.2a and b.

The computation was performed on Apple Mac, with 4Gb of RAM, and the number $N = 18$ was chosen because any further increase in the number of voids resulted in a large three-dimensional computation, which exceeded the amount of available memory. Although, increase in RAM can allow for a larger computation, it is evident that three-dimensional finite element computations for a meso-scale geometry have serious limitations. On the other hand, the analytical asymptotic formula can still be used on the same computer for significantly larger number of voids.

In the next subsection, we show such an example where the number of voids within the meso-scale cloud runs up to $N = 1{,}000$, which would simply be unachievable in a finite element computation in COMSOL 3.5 with the same amount of RAM available.

Table 10.1 Data for the voids $F^{(j)}$, $j = 1, \ldots, 18$

Void	Centre	ρ_j/R	Void	Centre	ρ_j/R
$F^{(1)}$	$(-50, 0, 0)$	0.0417	$F^{(10)}$	$(-72, 0, 0)$	0.0417
$F^{(2)}$	$(-50, 0, 22)$	0.0333	$F^{(11)}$	$(-72, 0, 22)$	0.0458
$F^{(3)}$	$(-50, 22, 0)$	0.0292	$F^{(12)}$	$(-72, 22, 0)$	0.0292
$F^{(4)}$	$(-50, 0, -22)$	0.0375	$F^{(13)}$	$(-72, 0, -22)$	0.0375
$F^{(5)}$	$(-50, -22, 0)$	0.0458	$F^{(14)}$	$(-72, -22, 0)$	0.0417
$F^{(6)}$	$(-50, 22, 22)$	0.0292	$F^{(15)}$	$(-72, 22, 22)$	0.0333
$F^{(7)}$	$(-50, 22, -22)$	0.025	$F^{(16)}$	$(-72, 22, -22)$	0.05
$F^{(8)}$	$(-50, -22, 22)$	0.0375	$F^{(17)}$	$(-72, -22, 22)$	0.0333
$F^{(9)}$	$(-50, -22, -22)$	0.0375	$F^{(18)}$	$(-72, -22, -22)$	0.0375

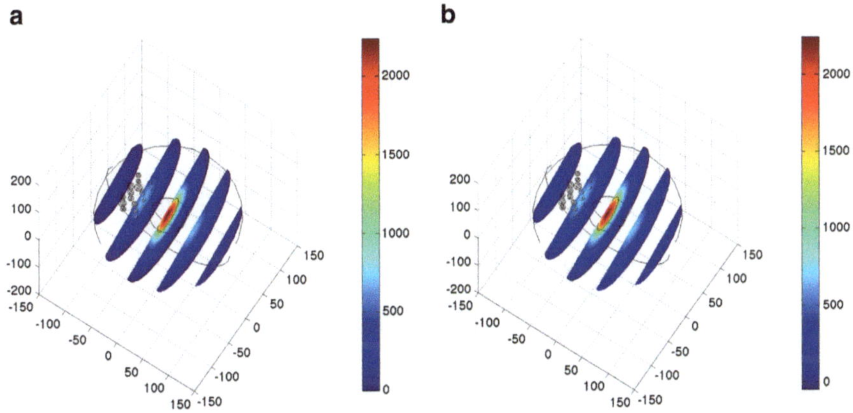

Fig. 10.2 Perforated domain containing 18 holes: (**a**) Numerical solutions produced in COMSOL; (**b**) Asymptotic approximation

10.7.3 Non-uniform Cloud Containing a Large Number of Spherical Voids

Here we consider the same mixed boundary value problem as in Sect. 10.7.1, but the cloud of voids is chosen in such a way that the number N may be large and voids of different radii are distributed in a non-uniform arrangement. For different values of N, the overall volume of voids is preserved—examples of the clouds used here are shown in Fig. 10.1.

The results are based on the numerical implementation of formula (10.83) in MATLAB.

The cloud ω is assumed to be the cube with side length $\frac{1}{\sqrt{3}}$ and the centre at $(3, 0, 0)$. Positioning of voids is described as follows. Assume we have $N = m^3$ voids, where $m = 2, 3, \ldots$. Then ω is divided into N smaller cubes of side length $h = \frac{1}{\sqrt{3}m}$, and the centres of voids are placed at

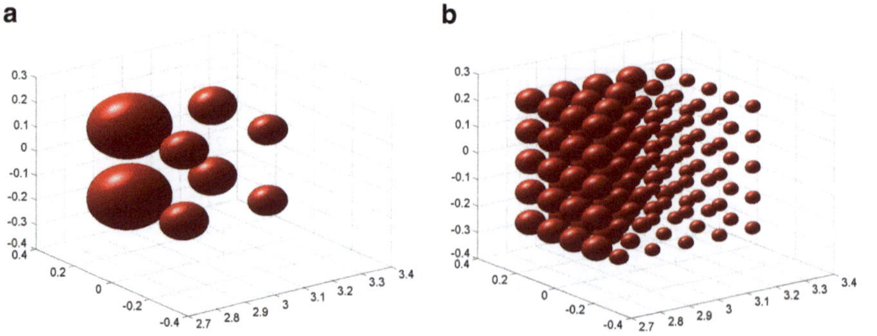

Fig. 10.3 The cloud of voids for the cases when (**a**) $N = 8$ and (**b**) $N = 125$

$$\mathbf{O}^{(p,q,r)} = \Big(3 - \frac{1}{2\sqrt{3}} + \frac{2p-1}{2}h, -\frac{1}{2\sqrt{3}} + \frac{2q-1}{2}h, -\frac{1}{2\sqrt{3}} + \frac{2r-1}{2}h\Big)$$

for $p, q, r = 1, \ldots, m$, and we assign their radii $\rho_{p,q,r}$ by

$$\rho_{p,q,r} = \begin{cases} \dfrac{h}{5} & \text{if } p > q\,, \\ \dfrac{\alpha h}{2} & \text{if } p < q\,, \\ \dfrac{h}{4} & \text{if } p = q\,, \end{cases}$$

where $\alpha < 1$, and it is chosen in such a way that the overall volume of all voids within the cloud remains constant for different N. An elementary calculation suggests that there will be m^2 voids with radius $\frac{h}{4}$ and equal number $\frac{m^2(m-1)}{2}$ of voids with radius $\frac{h}{5}$ or $\frac{\alpha h}{2}$.

Assuming that the volume fraction of all voids within the cube is equal to β, we have

$$\frac{4\pi h^3}{3}\Big(\frac{m^2(m-1)(8+125\alpha^3)}{2000} + \frac{m^2}{64}\Big) = \beta \frac{1}{3\sqrt{3}}\,,$$

and hence

$$\alpha^3 = \frac{16m}{m-1}\Big\{\frac{3}{4\pi}\beta - \frac{125 + 32(m-1)}{8000m}\Big\}\,. \tag{10.104}$$

In particular, if $N \to \infty$, the limit value α_∞ becomes

$$\alpha_\infty = \Big\{\frac{12}{\pi}\beta - \frac{8}{125}\Big\}^{1/3}\,. \tag{10.105}$$

In the numerical computation of this section, $\beta = \pi/25$.

Taking $R = 7$ and $\rho = 2$, we compute the leading order approximation of $u_N - v$, as defined in the asymptotic formula (10.83), along the line γ at the intersection

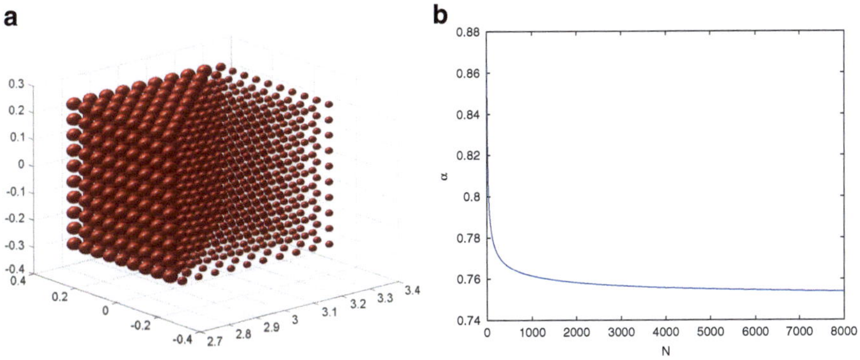

Fig. 10.4 (**a**) The cloud of voids for the case when $N = 1{,}000$, (**b**) The graph of α versus N given by formula (10.104) when $\beta = \pi/25$, for large N we see that α tends to 0.7465 which is predicted value present in (10.105)

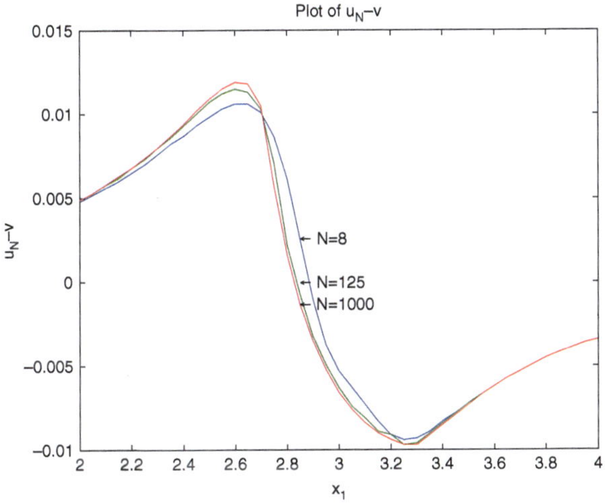

Fig. 10.5 The graph of $u_N - v$ given by (10.83), for $2 \le x_1 \le 4$ plotted along the straight line γ adjacent to the cloud of small voids

of the planes $x_2 = -1/(2\sqrt{3})$ and $x_3 = -1/(2\sqrt{3})$, for $N = 8{,}125{,}1{,}000$. Figure 10.3 shows the configuration of the cloud of voids for a) $N = 8$ and b) $N = 125$. For a large number of voids ($N = 1{,}000$), Fig. 10.4a shows the cloud and Fig 10.4b includes the graph of α versus N. The plot of $u_N - v$ given by (10.83) for $2 \le x_1 \le 4$ is shown in Fig. 10.5. The asymptotic correction has been computed along the straight line $\gamma = \{x_1 \in \mathbb{R}, x_2 = -1/(2\sqrt{3}), x_3 = -1/(2\sqrt{3})\}$. Dipole type fluctuations are clearly visible on the diagram. Beyond $N = 1{,}000$ the graphs are visually indistinguishable and hence the values $N = 8{,}125{,}1{,}000$, as in Figs. 10.3 and 10.4 have been chosen in the computations. The algorithm is fast and does not impose periodicity constraints on the array of small voids.

Bibliographical Remarks

Chapters 1–5 of Part I of the book address the asymptotics of Green's functions for boundary value problems for the Laplacian.

The analysis of uniform asymptotic approximations for Green's functions for Dirichlet problems in multi-dimensional domains with small perforations is included in Chap. 1, which is based on the papers [23, 24].

Chapter 2 incorporates the results of the paper [26], which deals with Neumann and mixed boundary value problems, with Neumann boundary conditions on the boundaries of small holes. The analysis of [26] includes uniform asymptotics of Green's kernels in two- and three-dimensional domains containing a small hole.

Chapters 3 and 4 address uniform asymptotics of Green's kernels in domains with several perforations and the numerical simulations. The material of these chapters is based on the results of [32]. The paper [25] shows other examples of uniform approximations of Green's functions in singularly perturbed domains, such as thin bodies, truncated cones and domains with small grooves on the exterior boundaries—this material is discussed in Chap. 5.

Part II of the book, incorporating Chaps. 6, 7 and 8 presents the asymptotic approach for uniform approximations of Green's kernels in vector problems of elasticity in two- and three-dimensional elastic bodies with small holes. Chapter 6 discussing the case of a domain with a single inclusion is based on the paper [33], and Chap. 7 addressing the case of multiply-perforated elastic bodies includes the results of [32].

In Part III, we consider the case when the number of perforations becomes large. A new method of meso-scale asymptotic approximations is introduced in Chaps. 9 and 10. Chapter 9 on meso-scale approximations for solutions of Dirichlet problems uses the results [27], and the case of mixed boundary value problems in multiply-perforated domains of Chap. 10 is discussed in [29].

V. Maz'ya et al., *Green's Kernels and Meso-Scale Approximations in Perforated Domains*, Lecture Notes in Mathematics 2077, DOI 10.1007/978-3-319-00357-3,

References

1. G. Allaire, Homogenization and two-scale convergence. SIAM J. Math. Anal. **23**, 1482–1518 (1992)
2. N.S. Bakhvalov, G.P. Panasenko, *Homogenization: Averaging Processes in Periodic Media* (Kluwer, Dordrecht, 1989), (translated from Russian: *Osrednenie Processov v Periodicheskih Sredah* (Nauka, Moscow, 1984))
3. R.W. Barnard, K. Pearce, C. Campbell, A survey of applications of the Julia variation. Ann. Univ. Mariae Curie-Sklodowska Sect. A **54**, 1–20 (2000)
4. G.A. Chechkin, Homogenization in perforated domains. In *Topics on Concentration Phenomena and Problems with Multiple Scales*, ed. by A. Braides, V. Chiadó Piat. Lecture Notes of the Unione Matematica Italiana, vol. 2 (Springer, Berlin, 2006), pp. 189–208
5. D. Cioranescu, F. Murat, *A Strange Term Brought from Somewhere Else*, Nonlinear Partial Differential Equations and their Applications, Collège de France Seminar, vol. II and III. Research Notes in Mathematics, vol. 60 and 70, pp. 98–138 and 154–178 (1982)
6. R. Courant, D. Hilbert, *Methods of Mathematical Physics*, vol. I (Interscience, New York, 1953)
7. M. Englis, D. Lukkassen, J. Peetre, L.E. Persson, On the formula of Jacques-Louis Lions for reproducing kernels of harmonic and other functions. J. Reine Angew. Math. **570**, 89–129 (2004)
8. G. Fichera, Il teorema del massimo modulo per l'equazione dell'elastostatica tridimensionale. Arch. Ration. Mech. Anal. **7**(1), 373–387 (1961)
9. R. Figari, A. Teta, A boundary value problem of mixed type on perforated domains. Asymptotic Anal. **6**, 271–284 (1993)
10. R. Figari, G. Papanicolaou, J. Rubinstein, The point interaction approximation for diffusion in regions with many small holes. Stochastic Methods in Biology. Lect. Note Biomater. **70**, 75–86 (1987)
11. G. Fremiot, J. Sokolowski, *Shape Sensitivity Analysis of Problems with Singularities. Shape Optimization and Optimal Design* (Cambridge University Press, Cambridge 1999), pp. 255–276, Lecture Notes in Pure and Appl. Math., vol. 216 (Dekker, New York, 2001)
12. D. Gilbarg, N.S. Trudinger, *Elliptic Partial Differential Equations of Second Order* (Springer, Berlin, 1983)
13. J. Hadamard, Sur le problème d'analyse relatif à l'équilibre des plaques élastiques encastrées. Mémoire couronnéen 1907 par l'Académie: Prix Vaillant, Mémoires présentés par divers savants à l'Académie des Sciences 33, No 4 (1908). In *Oeuvres de Jacques Hadamard*, vol. 2 (Centre National de la Recherche Scientifique, Paris, 1968), pp. 515–629
14. A. Hönig, B. Niethammer, F. Otto, On first-order corrections to the LSW theory I: Infinite systems. J. Stat. Phys. **119**, 61–122 (2005)

V. Maz'ya et al., *Green's Kernels and Meso-Scale Approximations in Perforated Domains*, Lecture Notes in Mathematics 2077, DOI 10.1007/978-3-319-00357-3,

15. A.M. Il'in, *Matching of Asymptotic Expansions of Solutions of Boundary Value Problems*, Translations of Mathematical Monographs (American Mathematical Society, Providence, 1992)
16. G. Julia, Sur une équation aux dérivées fonctionnelles analogue à l'équation de M. Hadamard. C.R. Acad. Sci. Paris **172**, 831–833 (1921)
17. G. Komatsu, Hadamard's variational formula for the Bergman kernel. Proc. Jpn. Acad. Ser. A. Math. Sci. **58**(8), 345–348 (1982)
18. V.A. Kondratiev, O.A. Oleinik: On the behavior at infinity of solutions of elliptic systems with a finite energy integral. Arch. Ration. Mech. Anal. **99**(1), 75–89 (1987)
19. V. Kozlov, V. Maz'ya, A. Movchan, *Fields in Multi-Structures, Asymptotic Analysis* (Oxford University Press, Oxford, 1999)
20. N.S. Landkof, *Foundations of Modern Potential Theory* (Spinger, Berlin, 1972)
21. V.A. Marchenko, E.Y. Khruslov, *Homogenization of Partial Differential Equations* (Birkhäuser, Boston, 2006). Russian Edition: Kraevye zadachi v oblastiakh s melkozernistoi granitsei (Naukova dumka, Kiev, 1974)
22. V. Maz'ya, *Sobolev Spaces* (Springer, Berlin, 1985)
23. V.G. Maz'ya, A.B. Movchan, Uniform asymptotic formulae for Green's kernels in regularly and singularly perturbed domains. C. R. Acad. Sci. Paris. Ser. I **343**, 185–190 (2006)
24. V.G. Maz'ya, A.B. Movchan, Uniform asymptotic formulae for Green's functions in singularly perturbed domains. J. Comput. Appl. Math. **208**(1), 194–206 (2007)
25. V. Maz'ya, A. Movchan, Uniform asymptotic approximations of Green's functions in a long rod. Math. Meth. Appl. Sci. **31**, 2055–2068 (2008)
26. V. Maz'ya, A. Movchan, Uniform asymptotics of Green's kernels for mixed and Neumann problems in domains with small holes and inclusions, in *Sobolev Spaces in Mathematics III*. Applications in Mathematical Physics (Springer, New York, 2009), pp. 277–316
27. V. Maz'ya, A. Movchan, Asymptotic treatment of perforated domains without homogenization. Math. Nachr. **283**(1), 104–125 (2010)
28. V. Maz'ya, A. Movchan, M. Nieves, *Green's kernels for transmission problems in bodies with small inclusions*, Operator Theory and Its Applications, In Memory of V. B. Lidskii (1924–2008), ed. by M. Levitin, D. Vassiliev, American Mathematical Society Translations, Series 2, vol. 231 (American Mathematical Society, Providence, RI, 2010), pp. 127–171
29. V. Maz'ya, A. Movchan, M. Nieves, Mesoscale asymptotic approximations to solutions of mixed boundary value problems in perforated domains. Multiscale Model. Simulat. **9**(1), 424–448 (2011)
30. V. Maz'ya, S. Nazarov, B. Plamenevskii, *Asymptotic Theory of Elliptic Boundary Value Problems in Singularly Perturbed Domains*, vols. 1–2 (Birkhäuser, Boston, 2000)
31. V. Maz'ya, B. Plamenevskii, *Estimates in L_p and in Hölder classes and the Miranda-Agmon maximum principle for solutions of elliptic boundary value problems in domains with singular points on the boundary.* In Elliptic boundary value problems ed. by V.Maz'ya et al. AMS Translations, Series 2, v. 123, 1–56 (1984)
32. V.G. Maz'ya, A.B. Movchan, M.J. Nieves, *Uniform asymptotic formulae for Green's tensors in elastic singularly perturbed domains with multiple inclusions*, Rendiconti della accademia nazionale delle scienze detta dei XL, Memorie di matematica e applicazioni, Serie V, Vol. XXX, Parte I, 103–158 (2006)
33. V.G. Maz'ya, A.B. Movchan, M.J. Nieves, Uniform asymptotic formulae for Green's tensors in elastic singularly perturbed domains. Asymptotic Anal. **52**(3/4), 173–206 (2007)
34. A.B. Movchan, N.V. Movchan, C.G. Poulton, *Asymptotic Models of Fields in Dilute and Densely Packed Composites* (Imperial College Press, London, 2002)
35. O.A. Oleinik, G.A. Yosifian, Boundary value problems for second order elliptic equations in unbounded domains and Saint-Venant's principle. Annali della Scuola Normale Superiore di Pisa, Classe di Scienze 4e série, tome **4**(2), 269–290 (1977)
36. S. Ozawa, Approximation of Green's function in a region with many obstacles. In *Geometry and Analysis on Manifolds*, ed. by T. Sunada, Lecture Notes in Mathematics, vol. 1339 (Springer, New York, 1988), pp. 212–225

37. B. Palmerio, A. Dervieux, Hadamard's variational formula for a mixed problem and an application to a problem related to a Signorini-like variational inequality. Numer. Funct. Anal. Optim **1**(2), 113–144 (1979)
38. G. Pólya, G. Szegö, *Isoperimetric Inequalities in Mathematical Physics* (Princeton University Press, Princeton, 1951)
39. E. Sánchez-Palencia, *Non-homogeneous Media and Vibration Theory.* Lecture Notes in Physics, vol. 27 (Springer, New York, 1980)
40. E. Sánchez-Palencia, *Homogenization Method for the Study of Composite Media. Asymptotic Analysis, II.* Lecture Notes in Math., vol. 985 (Springer, Berlin, 1983), pp. 192–214
41. E. Sánchez-Palencia, Homogenization in mechanics. A survey of solved and open problems. Rend. Sem. Mat. Univ. Politec. Torino **44**(1), 1–45 (1986)
42. E.M. Stein, *Singular Integrals and Differentiability Properties of Functions* (Princeton University Press, Princeton, 1970)
43. V.V. Zhikov, S.M. Kozlov, O.A. Oleinik, *Homogenization of Differential Operators* (Nauka, Moscow, 1993); English transl., *Homogenization of Differential Operators and Integral Functionals* (Springer, Berlin, 1994)
44. V.V. Zhikov, Averaging of problems in the theory of elasticity on singular structures. Izv. Ross. Akad. Nauk Ser. Mat. **66**(2), 81–148 (2002); English transl., Izv. Math. **66**(2), 299–365 (2002)

Subjects Index

V. Maz'ya et al., *Green's Kernels and Meso-Scale Approximations in Perforated Domains*, Lecture Notes in Mathematics 2077, DOI 10.1007/978-3-319-00357-3,

Author Index

V. Maz'ya et al., *Green's Kernels and Meso-Scale Approximations in Perforated Domains*, Lecture Notes in Mathematics 2077, DOI 10.1007/978-3-319-00357-3,

LECTURE NOTES IN MATHEMATICS Springer

Edited by J.-M. Morel, B. Teissier; P.K. Maini

Editorial Policy (for the publication of monographs)

1. Lecture Notes aim to report new developments in all areas of mathematics and their applications - quickly, informally and at a high level. Mathematical texts analysing new developments in modelling and numerical simulation are welcome.
   Monograph manuscripts should be reasonably self-contained and rounded off. Thus they may, and often will, present not only results of the author but also related work by other people. They may be based on specialised lecture courses. Furthermore, the manuscripts should provide sufficient motivation, examples and applications. This clearly distinguishes Lecture Notes from journal articles or technical reports which normally are very concise. Articles intended for a journal but too long to be accepted by most journals, usually do not have this "lecture notes" character. For similar reasons it is unusual for doctoral theses to be accepted for the Lecture Notes series, though habilitation theses may be appropriate.

2. Manuscripts should be submitted either online at www.editorialmanager.com/lnm to Springer's mathematics editorial in Heidelberg, or to one of the series editors. In general, manuscripts will be sent out to 2 external referees for evaluation. If a decision cannot yet be reached on the basis of the first 2 reports, further referees may be contacted: The author will be informed of this. A final decision to publish can be made only on the basis of the complete manuscript, however a refereeing process leading to a preliminary decision can be based on a pre-final or incomplete manuscript. The strict minimum amount of material that will be considered should include a detailed outline describing the planned contents of each chapter, a bibliography and several sample chapters.
   Authors should be aware that incomplete or insufficiently close to final manuscripts almost always result in longer refereeing times and nevertheless unclear referees' recommendations, making further refereeing of a final draft necessary.
   Authors should also be aware that parallel submission of their manuscript to another publisher while under consideration for LNM will in general lead to immediate rejection.

3. Manuscripts should in general be submitted in English. Final manuscripts should contain at least 100 pages of mathematical text and should always include

   - a table of contents;
   - an informative introduction, with adequate motivation and perhaps some historical remarks: it should be accessible to a reader not intimately familiar with the topic treated;
   - a subject index: as a rule this is genuinely helpful for the reader.

   For evaluation purposes, manuscripts may be submitted in print or electronic form (print form is still preferred by most referees), in the latter case preferably as pdf- or zipped psfiles. Lecture Notes volumes are, as a rule, printed digitally from the authors' files. To ensure best results, authors are asked to use the LaTeX2e style files available from Springer's web-server at:

   ftp://ftp.springer.de/pub/tex/latex/svmonot1/ (for monographs) and
   ftp://ftp.springer.de/pub/tex/latex/svmultt1/ (for summer schools/tutorials).

Additional technical instructions, if necessary, are available on request from lnm@springer.com.

4. Careful preparation of the manuscripts will help keep production time short besides ensuring satisfactory appearance of the finished book in print and online. After acceptance of the manuscript authors will be asked to prepare the final LaTeX source files and also the corresponding dvi-, pdf- or zipped ps-file. The LaTeX source files are essential for producing the full-text online version of the book (see http://www.springerlink.com/openurl.asp?genre=journal&issn=0075-8434 for the existing online volumes of LNM). The actual production of a Lecture Notes volume takes approximately 12 weeks.

5. Authors receive a total of 50 free copies of their volume, but no royalties. They are entitled to a discount of 33.3 % on the price of Springer books purchased for their personal use, if ordering directly from Springer.

6. Commitment to publish is made by letter of intent rather than by signing a formal contract. Springer-Verlag secures the copyright for each volume. Authors are free to reuse material contained in their LNM volumes in later publications: a brief written (or e-mail) request for formal permission is sufficient.

Addresses:

Professor J.-M. Morel, CMLA,
École Normale Supérieure de Cachan,
61 Avenue du Président Wilson, 94235 Cachan Cedex, France
E-mail: morel@cmla.ens-cachan.fr

Professor B. Teissier, Institut Mathématique de Jussieu,
UMR 7586 du CNRS, Équipe "Géométrie et Dynamique",
175 rue du Chevaleret
75013 Paris, France
E-mail: teissier@math.jussieu.fr

For the "Mathematical Biosciences Subseries" of LNM:

Professor P. K. Maini, Center for Mathematical Biology,
Mathematical Institute, 24-29 St Giles,
Oxford OX1 3LP, UK
E-mail : maini@maths.ox.ac.uk

Springer, Mathematics Editorial, Tiergartenstr. 17,
69121 Heidelberg, Germany,
Tel.: +49 (6221) 4876-8259

Fax: +49 (6221) 4876-8259
E-mail: lnm@springer.com

MIX
Papier aus verantwortungsvollen Quellen
Paper from responsible sources
FSC® C105338

If you have any concerns about our products,
you can contact us on
ProductSafety@springernature.com

In case Publisher is established outside the EU,
the EU authorized representative is:
Springer Nature Customer Service Center GmbH
Europaplatz 3, 69115 Heidelberg, Germany

Printed by Libri Plureos GmbH
in Hamburg, Germany